Vol. 32.	**lethods and Procedures.** By Frederick I.
Vol. 33.	**tions.** By D. D. Perrin
Vol. 34.	**Neutron Activation Analysis.** By D. De Soete, R. Gijbels, and J. Hoste
Vol. 35.	**Laser Raman Spectroscopy.** By Marvin C. Tobin
Vol. 36.	**Emission Spectrochemical Analysis.** By Morris Slavin
Vol. 37.	**Analytical Chemistry of Phosphorus Compounds.** Edited by M. Halmann
Vol. 38.	**Luminescence Spectrometry in Analytical Chemistry.** By J. D. Winefordner, S. G. Schulman, and T. C. O'Haver
Vol. 39.	**Activation Analysis with Neutron Generators.** By Sam S. Nargolwalla and Edwin P. Przybylowicz
Vol. 40.	**Determination of Gaseous Elements in Metals.** Edited by Lynn L. Lewis, Laben M. Melnick, and Ben D. Holt
Vol. 41.	**Analysis of Silicones.** Edited by A. Lee Smith
Vol. 42.	**Foundations of Ultracentrifugal Analysis.** By H. Fujita
Vol. 43.	**Chemical Infrared Fourier Transform Spectroscopy.** By Peter R. Griffiths
Vol. 44.	**Microscale Manipulations in Chemistry.** By T. S. Ma and V. Horak
Vol. 45.	**Thermometric Titrations.** By J. Barthel
Vol. 46.	**Trace Analysis: Spectroscopic Methods for Elements.** Edited by J. D. Winefordner
Vol. 47.	**Contamination Control in Trace Element Analysis.** By Morris Zief and James W. Mitchell
Vol. 48.	**Analytical Applications of NMR.** By D. E. Leyden and R. H. Cox
Vol. 49.	**Measurement of Dissolved Oxygen.** By Michael L. Hitchman
Vol. 50.	**Analytical Laser Spectroscopy.** Edited by Nicolo Omenetto
Vol. 51.	**Trace Element Analysis of Geological Materials.** By Roger D. Reeves and Robert R. Brooks

Trace Element Analysis
of Geological Materials

CHEMICAL ANALYSIS

A SERIES OF MONOGRAPHS ON ANALYTICAL CHEMISTRY AND ITS APPLICATIONS

Editors

P. J. ELVING   J. D. WINEFORDNER

Editor Emeritus I. M. KOLTHOFF

Advisory Board

Fred W. Billmeyer, Jr.
Eli Grushka
Barry L. Karger
Viliam Krivan

Victor G. Mossotti
A. Lee Smith
Bernard Tremillon
T. S. West

VOLUME 51

A WILEY-INTERSCIENCE PUBLICATION

JOHN WILEY & SONS

New York / Chichester / Brisbane / Toronto

Trace Element Analysis of Geological Materials

R. D. REEVES

R. R. BROOKS

Department of Chemistry, Biochemistry and Biophysics
Massey University
Palmerston North
New Zealand

A WILEY-INTERSCIENCE PUBLICATION

JOHN WILEY & SONS
New York / Chichester / Brisbane / Toronto

Copyright © 1978 by John Wiley & Sons, Inc.

All rights reserved. Published simultaneously in Canada.

Reproduction or translation of any part of this work beyond that permitted by Sections 107 or 108 of the 1976 United States Copyright Act without the permission of the copyright owner is unlawful. Requests for permission or further information should be addressed to the Permissions Department, John Wiley & Sons, Inc.

Library of Congress Cataloging in Publication Data:

Reeves, Roger D., 1940–
    Trace element analysis of geological materials.

    (Chemical analysis; v. 51)
    "A Wiley-Interscience publication."
    Includes index.
    1. Trace elements—Analysis. 2. Rocks—Analysis.
I. Brooks, R. R., joint author. II. Title.
III. Series.
QE516.T85R43   552'.06   78-8064

ISBN 0-471-71338-4

Printed in the United States of America

10 9 8 7 6 5 4 3 2 1

PREFACE

In recent years there have been increasing demands for trace element analysis of geological materials of all kinds—rocks, ores, sediments, soils, natural waters, and extraterrestrial materials. These demands have arisen in the course of work in a wide variety of fields, including mineral exploration, fundamental geochemical studies, soil science, agricultural and horticultural research, oceanography, archaeology, epidemiology, and environmental pollution. In this volume we have attempted to provide an introduction to trace element analysis that will be of use not only to those with formal training in analytical chemistry and geochemistry, but also to those who are being drawn into this field from many other branches of science. Portions of this book have been used in our recent courses in analytical chemistry and geochemistry, and should be useful for senior undergraduates and graduate students in chemistry and the earth sciences.

Because analytical data can be no more reliable than any of the steps that lead to their acquisition, it is obviously unwise to devote a great deal of time, care, and expense to the analytical measurements themselves if the original samples have not been properly collected or treated before analysis. The first part of the book therefore deals with sampling techniques and with physical and chemical methods of sample pretreatment. The book then outlines the basic principles underlying the most widely used methods of separating and determining the trace constituents of geological materials. A brief account is given of the uses that have been made of trace element data in some of the fields mentioned above, and of some applications of statistical techniques in the interpretation of geochemical trace element data.

In discussing the analytical methods themselves, we have endeavored to provide a balance that reflects the relative importance of the methods to geochemical trace analysis. Limited space is therefore devoted to gravimetric and volumetric techniques, which are of particular importance in the classical analysis of silicate rocks for major constituents, and which have been covered thoroughly elsewhere. A number of other analytical methods that are important in other fields but are rarely used for trace analysis of geological materials have also been neglected.

In a wide-ranging work of this kind, space limitations prevent the inclusion of detailed laboratory procedures. However, frequent reference has been made to specialist texts on individual analytical techniques and to papers illustrating applications to geological materials. A rather detailed bibliography has been provided in some chapters, such as those on atomic absorption and colorimetric methods (which are practised in almost every geochemical laboratory) and neutron activation analysis (which has been used for much of the definitive work on ultratrace concentrations). In other cases the reader has been referred to reviews covering a wide range of original literature. In this connection, the biennial reviews of *Analytical Chemistry* are recommended for the summaries of developments in particular analytical techniques (even-numbered years) and in particular fields of application (odd-numbered years). Among the latter the reviews on "Inorganic and Geological Materials" (commenced in 1975) are especially valuable to those interested in keeping abreast of progress in geochemical trace analysis.

We are grateful to several friends and colleagues for reading portions of the text and making useful suggestions. In particular, we wish to acknowledge the benefit of consultations with Professor B. I. Hayman, Dr. V. E. Neall, Dr. P. C. Rankin, Mr. J. R. Sewell, and Dr. S. Whineray. Special thanks are also due to Jenny Trow for her work in preparing many of the graphs and diagrams, and to Linda Smith and Nancy P. Reaburn for their patience and care in the typing of the manuscript.

R. D. REEVES
R. R. BROOKS

Palmerston North, New Zealand
July 1978

CONTENTS

CHAPTER 1.	GENERAL INTRODUCTION	1
CHAPTER 2.	SAMPLING AND STORAGE TECHNIQUES	4
	2.1 Introduction	5
	2.2 Rocks, Minerals, Soils, and Lake and Ocean Sediments	5
	2.3 Natural Waters	7
	2.4 Gases	10
	2.5 Biological Material	11
	2.6 Sampling from the Air	11
CHAPTER 3.	PHYSICAL METHODS OF SAMPLE PRETREATMENT	13
	3.1 Introduction	13
	3.2 Crushing and Pulverizing Rocks, Ores, and Minerals	13
	3.3 Crushing of Soils	16
	3.4 Sieving	16
	3.5 Sample Splitting	18
	3.6 Separation of Minerals	19
	3.7 Filtering of Natural Waters	25
	3.8 Other Methods of Sample Pretreatment	26
	3.9 Errors in Sampling Solid Geological Materials for Trace Element Analysis	26
CHAPTER 4.	CHEMICAL METHODS OF SAMPLE PRETREATMENT	29
	4.1 Decomposition of Solid Samples	29
	4.2 Separation and Concentration Techniques	36
CHAPTER 5.	ABUNDANCE DATA AND STANDARD ROCKS	76
	5.1 Introduction	76
	5.2 Abundance Data	79
	5.3 Standard Rocks	81

CONTENTS

CHAPTER 6. GRAVIMETRIC AND TITRIMETRIC METHODS OF ANALYSIS — 90
- 6.1 Introduction — 90
- 6.2 Gravimetric Techniques — 91
- 6.3 Titrimetric Techniques — 92

CHAPTER 7. SOLUTION ABSORPTIOMETRY — 99
- 7.1 Introduction — 99
- 7.2 Instrumentation — 104
- 7.3 Aspects of Absorptiometric Analysis — 109
- 7.4 Applications to the Determination of Various Elements — 116

CHAPTER 8. MOLECULAR FLUORIMETRY — 126
- 8.1 Introduction and Theory — 126
- 8.2 Instrumentation — 130
- 8.3 Applications — 132

CHAPTER 9. EMISSION SPECTROCHEMICAL ANALYSIS — 138
- 9.1 Introduction — 138
- 9.2 The Origin of Atomic Spectra — 138
- 9.3 Sources and Electrodes — 140
- 9.4 Optical Spectrographs — 143
- 9.5 Photographic Measurement of Spectral Line Intensities — 146
- 9.6 Qualitative and Quantitative Analysis with the D.C. Arc — 150
- 9.7 Direct-Reading Spectrometers — 152
- 9.8 Spectrochemical Analysis of Liquids — 153
- 9.9 The Laser Microprobe — 155
- 9.10 Comparative Evaluation and Recent Developments — 156

CHAPTER 10. ATOMIC ABSORPTION SPECTROPHOTOMETRY — 160
- 10.1 Theory — 161
- 10.2 Instrumentation — 165
- 10.3 Aspects of Atom Production — 175
- 10.4 Analytical Techniques — 183
- 10.5 Applications to the Analysis of Geological Materials — 191

CHAPTER 11. FLAME EMISSION AND ATOMIC FLUORESCENCE SPECTROMETRY — 214

- 11.1 General Introduction — 214
- 11.2 Theoretical and Experimental Considerations — 215
- 11.3 Instrumentation — 217
- 11.4 Practice and Applications — 219
- 11.5 Atomic Fluorescence Spectrometry — 225

CHAPTER 12. X-RAY EMISSION SPECTROMETRY — 232

- 12.1 Introduction — 232
- 12.2 X-Ray Spectra — 233
- 12.3 Instrumentation — 238
- 12.4 X-Ray Spectrochemical Analysis — 244
- 12.5 Other X-Ray Emission Techniques — 246

CHAPTER 13. RADIOMETRIC AND RADIOACTIVATION METHODS — 250

- 13.1 Radioactivity and Its Measurement — 250
- 13.2 Trace Analysis Using Natural Radioactivity — 266
- 13.3 Neutron Activation Analysis — 270
- 13.4 Other Forms of Activation Analysis — 291
- 13.5 Radiochemical Isotope Dilution — 293

CHAPTER 14. ELECTROANALYTICAL METHODS — 303

- 14.1 Potentiometry with Ion-Selective Electrodes — 303
- 14.2 Controlled Potential Methods — 309

CHAPTER 15. MASS SPECTROMETRY AND SPARK-SOURCE MASS SPECTROGRAPHY — 322

- 15.1 Introduction and Theory — 322
- 15.2 Instrumentation — 324
- 15.3 Mass Spectrometric Isotope Dilution — 326
- 15.4 Spark-Source Mass Spectrography — 330

CHAPTER 16. PHYSICAL AND CHEMICAL FIELD TESTS FOR TRACE ELEMENTS — 336

- 16.1 Physical Field Tests — 336
- 16.2 Chemical Field Tests — 337
- 16.3 Mobile Geochemical Laboratories — 338

CHAPTER 17. USES OF DATA ON TRACE ELEMENTS IN GEOLOGICAL MATERIALS 341

17.1 Geochemical Studies 341
17.2 Mineral Exploration 347
17.3 Plant, Animal, and Human Health 354
17.4 Environmental Chemistry 358
17.5 Archaeology 363

CHAPTER 18. STATISTICAL INTERPRETATION OF GEOCHEMICAL DATA 374

18.1 Introduction 374
18.2 Simple Tests of Significance 379
18.3 Bivariate Analysis 384
18.4 Multivariate Analysis 391
18.5 Accuracy and Precision in Trace-Element Analysis 398

Appendix I. Table of Values for F Test 403
Appendix II. Table of Values for t test 406
Appendix III. Table of Values for χ^2 Test 407
Appendix IV. Table of Significance Values for r 408

INDEX 409

CHAPTER

1

GENERAL INTRODUCTION

During the last 20 years there has been an enormous increase of interest in the role played by elements present in low concentrations in geological and biological systems. The growth of research in these fields owes a great deal to the development and improvement of methods of chemical analysis in this period. In turn, many of the advances in analytical techniques and instrumentation have been stimulated by the special requirements of fields as diverse as geochemistry, mineral exploration, soil science, agriculture and horticulture, environmental science, clinical science, toxicology, forensic science, and archaeology. These special requirements have included the determination of large numbers of elements at trace or ultratrace levels, the use of very small samples, and the processing of very large numbers of samples in a short time.

The definition of terms such as *trace* and *ultratrace* is to some extent arbitrary. The term *trace element* is used by some workers in geochemistry to denote any element with an average abundance in the earth's crust of less than 1000 μg/g (0.1%). With such a restrictive definition, aluminum, iron, calcium, sodium, potassium, and magnesium, for example, would never be regarded as trace elements in dealing with geological materials. We have preferred to regard a trace element as one present in a given sample at a concentration below 1000 μg/g, regardless of its average crustal abundance. With this definition, even elements such as those noted above can become trace elements when they occur as constituents of "fresh" waters or as impurities in certain minerals. The term *ultratrace* may be used to imply a concentration below 1 μg/g in the sample in question.

Although major constituents of very small samples do not fall within the definition above, some methods for their determination have been described because some of the techniques used are similar to those required for trace constituents of larger samples. It is also noted that major and minor elements in large samples are now often determined by instrumental methods after considerable dilution to bring their concentrations into the trace-element range.

A knowledge of trace-element concentrations in geological materials such as minerals, rocks, ores, soils, sediments, and fresh and saline waters, is important in a number of scientific fields. The first of these is geochemistry itself, the progress of which has always relied heavily on developments in chemical analysis. Geochemistry is primarily concerned with elucidating the behavior of the chemical elements through the natural processes occurring in the earth and in extraterrestrial bodies. The collection of data on the abundance of elements in various rock types is therefore fundamental to the development of theories of distribution and migration and can provide clues about the nature of the physical and chemical environment of the rock throughout its geological history. For many purposes the study of trace elements is more rewarding than that of macroconstituents, because their concentrations can range over several orders of magnitude, compared with perhaps a factor of three or four for the macroconstituents.

In the field of mineral exploration the development of rapid instrumental methods for multielement trace analysis has facilitated surveys of rock samples, stream sediments, soils, and vegetation on an unprecedented scale. In soil science, agriculture, and horticulture, interest in trace elements has been stimulated by the discovery of the role of several essential elements in plant and animal nutrition. It has also been possible to establish the part played by deficiencies or excesses of certain elements in contributing to disorders of plants, animals, and humans. In archaeological investigations trace analysis can help to indicate the origin of materials such as pottery fragments, metal implements and ornaments, and artifacts made from stone or volcanic glass.

Trace-element analysis is also one of the cornerstones of environmental science. Many elements have harmful environmental effects even when they are present in waters at 1 $\mu g/ml$ concentrations or in air at 1 $\mu g/m^3$ levels. Such levels may arise from natural processes or from human activities. The chemical monitoring of the environment, both to measure trace elements and to establish their chemical form, is therefore of increasing concern.

The selection of an appropriate analytical method is dictated by many factors. Methods differ widely in their sensitivity, specificity, precision, accuracy, ease of operation, speed, and cost. Compromise among these factors is largely determined by the purpose for which the data are to be used and to some extent by the nature of the sample. For example, a definitive study of the concentration of platinum metals in a meteorite calls for a method, such as neutron-activation analysis, with very high sensitivity and specificity and moderately high precision and accuracy. Speed, cost, and the technical requirements of the method are of lesser importance. On the other hand, in a survey of copper, lead, and zinc levels in stream sediments aimed at establishing a geochemical anomaly, rapid analysis of large numbers of samples, ease of operation, and low cost may be prime considerations; these considerations point to a method such as atomic absorption or direct-reading emission spectrometry. Sensitivity and precision are secondary features, and absolute accuracy may be relatively unimportant.

In many projects involving geological materials detailed statistical analysis is required once the chemical data have been obtained. Without the statistical treatment there is a danger that the observer will make qualitative interpretations colored by some unconscious bias. With the widespread availability of computing facilities, it is now possible to use some relatively sophisticated statistical techniques for interpreting geochemical data. The last part of this book is concerned with some elementary statistical parameters, such as standard deviations and correlation coefficients, and indicates how more elaborate statistical methods may be used in geochemical investigations.

CHAPTER

2

SAMPLING AND STORAGE TECHNIQUES

2.1 INTRODUCTION

Proper sampling and storage of geological samples is a vital prerequisite for the analytical determination of an element or elements in the material under study. Surprisingly, this vital step is often carried out haphazardly. In the case of homogeneous materials such as natural waters and gases, almost any size of sample will suffice, but when dealing with heterogeneous materials such as rocks, two obstacles must be surmounted before a satisfactory sample can be obtained. First, the sample selected must be representative of the *formation* under study; second, the sample size must be such that it takes into account the *grain size* of rocks from this formation.

The analyst can seldom obtain a "true" value for elemental abundances in heterogeneous samples because to do so would necessitate sampling the entire formation. For most geological units this is hardly possible. Griffiths[1,2] has defined "target" populations of interest in dealing with geological formations. The *existent population* is defined as the volume of rocks of interest existing in a specified volume of the earth's crust. In general, the existent population cannot be sampled randomly, because only parts of it are available to the field worker.

The existent population is related to the *hypothetical population*, which is the volume of rocks of interest that occurred in the specified volume of the earth's crust at some previous date. In other words it is the existent population minus additions during geological time plus losses during this period. Since the hypothetical population is difficult to deduce from the existent population, the latter is the usual target of the geologist.

The usual procedure of the field geologist is to infer the composition of the existent population by sampling the *available population*, that part of the existent population available by sampling at outcrops, quarries, roadcuts, and so on, or by drilling.

2.2 ROCKS, MINERALS, SOILS, AND LAKE AND OCEAN SEDIMENTS

2.2.1 Rocks and Minerals

Rocks and mineral samples should preferably be collected by a geologist or field worker who is familiar with the particular formation under study. Although there is a temptation to sample at outcrops, quarries, or mines, such localities are likely to provide biased samples. Quarries very often represent sites selected for providing particularly uniform and homogeneous material, whereas mines, by definition, represent areas of concentration of specific ore materials. Even outcrops may not be suitable for selection, because they may represent harder parts of the formation more resistant to weathering. In general, drilling or channeling provides satisfactory samples, although there are many cases where there is no option but to sample outcrops and other exposures of the existent population.

It is essential to take a sufficiently large sample of the material concerned. A return visit may prove to be expensive, and it is far better to leave with too much material rather than with too little.

Rocks or minerals should be selected so that all the sample is fresh and does not contain an outer layer of altered material. The minimum sample size is governed by the grain size of the rock or mineral. The size of grain can usually be determined approximately by use of a hand lens. A good general rule is to collect 1 kg of sample for every millimeter of grain size down to 5 mm, for example, 20 kg for rocks of grain size 20 mm diameter. When the grain size is below 5 mm, a proportionately greater mass of rock should be taken, with a minimum of 500 g for the finest grade formations.

Some field workers prefer to take composites of a large number of chip samples in the mistaken belief that this affords a random sample. The only merit of such a procedure is that it results in lower analytical costs, but far more information is obtained by analyzing each rock chip separately.

There are various kinds of samples used in geological investigations. The commonest of these is the *spot sample*, selected randomly. Such samples are, however, only representative of the structure as a whole if it is a homogeneous formation. Another possibility is to take a number of random spot samples and either analyze them separately or make a composite of them. One way of ensuring a certain degree of randomness is to select sites on a symmetrical pattern that is designed to suit the particular environmental conditions.[3]

A useful self-weighting method of obtaining representative material from rocks is to use *channel sampling*. The surface is first cleared of soil and vegetation, and a selected channel (10–20 cm wide) is marked with chalk across the formation in such a way as to cut across layers in the rock. The channel is then cut out to a predetermined depth by a hammer and chisel or machine drill.

One problem of the channel sample is that it generally depends on the existence of outcrops or near-surface structures. Since outcrops are usually themselves biased, this affects the reliability of the channel sample.

A better approach to random sampling involves use of auger holes or drill holes, since these are easier to select in a random manner but they are also more expensive than channel samples or spot samples. It is apparent that, in each case, questions of cost and time have to be balanced against the need to obtain a truly representative sample.

2.2.2 Soils

Soil sampling presents a different problem from rock or mineral sampling. Unlike rocks, soils usually have well-differentiated horizontal layers known as *horizons*. Elemental concentrations can vary appreciably within the soil profile. In general, there is an upper humic A horizon. Some elements are leached from the A horizon to the lower B horizon, which overlays the C horizon. The C horizon is usually the weathered layer derived from the original rock. In soil surveys (particularly in mineral exploration), it is customary to select the same horizon at each sampling site. This presupposes an adequate pedological knowledge on the part of the sampler and also presupposes that the particular soil horizon is always available. Some workers compromise by sampling at a predetermined depth at each site, but this procedure does not always result in the same horizon being sampled. A safer procedure, although not always practicable, is to dig a small pit at each site and hence select the correct horizon.

A useful tool in soil sampling is a simple soil auger made by welding a metal "T" piece on to a carpenter's wood auger from which the leading screw point has been removed. These can be used to take samples up to 2 m deep. Power augers can also be used and can operate up to 10 m deep.

As in the case of rock samples, it is essential to take samples on a random basis, and this is probably best achieved by means of a symmetrical grid pattern. Extreme care must be taken in soil sampling because the reliability of analytical data for soils can depend more on the soundness of the sampling program than on the subsequent physical treatment and chemical analysis.[4]

2.2.3 Unconsolidated Sediments

The collection of unconsolidated specimens from lakes and oceans presents technical problems and usually requires the use of specialized equipment such as bailer devices for lakes[5] or box or piston corers for ocean sediments. In ocean work there is also the difficulty of precise identification of the location of the sample site. One of the commonest sampling devices, the piston corer, depends on its own momentum to sink into the sediments but in doing so compresses the sediment profile by an amount of up to 50%.

2.2.4 Storage of Solid Samples

The storage of heterogeneous solid samples presents relatively few problems. Samples of rocks should preferably be stored in well-labelled plastic containers, but on no account should the rock sample itself be marked with paint as this can cause contamination problems during the analytical procedure. Powdered rocks or soils should be kept in stoppered plastic or glass containers. Finely powdered rock samples can oxidize rapidly in contact with air and it is therefore essential to have an airtight seal on the container. Plastic containers are preferable to glass bottles, because they are unbreakable and are much less likely to contaminate the material.

2.3 NATURAL WATERS

The sampling and storage of homogeneous samples such as natural waters present entirely different problems from the handling of rocks. The collection of representative water samples is relatively simple, but storage presents the greater problems.

The simplest way to collect surface water samples is to use a polyethylene bottle, taking care to rinse well with the sampling material. If shallow subsurface samples are needed, the polyethylene bottle can be affixed to a weighted line, lowered to the appropriate depth, and the stopper removed by an attached string or cable.

For samples at greater depths, Nansen or Niskin bottles are often used. The Nansen bottle[6] comprises a brass bottle mounted in a frame and attached at both ends to the weighted lowering cable. The bottle is inverted and thereby traps a volume of air within it. A small weight (messenger) travels down the cable and on contact with the frame releases one of the attachments. The frame then rotates through 180°, and the bottle expels the air and is filled with seawater representative of that depth. The sample is automatically sealed within the bottle. The Nansen bottle is made of brass, which can cause contamination problems. These can be overcome to some extent by lining the bottle with polyethylene or teflon. The Niskin sampler[7] is a bellows-type sampler incorporating a spring-activated metal frame and a disposable plastic container.

The Van Dorn sampler[8] is one of the simplest, cheapest, and most satisfactory devices. It consists of a perspex tube open at both ends and attached

TABLE 2.1 Adsorption of Ions from Waters to Container Walls

Element(s)	Initial concentration (μg/l)	pH	Aging period	Polypropylene	Polyethylene	Borosilicate glass	Ref.
Ag	0.5	4.5	4 days	100	20	12	11
	0.5	2.0	36 days	100	20	25	11
	Trace	8.1	60 days	—	20	2	9
	Trace	1.5	60 days	—	2	—	9
	0.3	8.1	6 months	—	—	Negligible	12
Au	0.01	8.1	21 days	—	75	—	13
Ba, Ru, Y, Zr	Trace	Various	Various	Negligible	—	Negligible	10
Cd	1.0	6.0	20 days	—	Negligible	20	11
	1.0	2.0	20 days	—	Negligible	Negligible	11
Co	Trace	8.1	6 months	—	—	Negligible	12
	Trace	8.1	20 days	—	15	7	9
	Trace	1.5	60 days	—	Negligible	—	9
Cr, Se	Trace	8.1	6 months	—	—	Negligible	12
Cs	Trace	8.1	6 months	—	—	Negligible	12
	Trace	8.1	75 days	—	Negligible	Negligible	9
	Trace	1.5	75 days	—	Negligible	—	9
Fe	Trace	8.1	55 days	—	90	70	9
	Trace	1.5	55 days	—	Negligible	—	9
Hg	50	6.0	4 hours	—	40	—	14
	50	6.0	7 days	—	100	—	14
	20,000	0	51 days	1.9	9.7	0.80	15
	20	0	51 days	12.8	13.8	2.8	15
	Trace	5.5	1 week	65	82	35	16
	Trace	1.0	2 weeks	Negligible	Negligible	Negligible	16

Element							Ref
In	Trace	8.0	20 days	—	90	20	9
	Trace	1.5	20 days	—	Negligible	—	9
Ni	100	5.0	30 days	15	15	15	11
	100	2.0	30 days	15	15	15	11
Pb	10	6.0	4 days	85	76	95	11
	10	2.0	4 days	5.0	5.0	1.0	11
	10	2.0	24 days	60	60	1.0	11
	Trace	7.4	—	50	—	—	17
	Trace	3.4	—	Negligible	—	—	17
Rb	Trace	Various	Various	Negligible	—	Negligible	10
	Trace	8.1	50 days	—	1.0	10	9
	Trace	1.5	50 days	Negligible	Negligible	—	9
Sb	Trace	Various	Various	—	Negligible	Negligible	10
	Trace	8.1	75 days	—	Negligible	Negligible	9
	Trace	1.5	75 days	—	—	—	9
	0.3	—	6 months	—	—	Negligible	12
Sc	Trace	8.1	60 days	—	70	50	9
	Trace	1.5	60 days	—	35	—	9
Sr	Trace	Various	Various	Negligible	Negligible	Negligible	10
	Trace	8.1	75 days	—	Negligible	Negligible	9
	Trace	1.5	75 days	—	—	—	9
U	Trace	8.1	50 days	—	10	20	9
	Trace	1.5	50 days	—	10	—	9
Zn	100	5.0	60 days	—	Negligible	20	11
	100	2.0	60 days	—	Negligible	Negligible	11
	10	8.1	6 months	—	—	Negligible	12
	Trace	8.1	75 days	—	Negligible	Negligible	9
	Trace	1.5	75 days	—	Negligible	—	9

vertically to the lowering cable. As the cable is lowered, the water passes smoothly through the tube. When a messenger is lowered it releases two plumber's force cups attached together by a rubber band passing through the tube. The cups then seal both ends of the tube and trap the water sample within it. As with Nansen and Niskin bottles, sampling at various depths can be carried out by a series of vessels and messengers. As each messenger activates a bottle it also releases another messenger that travels down the cable to the next container.

The storage of natural waters presents considerable problems related to adsorption of ions on to the walls of the container, or contamination of the sample by ions leached from the vessel. Much of the literature on this subject is conflicting. Robertson[9] studied the adsorption of 11 elements in seawater to various containers and found that the optimum conditions were acidification to pH 1.5 and storage in polyethylene containers. On the other hand, Eichholz et al.[10] working with different ions, concluded that borosilicate containers were preferable to those made from polyethylene. Struempler[11] tested borosilicate, polypropylene, and polyethylene containers and found that no single container type was satisfactory for all ions. The conclusions of several workers are summarized in Table 2.1.

Conflicting conclusions are reached because the adsorption of ions on containers depends largely on the previous history of the container and the manner in which it was cleaned. Glass and most plastics are supercooled liquids that have distorted and broken bonds and therefore have a greater surface energy than crystalline substances. This leads to adsorption of ions from solution and the formation of bonds between the container and the ions. There appears to be general agreement that acidification at a pH of approximately 2 will greatly reduce adsorption of nearly all ions. It also appears that polyethylene is preferable to polypropylene, because the latter often cannot be cleaned of traces of such elements as cadmium and zinc. Moreover, since deep-freezing can be used as a method of preserving natural waters for an indefinite period, the only suitable container is polyethylene because glass and polypropylene containers fracture when deep-frozen after filling with water.

In deciding which type of container to use, it is advisable to consider each case on its merits before making a decision. In some cases it may even be necessary to carry out tests beforehand. Usually however, acidification to pH 2 and storage in polyethylene containers will present few problems.

2.4 GASES

Natural gases are collected much less frequently than rocks, soils, or waters and unlike other geological materials usually require specialized treatment or equipment specific for individual cases. For this reason and because gas analysis is outside the scope of this volume, the reader is referred to Wager and Brown[18] for a listing of these specialized procedures.

2.5 BIOLOGICAL MATERIAL

The analysis of biological material is frequently required in geological studies, for example, in environmental work and in the field of biogeochemical prospecting.[19-21] Vegetation samples should be collected with due regard to the physiology of the species selected. In biogeochemical prospecting, for example, it may be necessary to select a particular plant organ and perhaps be even more selective in using only organs of a certain age.[19,21] In the field twigs or leaves should be washed thoroughly in running water and then stored in brown paper bags. On return to the laboratory, the samples should be dried for 12 hr in a drying oven at 105°C. If samples cannot be washed in the field, they should be stored in polyethylene bags and upon return to the laboratory should be washed in running tap water, rinsed in distilled water, and again stored in brown paper bags before drying at 105°C.

Animal material should be deep-frozen upon return to the laboratory and kept in that condition until required for analysis. Samples may also be freeze-dried and can then be stored at room temperature for an indefinite period.

2.6 SAMPLING FROM THE AIR

In geochemical prospecting, the cost of taking a sample is often higher than the cost of analyzing it. Therefore, there has been an increase recently in the use of helicopters to improve the productivity of sampling parties.[22-24] In such cases parties can be landed and then transported to the next site a few minutes later. In very rugged country the helicopter may be used to sample materials directly with the aid of suitable equipment or indirectly by an operator suspended in a specially designed bucket.

Limited success has been obtained in sampling stream sediments from the air. A tubular missile attached to a line is dropped into stream beds and sediments trapped within the tube are retrieved. Although the helicopter is the obvious choice for sampling from the air, some success has been obtained with fixed-wing aircraft in Alaska,[25] where soils, sediments, rock fragments and vegetation samples were collected in flight.

References

1. J. C. Griffiths, in *Sedimentary Petrography*, Vol. 1, 4th ed. (ed. H. B. Milner), Allen and Unwin, London, 1962, p. 565.

2. J. C. Griffiths, *Scientific Method in Analysis of Sediments*, McGraw-Hill, New York, 1967.
3. W. G. Cochran, *Sampling Techniques*, Wiley, New York, 1953.
4. M. G. Cline, *Soil Sci.*, **58**, 275 (1944).
5. H. E. Hawkes and J. S. Webb, *Geochemistry in Mineral Exploration*, Harper Row, New York, 1962.
6. M. Knudsen, *J. Cons. Int. Explor. Mer*, **4**, 192 (1929).
7. S. L. Niskin, *Deep Sea Res.*, **9**, 501 (1962).
8. W. G. Van Dorn, *Trans. Am. Geophys. Union*, **37**, 682 (1956).
9. D. E. Robertson, *Anal. Chim. Acta*, **42**, 533 (1968).
10. G. G. Eichholz, A. E. Nagel, and R. B. Hughes, *Anal. Chem.*, **37**, 863 (1965).
11. A. W. Struempler, *Anal. Chem.*, **42**, 2251 (1973).
12. D. F. Schutz and K. K. Turekian, *Geochim. Cosmochim. Acta*, **29**, 259 (1965).
13. R. W. Hummel, *Analyst*, **82**, 483 (1957).
14. R. V. Coyne and J. A. Collins, *Anal. Chem.*, **44**, 1093 (1972).
15. M. R. Greenwood and J. W. Clarkson, *Am. Ind. Hyg. Ass. J.*, **31**, 251 (1970).
16. Y. K. Chau and H. Saitoh, *Environ. Sci. Technol.*, **4**, 839 (1970).
17. O. T. Høgdahl, *Semi-Ann. Prog. Rep. Cent. Inst. Res. Oslo*, **7**, 1 (1965).
18. L. R. Wager and G. M. Brown, in *Methods in Geochemistry*, (eds. A. A. Smales and L. R. Wager), Interscience, New York, 1960, p. 4.
19. R. R. Brooks, *Geobotany and Biogeochemistry in Mineral Exploration*, Harper Row, New York, 1972.
20. D. P. Malyuga, *Biogeochemical Methods of Prospecting*, Consultants Bureau, New York, 1964.
21. H. V. Warren, R. E. Delavault, and J. A. C. Fortescue, *Bull. Geol. Soc. Am.*, **66**, 229 (1955).
22. W. Dyck, *Geol. Surv. Can. Pap.*, **72-42**, 1 (1973).
23. R. J. Allan, E. M. Cameron, and C. C. Durham, *Geol. Surv. Can Pap.*, **72-50**, 1 (1973).
24. A. A. Levinson, *Introduction to Exploration Geochemistry*, Applied Publishing, Calgary, 1974.
25. C. L. Sainsbury, K. J. Curry, and J. C. Hamilton, *U.S. Geol. Surv. Bull.* **1361**, 1 (1973).

CHAPTER

3

PHYSICAL METHODS OF SAMPLE PRETREATMENT

3.1 INTRODUCTION

After sampling the material the next step is physical pretreatment before analysis. In the case of rocks and minerals this usually involves mechanical fracture and subsequent breakdown by grinding to a particle size appropriate to the method of analysis that is envisaged. For natural waters the usual pretreatment step is filtration. The same may apply to gases, although these will not be discussed further as gas analysis is outside the scope of this volume.

Although reduction of rocks and minerals to a small particle size is a basic prerequisite for most methods of chemical analysis, there is surprisingly little general agreement about the best sequence of steps that should be followed. There is, however, general agreement that serious errors can arise during this stage of the treatment. For example, Ingamells et al.[1] have shown that improper physical methods of treatment (i.e., grinding, splitting, etc.) can result in errors as high as 100%, which greatly exceed the error of the analytical method. Problems and errors arising from sampling and pretreatment of rocks and minerals have also been studied by Nicholls.[2] The statistical basis of calculation of such errors has been discussed by Garrett[3,4] and Michie.[5] For a more general treatment of the problem, the reader is referred to other works.[6-9]

3.2 CRUSHING AND PULVERIZING ROCKS, ORES, AND MINERALS

A bewildering selection of automatic and manual crushers and pulverizers is commercially available with prices ranging from over $1000 to less than $20. Economics will dictate which instruments are selected. Small-scale laboratory operations can be carried out with a minimum of equipment involving perhaps only a hammer, percussion mortar, agate mortar and pestle, and a selection of nylon sieves. Large-scale operations can involve an array of automatic jaw crushers, roll crushers, ball mills, rod mills, hammer mills, vibrating sieves, and rotary sample splitters.

The rock sample should first be cleaned by washing, if necessary, and care should be taken to avoid contamination from any identifying paint marks. The specimen may then be reduced to pea size fragments by use of a hydraulic rock crusher or a heavy hammer. At this stage the reduction to fine grain size may be effected by a manually operated percussion mortar and pestle. This comprises a block, cylinder, and pestle made of hardened steel. The cylinder rests upon the block and contains the rock sample which is then struck by a hammer-operated pestle. The pestle is rotated within the cylinder during use and has an eccentric head to prevent the material caking. The contents of the mortar are sieved at frequent intervals. By use of this unit, it is usually possible to reduce the sample to about -20 mesh size (i.e., smaller than 20 mesh, see p. 17) before proceeding to the next step. The final step can be grinding to -100 or -200 mesh with an agate mortar and pestle. This can be done either manually or with an automatic unit.

If appreciable quantities of material are to be crushed, the use of a percussion mortar followed by an agate mortar is extremely laborious. Even an automatic agate mortar and pestle is too slow for most cases. By far the speediest method of crushing is to use one of the modern high-speed pulverizers, which can reduce pea-size rock samples to -200 mesh powder in approximately 1 or 2 minutes. These pulverizers contain disks or rings of hardened chrome steel and are not suitable if trace quantities of chromium or cobalt (which are components of the steel) are to be measured in the sample. Tungsten carbide rings, however, can be used instead.

A good compromise between manual and large-scale automatic operations is to use a hammer mill for large numbers of small samples. These mills are quite inexpensive and can reduce 10 g of rock to -60 mesh size in a few minutes.

Throughout the crushing and grinding stages there is a serious risk of contamination from the apparatus used. This contamination arises mainly from iron and is not significant if iron is a macroconstituent but it can be serious if iron is to be determined at low concentrations. Some workers[9] recommend that iron contamination should be removed with a magnet, but this is a dangerous procedure as the operation is equally likely to remove magnetite or other highly magnetic minerals.

Problems can also arise from cross-contamination of equipment. A classical example of this is shown by Jeffery[7] who ground 5 g of a tungsten mineral to 150 mesh size with a mechanical mortar and pestle. The material was discarded and the equipment cleaned and dried. Eleven separate 5-g amounts of quartz were then ground in the same unit, which was cleaned between each operation. Each sample was then analyzed for tungsten. The data are summarized in Table 3.1. Even after eleven separate grindings, the quartz still contained 3.1 μg/g tungsten instead of its original content of

TABLE 3.1 Contamination of Quartz after Grinding in an Agate Mortar

Material	W content (μg/g)
Unground quartz	0.7
First ground quartz sample	1300
Second	61
Third	31
Fourth	22
Fifth	16
Sixth	7.4
Seventh	3.8
Eighth	2.9
Ninth	3.1
Tenth	2.5
Eleventh	3.1

Source: Jeffery.[7]

0.7 μg/g. The experiment clearly demonstrated the necessity of keeping separate equipment for grinding ores and not using it for ordinary rocks.

Although the "memory effect" of previously ground material is a serious problem for agate mortars and pestles, there can also be contamination due to loss of the agate itself. For example, when 200 g of quartz sand was ground with an agate mortar and pestle for a period of 19.5 hr, it was found that the loss of weight of the mortar was 0.189 g for 635 g (0.03%) and for the pestle was 0.102 g for 268 g (0.04%).[9] Although there can be contamination from agate mortars and pestles, this material is still perhaps the most suitable for small-scale final grinding of rocks, ores, and minerals.

As a general rule it is not advisable to grind rock material to a mesh size finer than that necessary for the analytical method that is contemplated. This is because the finer the mesh size, the greater the risk of contamination from the grinder (because of the longer grinding times), the greater the chance of oxidation of the material (e.g., Fe^{2+} to Fe^{3+}), and the greater the chance of loss or gain of water. The importance of mesh size cannot be overemphasized. For example, Mazzucchelli[10] has shown that the great nickel deposits of Western Australia remained undiscovered for a long time because geologists has previously chosen fine mesh material for analysis. The nickel is concentrated in the ferruginous soil fragments coarser than 80 mesh size.

Much of the contamination from grinding and crushing apparatus can be avoided by a procedure recommended by Hawley and MacDonald.[11] Coarse fragments of a silicate rock are heated to 600°C in a muffle furnace and quenched in distilled water. The material then is pulverized easily with an agate mortar and pestle. This procedure should not be used if readily volatile or oxidizable elements such as arsenic, cadmium, ferrous iron, lead, or selenium are to be determined.

Certain flakey minerals, such as mica present problems during grinding. There is a tendency for them to accumulate in the latter stages of grinding and pulverization, and for this reason it is very important that *all* the rock specimen should be ground to the final mesh size. If any residual material is thrown away, it will almost certainly result in erroneous data for some elements. An example of this is thallium, which is found almost exclusively in the biotite fraction of rocks such as granite and pegmatites. Micaceous samples may also be pulverized by use of a blender with swiftly rotating (15,000 rpm) arms.[12]

3.3 CRUSHING OF SOILS

In the reduction of soils to fine particles a somewhat different procedure is required from that used for rocks, because the whole of the sample is not required for analysis. It is usual to analyze the so-called *fine soil*, which may be defined as material of less than 2 mm diameter. Rocks and nodules are usually removed. The standard procedure is to air dry the soils, disintegrate the sample gently with a hardwood or porcelain mortar and pestle, pass the material through a 2-mm sieve, and then further grind the fine material to the required mesh size. Unlike rocks, soils usually contain a large organic fraction that may have to be removed by ignition at 500–600°C when certain analytical procedures (such as emission spectrography) are to follow.

3.4 SIEVING

Fractionation of material by sieving is an important operation in sample treatment. Sizing assists in preparation of a homogeneous sample and speeds up the crushing and pulverization processes, if finer material is removed continuously. A great deal of confusion exists about nomenclature of sieve sizes because of the various standards that exist side by side. Most sieves conform to the "U.S.," "Tyler," or "British" standards and have designated "mesh" sizes. In all three series "mesh" is defined as the number of wire meshes per lineal inch. Since a mesh can be defined as the diameter of one wire plus the width of the opening, it is obvious that a given mesh number can have different openings if different wire thicknesses are used. Herein lies the difference in the three series, as summarized in Table 3.2.

The German "DIN" series is included in this table for the sake of completeness, but this standard is related to the number of meshes per square meter

TABLE 3.2 Standard Sieve Sizes

Mesh (or DIN)	Opening width (μm)			
	Tyler series	U.S. series	British series	German DIN
5	4000	4000	3353	1200
10	1680	2000	1676	600
12	1410	1680	1405	490
16	1000	1190	1003	385
20	840	840	—	300
25	—	710	599	—
35	420	500	—	—
60	250	250	251	102
100	149	149	152	60
170	88	88	89	—
200	74	74	76	—
300	—	—	53	—
400	37	37	—	—

and is therefore not comparable with the others. It is obvious from the table that for mesh numbers greater than 35, the three main series are almost identical and are freely interchangeable. In general therefore it is seldom necessary to be concerned about which standard is involved when mesh sizes are quoted in the literature.

Although metal sieves with steel, brass, or bronze meshes are frequently used for sizing samples, serious contamination from elements such as copper, iron, and zinc can result. It is much safer to use sieves composed of nylon mesh set in perspex.

The problem often arises as to the mesh size to which the sample should be reduced. There is no universal criterion in this case because the required mesh size will depend largely on the analytical procedure that is to follow. Methods involving nondestructive analysis of powders usually require much finer material than methods requiring a dissolution of the sample. For example, fluorescent X-ray spectrography usually requires a powder of 300–400 mesh size, whereas for chemical attack, -100 mesh (i.e., smaller than 100 mesh) is usually satisfactory.

When different rocks have been ground to pass a sieve of a predetermined mesh size, there will still be large differences in the size of the various screened particles, and this will be mainly a function of the rock type. This is illustrated in Table 3.3, which shows screen analysis of three rock types sieved to pass a 72 mesh sieve.

TABLE 3.3 Screen Analysis of Three Rock Types Ground to Pass a 72-Mesh Sieve

Mesh (British Std.)	Percent retained		
	Quartzite	Granite	Diorite
85	0.06	1.92	0.03
100	20.64	27.27	21.53
120	12.60	10.95	11.02
150	12.00	11.65	10.69
170	6.16	6.99	7.16
200	4.22	6.26	13.15
240	8.68	10.17	17.91
300	3.53	4.23	5.22
350	2.64	2.57	1.23
−350	29.47	17.99	11.86

Source: Jeffery.[7]

3.5 SAMPLE SPLITTING

In theory, once a rock specimen has been reduced to a fine mesh size, it should only be necessary to shake the material in a large bottle to obtain a perfectly homogeneous sample. Certainly prolonged shaking is necessary since there is a continuing alteration of the composition of successive sievings of the fine material during the final grinding and sieving operations. In practice, shaking in a bottle may introduce a set of new problems due to the different densities of the components of the sample. There is a tendency for the heavier constituents to settle at the bottom of the container. The finer the material, the less is the danger that it will be segregated. Although some workers[13] have suggested that, provided the rock is ground to −200 mesh, there will be no segregation whatsoever, it is unwise to assume that this is necessarily the case, and it is advisable to use some suitable procedure for ensuring homogeneity of the sample.

The classical method of achieving sample homogeneity is the so-called "coning and quartering" technique. The sample is poured out in a thin stream to form a cone overlying a cardboard frame that divides it into four quarters. Two diagonally opposite quarters are selected, combined and again subjected to coning and quartering until the required sample size is obtained. Another method is the use of a laboratory riffle.[14] The sample is fed into a hopper, which divides it into a dozen or so separate streams via a system of chutes. Alternate streams are recombined to give two halves collected in trays at the bottom of the instrument. The material can be progressively halved until the required sample size is obtained. A rather more sophisticated instrument is a rotary-type sample splitter. A constant stream of fine powder falls vertically into a series of collecting vessels, which rotate around the circumference of a disk below the hopper.

Although the above instruments are inexpensive and readily available, some analysts content themselves with merely shaking the sample in a bottle and ignore the danger of segregation of the constituents. Unless the particle size is large (<50 mesh) the risk of segregation is probably small and the resultant error may well be lower than the subsequent analytical error. Nevertheless, sample splitting is a speedy and inexpensive operation and should be used whenever possible.

3.6 SEPARATION OF MINERALS

3.6.1 Preliminary Treatment

Separation of minerals from a rock requires specialized treatment. Before commencing any of the usual separation procedures, the sample should first be examined microscopically, either in thin section or with a reflecting microscope. The purpose of this step is to establish the approximate proportion of the minerals that are to be investigated, and to establish the presence of any constituents likely to interfere in the separation stage. The larger minerals can usually be hand picked with the aid of a magnifying glass or low-powered microscope, but the smaller-sized minerals (<1 mm) require specialized techniques.

As a general rule, the rock should be ground to the size of interfering inclusions in the mineral to be studied. For example, 5-mm felspars with 1-mm inclusions should be ground to at least -20 mesh. It is essential not to grind to too fine a mesh size, otherwise there will be a large amount of "fines" which may seriously impede subsequent physical methods of separation.

Because losses will be high, it is important to crush a sufficiently large amount of rock. For example less rock will be needed to obtain easily separable magnetite than for minerals such as felspars, which are difficult to recover.

As a general rule, rocks should be crushed to -80 to $+200$ mesh. The material should then be washed vigorously in tap water to remove fines, rinsed in distilled water, rinsed in acetone, and then dried. Sometimes chemical methods may be needed (for example, the use of dilute hydrochloric acid to remove limonitic staining). Such methods should however be avoided if possible as it is extremely difficult to be selective when chemical methods are used.

3.6.2 Magnetic Separations

Magnetic methods of separation are usually quite efficient since most minerals differ in their magnetic susceptibilities. A simple bar magnet may be used to remove the most magnetic constituents in a mixture (magnetite, titanoferrite, pyrrhotite), but is not very selective insofar as other constituents are removed along with the highly magnetic minerals.

The isodynamic type of electromagnetic separator is a highly efficient instrument and is particularly effective if the sample can be sized to grains of approximately equal size. The material is fed via a hopper to a vibrating chute that is inclined both downward and sideways. The material passes through the field of a powerful electromagnet and the particles are balanced by the magnetic field acting against the gravitational force. These two forces can be altered by varying the current and the sideways angle of the chute. The particles passing along the chute are divided into two streams depending on their magnetic susceptibilities and are collected in containers affixed to the end of the chute. Magnetite is removed before the material is fed into the separator as it tends to cause clogging. The effectiveness of the isodynamic separator has been demonstrated for separation of sulfide minerals.[15] The instrumental parameters have been summarized by Hess.[16] The main advantage of the isodynamic separator is that it can be used to separate not only the moderately magnetic minerals, but can also be used for weakly magnetic materials and even for diamagnetic substances that are repelled by a magnetic field. The usual tilt of the instrument is away from the strongly magnetic side, so that the more magnetic materials are pulled up-slope. When separating diamagnetic materials it is also possible to reverse the direction of tilt so that the diamagnetic grains are moved up-slope.

3.6.3 Electrostatic Separations

Electrostatic separators depend on the principle of inducing an electrostatic charge upon a mineral. The extent of this charge is governed by the conductivity of the material. Examples of good conductors are native metals, graphite, and sulfides. Poor conductors are represented by such minerals as quartz, calcite, barytes, and fluorspar.

The sand or finely crushed rock or ore is thoroughly dried and is dropped on to an electrically charged rotating iron cylinder. The charged minerals are then repelled from the iron cylinder to different degrees. The various streams of particles are then collected at different distances from the cylinder and represent separate minerals of varying degrees of purity.

3.6.4 Separation by Planning

The prospector's gold pan has for centuries been one of the most frequently used methods of mineral separation (see Chapter 17). Its use of course has been restricted mainly to stream sediments but not necessarily only for gold prospecting. Most heavy minerals can be separated by use of the gold pan, and although it has often been considered as being a qualitative rather than quantitative tool, an investigation by the U.S. Geological Survey[17] has shown that quite high recoveries could be obtained. For example a single panning of the -120 mesh fraction of a stream sediment sample gave a recovery of 100% for garnet, 70% for ilmenite, 75% for monazite, 20% for sillimanite, and 100% for zircon. In general, the efficiency of extraction is a function of specific gravity. For example 90% of minerals of density greater than 5.2 g/cm^3 and only 1% of those of density below 2.8 g/cm^3 can be separated via panning. The semiquantitative data afforded by the gold pan have also been used extensively for geochemical exploration in France,[18] and have resulted in important mineral discoveries.

In the laboratory, the gold pan can be replaced by a "superpanner,"[19] or "micropanner."[20] In these panning devices, the material is placed at the head of a sloping surface and water (or other liquids) flows over it. The minerals are fractionated with the lighter material passing down the slope faster than the heavier minerals. The surface is shaken and rocked during operation.

3.6.5 Separation by Heavy Liquids

Minerals may be separated by allowing them to sink or float in a dense liquid. The process is often assisted by centrifugation. The separating medium may be a liquid at room temperatures or may be a melt of a solid with relatively low melting point ($<100°C$). Table 3.4 lists a number of such compounds which have been used for mineral separations (for a fuller list see Refs. 21 and 22).

Ideally, a separating medium should be inexpensive, easily available, transparent, liquid at room temperatures, noncorrosive, chemically inert, odorless, nonviscous, and easily concentrated or diluted. Perhaps the reagent that meets these criteria most closely is bromoform. Unfortunately it is still relatively expensive ($12/kg) and is only dense enough to remove the lighter minerals such as quartz and felspar. However, this usually is all that is required and bromoform is therefore the most common heavy liquid used in mineral separation. Bromoform may be diluted with ethanol or acetone and is easily recovered by shaking the mixture with water. The ethanol and acetone

TABLE 3.4 Properties of Heavy Liquids and Fused Salts for Mineral Separations

Compound	Formula	Density (15–25°C or at m.p.)	m.p. (°C)	Cost ($US) kg	Cost ($US) liter	Diluent	Comments
Acetylene tetrabromide	$C_2H_2Br_4$	2.96	0.1	35	103	Chloroform	—
Antimony tribromide	$SbBr_3$	4.14	94.2	74	306	Water	—
Antimony trichloride	$SbCl_3$	3.06	73.4	17	52	Water	Decomposes easily very poisonous
Barium mercuric iodide	$BaI_2 \cdot HgI_2$	3.50	Aq. soln.	35	122	Water	Decomposes slightly
Bromoform	$CHBr_3$	2.87	9.0	12	34	Acetone	Poisonous
Cadmium borotungstate (Clerici's solution)	$Cd_5(BW_{12}O_{40})_2 \cdot 18H_2O$	3.36	Aq. soln.	35	117	Water	(remove carbonates)
Carbon tetrabromide	CBr_4	3.42	48.4	130	444	Chloroform	—
Carbon tetrachloride	CCl_4	1.60	−23.0	2	3	Chloroform	—
Lead tetrachloride	$PbCl_4$	3.18	−15.0	100	318	Chloroform	Explodes at 100°C
Mercuric nitrate	$Hg(NO_3)_2$	3.40	79.0	55	187	Water	—
Mercuric potassium iodide	K_2HgI_4	3.18	Aq. soln.	39	124	Water	Corrosive
Mercurous nitrate	$Hg_2(NO_3)_2$	4.30	70.0	55	236	Water	—
Methylene iodide	CH_2I_2	3.32	5.0	70	232	Benzene	Light sensitive
Stannic bromide	$SnBr_4$	3.35	30.0	50	167	Acetone	—
Stannic chloride	$SnCl_4$	2.27	−33.0	13	30	Water	Corrosive and very poisonous
Thallous formate	$TlCOOH$	3.39	94.0	105	355	Water	Corrosive and very poisonous
Thallous formate + thallous malonate	$TlCOOH + CH_2(COOTl)_2$	4.25	Aq. soln.	90	382	Water	

dissolve in the water and the bromoform may be recovered by decanting off the aqueous layer and filtering the lower layer through a filter paper soaked in bromoform.

The most generally useful heavy liquids and their densities in grams per milliliter are bromoform, 2.87; methylene iodide, 3.32; and Clerici's solution, 4.25. Although comprehensive discussion of methods of separation is outside the scope of this work (for further details see Refs. 23 and 24), mention will be made of a few techniques.

Figure 3.1 illustrates a simple centrifuge tube (*a*) for separating heavy minerals. The sample and liquid are centrifuged in the bulb-shaped tube, and the constriction is then sealed with the stopper. The upper layer can now be removed by simple inversion followed by subsequent removal of the lower layer after removal of the stopper.

One of the simplest devices for mineral separation is the Fraser[25] tube (*b*). The liquid is first placed in the tube held in position A with the wider end vertical. The sample is now added through this wide end, and the lighter minerals rise to the top and collect at the liquid surface under the stopper. The tube is inverted to position B and the heavy minerals are thereby rejected and collected on a watch glass or microscope slide.

Another simple device is the Spaeth sedimentation glass[26] also illustrated in Fig. 3.1. The conical vessel (*c*) contains the sample and heavy liquid. The heavy minerals concentrate inside a cup built into the stopcock. Turning of the stopcock isolates the heavy minerals from the system.

If fused salts are to be employed for mineral separations, a slightly different procedure is required. The samples and salt are fused in a small test tube held on a bunsen burner. After agitation, the system is allowed to cool and the tube is broken. The cylindrical melt is then cut into appropriate sections and dissolved in some suitable solvent.

Because of the very high cost of nearly all heavy liquids and fusible salts, most operations have to be carried out on a small scale. When larger amounts of materials are involved, it is possible to use a three-stage process beginning with preliminary concentration by panning, removal of lighter minerals such as quartz and felspar by use of bromoform, and then a final separation with the requisite heavy liquid or salt. The reader is referred to standard texts[21-24] for further details of laboratory procedures for separating minerals with liquids or fused salts.

3.6.6 Other Methods of Separation

There are several other methods of separating minerals involving modification and combination of the basic procedures outlined above. The reader is referred to Krumbein and Pettijohn[24] for fuller details. Mention may also

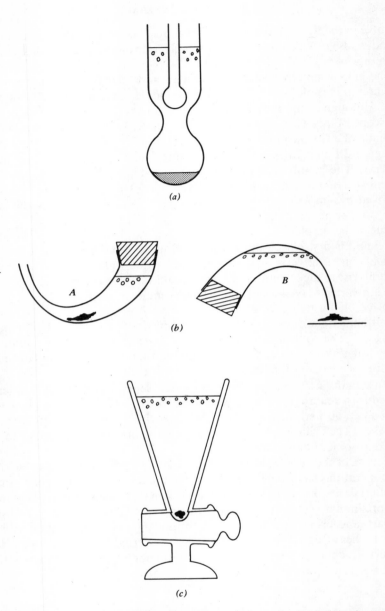

Fig. 3.1. Simple apparatus for mineral separations using dense liquids: (*a*) Simple centrifuge tube; (*b*) Fraser tube[25]; (*c*) Spaeth sedimentation glass.[26]

3.7 FILTERING OF NATURAL WATERS

If natural waters are analyzed without filtering, erroneous results almost certainly occur. Nearly all natural waters contain suspended or colloidal material. Suspended inorganic matter results from rock weathering and organic detritus results from the decay of organisms. This suspended matter can cause a dilemma for a hydrogeochemist involved in mineral exploration, because filtered material may be composed of colloids or clays upon which economic metals may be adsorbed,[27] and important anomalies may be missed if the water is filtered. The ideal procedure would therefore seem to be to analyze both the filtered water and the material suspended in it.

An ideal filter should have a reproducible and uniform pore size, the filtration rate should be high, the suspended matter should be retained on the surface of the filter so that it may be easily removed if needed for study, the filter should not adsorb trace elements, it should have a low ash content to prevent contamination if the filter is analyzed, and it should have no loose fibers to contaminate the filtrate.

The above conditions are most easily met by the use of commercial "millipore" microfilters that have pore sizes from 10 nm to 5 μm. On ashing, the residue is only about 0.0001% of the original mass. The filters are used with suction apparatus and are supported on Buchner funnels. A suitable pore size for most natural waters is 0.45 μm. The importance of filtration is shown by the data in Table 3.5,[28] which gives values for the concentration of four elements in soluble and particulate (>0.45 μm) form in seawater.

TABLE 3.5 Concentrations (μg/l) of Trace Elements in Soluble S) and Particulate (P) Form in Seawater Collected off the Los Angeles Basin

Sample No.	Copper		Iron		Lead		Zinc	
	S	P	S	P	S	P	S	P
1	3.0	1.3	11.7	32.6	5.2	13.0	0.8	2.1
2	2.8	0.2	10.8	3.8	3.8	1.4	13.8	0.3
3	1.3	0.3	1.3	5.0	2.6	1.6	3.4	0.5
4	1.7	0.5	0.6	1.3	0.2	0.4	0.4	0.7
5	6.6	1.6	3.2	37.4	1.8	4.0	28.2	13.0
6	4.0	0.9	1.8	6.5	4.2	1.7	16.4	2.0
7	4.8	0.6	2.5	11.0	4.0	0.7	22.0	2.2
8	1.9	0.6	0.9	4.4	2.3	1.8	4.8	3.0
9	2.6	6.4	1.1	4.4	1.2	2.6	6.4	1.6

Source: Brooks et al.[28]

3.8 OTHER METHODS OF SAMPLE PRETREATMENT

A number of other methods of physical pretreatment may be used for specific cases. These include special methods for soils and clays, drilling techniques for meteorites and other rocks whose mineral separation is difficult, and separation of minerals on the basis of their dielectric constants. The reader is referred to standard texts[6,9,21,23,24] for further discussion of such procedures.

3.9 ERRORS IN SAMPLING SOLID GEOLOGICAL MATERIALS FOR TRACE-ELEMENT ANALYSIS

There are several potential sources of error in the sampling of solid geological materials for trace analysis. The first, noted in Chapter 2, is associated with selection of a representative sample of the material from its original location. Errors may then be introduced during reduction of the sample to a finely divided powder, stored in smaller portions for subsequent analysis. At this stage errors can be introduced by contamination from grinding equipment and by insufficient homogenization during procedures such as riffling or coning and quartering.

Even if the sample is reduced to a very fine powder (e.g., -200 mesh) without contamination, sampling errors can arise during storage and handling. For example, differentiation of the sample can occur, with minerals of different density becoming stratified. Concern has been expressed about the possible inhomogeneity of the U.S. Geological Survey standard rock G-1 (p. 76) as a result of surveys of lead and tungsten analyses.[29,30] Vistelius[31] concluded that some stratification of the constituents of the standard rocks G-1 and W-1 occurred before they were subdivided and samples issued to various laboratories, but Chayes[32] has asserted that the observed dispersion of the results, at least for major constituents, was probably due to systematic interanalyst bias. The G-1, W-1 program was not, in fact, designed to allow the various sources of error to be identified unequivocally.

The final, and sometimes most serious, sampling error can occur at the stage when a small subsample is selected for analysis. The problem arises from the nonuniform distribution of some trace elements among the various minerals of a powdered rock sample. Wilson[33] observed that trace elements not associated with major elements, but forming characteristic minerals, are subject to a much greater sampling error than those contained in major mineral species. The effect of sample inhomogeneity on this scale is obviously more serious in analytical methods, such as emission spectrography, that use very small samples.

Where a trace element exists largely in "special grains" of low abundance in the rock, there is a risk that these grains will be missed completely if the sample taken for analysis is too small. A situation of this kind arises in the case of chromium in G-1. A large part of the chromium is contained in chromite grains[34] and only about one in twenty 10-mg subsamples of 170-mesh G-1 contains one of these grains. The background value for chromium in G-1 is about 10 μg/g, but a 10-mg subsample containing one chromite grain would show about 200 μg/g. An analyst would probably reject such an isolated high value as erroneous, but in fact the frequency of such high values constitutes valuable information about the inhomogeneity of the sample. A set of determinations from which one or two high values are rejected therefore may show high precision, but the mean of the concordant results may be very inaccurate because of the sampling error.

Ingamells and Schwitzer[34] have proposed use of a *sampling constant* for any given element in a given geological sample in a given state of subdivision. The sampling constant K_s has the units of mass and is defined as the subsample mass necessary to ensure a relative standard deviation R of 1% in a single determination. Its square root is numerically equal to the expected R value for results obtained for 1-g subsamples in a procedure that is *free from analytical error*.

The relationship can be expressed as follows:

$$K_s = R^2 w$$

where w is the subsample mass in grams. For chromium in 80-mesh G-1, K_s was estimated to be about 2000. In this case the use of a 1-g subsample (which contains on average only about five grains of chromite) gives an R value of approximately 45%, and even the use of a 20-g subsample leads to an R value of approximately 10%.

Sampling constants can be estimated from replicate determinations of an element if an independent estimate of analytical error can be made and subtracted from the total error. It has been proposed that sampling constants be calculated for all constituents of a number of reference samples as part of the certification of such materials.[34]

For further discussion of this topic the reader is referred to the work of Wilson[33] and of Ingamells and Schwitzer.[34] It should be clear from the above that in reporting trace-element analyses of nonuniform geological materials, it is important for an analyst to state the subsample mass and to take great caution in rejecting occasional anomalously high values.

References

1. C. O. Ingamells, J. C. Engels, and P. Schwitzer, *Proceedings of the 24th International Geological Congress, Montreal*, Sec. B, (1972), p. 405.

2. G. D. Nicholls, *Trans. Inst. Min. Metall.*, Sec. B, **80**, B299 (1971).
3. R. G. Garrett, *Econ. Geol.*, **64**, 568 (1969).
4. R. G. Garrett, *Econ. Geol.*, **68**, 282 (1973).
5. U. McL. Michie, *Econ. Geol.*, **68**, 281 (1973).
6. L. R. Wager and G. M. Brown, in *Methods in Geochemistry* (eds. A. Smales and L. R. Wager), Interscience, New York, 1960, p. 4.
7. P. G. Jeffery, *Chemical Methods of Rock Analysis*, Pergamon, Oxford, 1970.
8. A. A. Levinson, *Introduction to Exploration Geochemistry*, Applied Publishing, Calgary, 1974.
9. W. F. Hillebrand and G. E. F. Lundell, *Applied Inorganic Analysis*, Wiley, New York, 1953.
10. R. H. Mazzucchelli, *J. Geochem. Explor.*, **1**, 103 (1972).
11. J. E. Hawley and G. MacDonald, *Geochim. Cosmochim. Acta*, **10**, 197 (1956).
12. S. Abbey and J. A. Maxwell, *Chem. Can.*, **12**, 37 (1960).
13. G. F. Smith, L. V. Hardy, and E. L. Gard, *Ind. Eng. Chem., Anal. Ed.*, **1**, 228 (1929).
14. H. H. Willard and H. Diehl, *Advanced Quantitative Analysis*, Van Nostrand, New York, 1943.
15. A. M. Gaudin and H. R. Spedden, *Miner. Technol.*, 563 (Jan. 1943).
16. H. H. Hess, Instrument Instruction Booklet of S. G. Frantz Co. Inc., Trenton, N.J., 1966.
17. P. K. Theobald Jr., *U.S. Geol. Surv. Bull.*, **1071-A**, 1 (1957).
18. J. Guigues and P. Devismes, *Mem. BRGM.*, **71**, 1 (1969).
19. H. E. T. Haultain, *Trans. Can. Inst. Min. Metall.*, **40**, 229 (1937).
20. L. D. Muller, *Trans. Inst. Min. Metall. Lond.*, Part 1, **68**, 1 (1958).
21. H. B. Milner, *Sedimentary Petrography*, Vol. 1, Allen and Unwin, London, 1962.
22. J. D. Sullivan, *U.S. Bur. Mines. Tech. Pap.*, **381**, 5 (1927).
23. G. Müller, *Methods in Sedimentary Petrology*, E. Schweitzerbartsche Verlagsbuchhandlung, Stuttgart, 1967.
24. W. C. Krumbein and F. J. Pettijohn, *Manual of Sedimentary Petrography*, Appleton-Century-Crofts, New York, 1966.
25. F. J. Fraser, *Econ. Geol.*, **23**, 99 (1928).
26. A. O. Woodford, *Econ. Geol.*, **20**, 103 (1925).
27. J. D. Hem, *U.S. Geol. Surv. Water Supply Pap.*, **1473**, 1 (1970).
28. R. R. Brooks, B. J. Presley, and I. R. Kaplan, *Talanta*, **14**, 809 (1967).
29. F. J. Flanagan, *U.S. Geol. Surv. Bull.*, **1113**, 113 (1960).
30. D. H. F. Atkins and A. A. Smales, *Anal. Chim. Acta*, **22**, 462 (1960).
31. A. B. Vistelius, *Math. Geol.*, **2**, 1 (1970).
32. F. Chayes, *Math. Geol.*, **2**, 207 (1970).
33. A. D. Wilson, *Analyst*, **89**, 18 (1964).
34. C. O. Ingamells and P. Schwitzer, *Talanta*, **20**, 547 (1973).

CHAPTER

4

CHEMICAL METHODS OF SAMPLE PRETREATMENT

4.1 DECOMPOSITION OF SOLID SAMPLES

4.1.1 Introduction

Many methods of analysis, including flame emission and atomic absorption spectrophotometry, solution absorptiometry and fluorescence, and the various electrochemical methods, generally require the sample to be in the form of a solution. In only a few cases can suitable solutions containing the required elements be prepared by extracting the sample with a solvent (e.g., carbon disulfide or pyridine for elemental sulfur, water for various ionic traces in specimens of minerals such as sylvite, rock salt and nitre, and various aqueous extractant solutions for the removal of "exchangeable" cations from soil samples). More often it is necessary to attack the sample by more drastic methods such as decomposition by aqueous acid mixtures or fusion with one of a variety of alkali fusion mixtures.

The method used in any given case is determined by such factors as the nature of the sample, the technique to be employed for the analysis itself, and the use to be made of the analytical data. The monograph by Dolezal et al.[1] is a valuable source of information on inorganic decomposition techniques, and the review by Tölg[2] contains much useful material on sample preparation, with particular attention to the requirements of extreme trace analysis, that is, determination of constituents at concentrations near one part in 10^9.

4.1.2 Acid Dissolution

Many minerals are soluble in inorganic acids or mixtures of acids. Complete dissolution of the sample is not always necessary, or even desirable. For example, in geochemical exploration it may be sufficient to dissolve the traces of metals that have been introduced into the rocks by hydrothermal or other genetic processes. The background amounts of such metals in the silicate lattice may be of little significance for this purpose. Similarly, in the analysis of carbonatites, decomposition of carbonate minerals alone may be required, with siliceous matter remaining unattacked.

Digestion of solids with acids has the advantage of introducing only the minimum amount of foreign cations, in the form of impurities in the acids used. In the analysis of anionic constituents great care must be taken in the choice of the digesting acid; this must be more volatile than the conjugate acid of the anion being determined. The acids or mixtures most commonly used for various purposes are noted below.

Hydrochloric Acid

Dilute hydrochloric acid, either cold or hot, is satisfactory for the decomposition of most carbonates and carbonatites and is often useful for the analysis of the carbonate fractions of more complex materials. (Carbonate minerals of the scapolite family are not significantly attacked and should be considered as aluminosilicates for the purpose of dissolution.) Some oxide and hydroxide ores, such as those of iron and the alkaline earth metals, are also attacked by hydrochloric acid, as are some silicates (e.g., wollastonite) and some sulfides. The addition of oxidizing agents such as bromine or perchloric acid renders oxide ores such as those of copper, zinc, lead, uranium, and molybdenum, susceptible to attack in a hydrochloric acid medium.

Nitric Acid

Nitric acid solutions are effective in attacking ores of metals such as cobalt, copper, manganese, nickel, and lead. Carbonates and sulfides are decomposed by concentrated nitric acid, and boiling nitric acid has been recommended for leaching sulfides of elements such as lead and zinc from a silicate matrix. Hydrochloric acid is sometimes added to nitric acid to assist the decomposition of carbonates, and the dissolution of sulfides may be facilitated by making the medium more oxidizing, for example by addition of bromine to nitric acid, or by the use of aqua regia (three parts concentrated hydrochloric acid to one part concentrated nitric acid).

Hydrofluoric Acid

The decomposition of materials containing silicon as a major constituent (e.g., silicate rocks and minerals, lavas, and volcanic glasses) is often achieved with hydrofluoric acid, either alone or in conjunction with nitric, perchloric or sulfuric acids. The ability of hydrofluoric acid alone to decompose a wide variety of silicate minerals has been studied by Langmyhr and Sveen.[3] Many silicates are satisfactorily decomposed at 85 to 100°C. For more resistant minerals, attacked at temperatures above the boiling point of the hydrofluoric acid/water azeotrope (38 % HF, b.p. 112°C), use can be made of a sealed bomb lined with teflon, platinum, or other resistant material.[3-7] The powdered sample is usually moistened with water or aqua regia[7] before the addition of hydrofluoric acid. The bomb technique is useful when it is desired to retain all the silica in solution as part of a comprehensive silicate analysis scheme. Any precipitated fluorides remaining after the digestion can be taken into solution by adding boric acid.

If silicon itself is not being determined, it is often satisfactory to use an open vessel of platinum, teflon, or polypropylene for digestion by hydrofluoric acid or a hydrofluoric/nitric acid mixture. The presence of nitric acid helps to decompose small amounts of carbonate, to oxidize any organic matter and sulfides, and to convert several elements into higher valence states. Addition of a less volatile acid such as perchloric or sulfuric acid, ensures virtually complete removal of the silicon and fluorine from solution as the mixture is evaporated. The residue can be dissolved in water or dilute acid (usually hydrochloric acid). Although perchloric acid is a little less effective than sulfuric acid in the removal of the residual fluorine,[8] the evaporation, crystallization, and dissolution of residue usually proceed more smoothly when perchloric acid is used. A further advantage of perchloric acid is the smaller likelihood of interferences when flame spectrometric methods of analysis are used.

Elements that can be taken into solution following hydrofluoric acid attack include the alkali and alkaline earth metals, iron, aluminum, manganese, titanium, and most of the transition elements that commonly occur in trace amounts in silicate rocks. Some minerals, such as zircon, rutile, and corundum, are relatively resistant to attack by hydrofluoric acid media and may remain as an insoluble residue unless the bomb technique is used. Where the solution contains sulfuric acid, some sparingly soluble sulfates, such as barium sulfate, may also remain.[9]

The decomposition of sulfide ores and minerals can also be carried out in a bomb lined with teflon,[10] using a mixture of hydrofluoric, nitric, and hydrochloric acids. This procedure is appropriate when a complete digestion of both sulfide species and gangue is required.

4.1.3 Fusion Mixtures

An alternative general technique for dissolving solid geological samples involves taking the powdered sample, mixing it with approximately 5 to 20 times its mass of an inorganic fusion reagent and heating the mixture for 10 to 120 min to form a homogeneous melt. After cooling, the solidified melt is extracted with water or acid. Widely used reagents include alkali metal carbonates, hydroxides, peroxides, bisulfates and pyrosulfates, nitrates, fluorides and borates. Melting points of the compounds used range from about 200°C ($NaHSO_4$, $KHSO_4$) to almost 1000°C (NaF). Many mixtures (e.g., sodium and potassium carbonates) have lower melting points than either of the pure components.

Bisulfates and Pyrosulfates

Acidic fusion reagents include the bisulfates and pyrosulfates of sodium and potassium. When bisulfate is used, conversion to pyrosulfate occurs at an early stage of the heating

$$2KHSO_4 \longrightarrow K_2S_2O_7 + H_2O$$

and the liberation of water vapor may be accompanied by "spitting" of the mixture. There is little attack on metal oxides until the conversion into pyrosulfate has occurred.

Silicate minerals are not significantly decomposed by pyrosulfate fusion. This treatment is best reserved for certain minerals such as rutile and ilmenite, which are resistant to attack by hydrofluoric acid at low temperatures and pressures, and which in some cases form a significant part of the residue from digestion of silicate rocks with that reagent. Various other silicates (e.g., zircon, tourmaline) and oxides (e.g., chromite, cassiterite) which may remain after a conventional hydrofluoric acid attack are more satisfactorily fused with sodium peroxide or sodium carbonate, than with pyrosulfate.

Because of the resistance of silica to attack by pyrosulfates, silica vessels are preferred to those of platinum for this fusion. Platinum undergoes significant attack by acid sulfate fusion mixtures, although its use is sometimes recommended.

Carbonates

Most silicate rocks decompose completely on prolonged heating at about 1000°C with anhydrous sodium carbonate or with an equimolar mixture of sodium and potassium carbonates. A short period at 1200°C is recommended for particularly resistant minerals such as zircon and chromite. The fusion may be carried out in vessels of platinum or noble-metal alloys of greater mechanical strength or lower cost (e.g., platinum-iridium, palladium-gold).

Where significant amounts of sulfides or carbonaceous material are present, the medium may become sufficiently reducing for some elements to be converted into the metal, with a danger of subsequent alloying with the metal crucible. Although in some cases this problem can be overcome by including an oxidizing agent such as potassium nitrate in the fusion mixture, roasting of the sample before carbonate fusion is a better alternative.

Hydroxides

Molten sodium and potassium hydroxides are effective in decomposing silicates, but are less satisfactory for a number of accessory minerals. Lower temperatures can be used than for carbonate fusions. Prefusion of the hydroxide is advisable to remove moisture before the sample is added. Platinum vessels are attacked by molten alkali-metal hydroxides, and the use of crucibles of iron, nickel, or zirconium is recommended. These materials are attacked to some extent, but are satisfactory if the metal in question is not being determined, and if traces of it do not interfere with later stages of the analysis.

Sodium Peroxide

Certain minerals such as chromite and cassiterite, which do not decompose readily in a number of fusion mixtures, can be taken into solution with sodium peroxide. High-temperature fusion ($>500°C$) is often unnecessary, as many minerals in silicate rocks can be decomposed by sintering the powdered sample with sodium peroxide in a platinum crucible[11] at about 480°C. Above 500°C platinum is attacked, and zirconium vessels are more suitable. Various aspects of the use of sodium peroxide fusions have been discussed by Belcher.[12]

The cooled melts resulting from hydroxide or peroxide fusion are often extracted with water. The strongly alkaline solutions contain silicate and aluminate, whereas metals such as iron that have hydroxides insoluble in excess alkali remain in the form of a hydroxide residue.

Borates

Borax (sodium tetraborate, $Na_2B_4O_7$), alone or in combination with boric oxide, boric acid, or sodium carbonate, has been used extensively to attack aluminosilicates and other materials containing large amounts of alumina. Cooled melts containing borate and carbonate are easily broken up on the addition of dilute hydrochloric acid.

Lithium tetraborate has been suggested as a flux in pretreatment of samples for emission spectroscopy,[13] and a combination of boric acid and lithium fluoride has been used to decompose silicate rocks and minerals.[14] This mixture forms lithium tetraborate on heating to 800 to 850°C in a platinum vessel, the fluoride removing silicon as the volatile silicon tetrafluoride. Boron and excess fluoride are eliminated by adding concentrated sulfuric acid and heating. Residual matter includes insoluble sulfates and those elements that form precipitates in acid solution (e.g., tungsten, niobium, tantalum). This fusion mixture is satisfactory for minerals such as zircon, sillimanite, rutile, corundum, and garnet, which are not significantly attacked by hydrofluoric/sulfuric acid mixtures.[9]

Fusions with lithium metaborate, $LiBO_2$, in vessels of platinum or graphite have been studied in some detail.[15–17] After being heated at 950°C for 10 to 15 min, the melt is cooled and dissolved in nitric acid. The digestion is carried out at room temperature to prevent the formation of colloidal or insoluble hydrated silicic acid polymers. Lithium metaborate decomposes silicates and minerals such as chromite and ilmenite[15] and has been used to prepare solutions from various igneous and volcanic rocks, sulfides, phosphate rock, and lignite ash for analysis by spectrophotometry, emission spectrography, flame emission, and atomic absorption.

Fluorides

Fusions have been carried out in platinum vessels with alkali-metal fluorides and with ammonium fluoride. Not all silicates are attacked, and the cooled melt must generally be heated with sulfuric acid to convert fluorides into sulfates and expel excess fluoride as hydrofluoric acid. A final fusion with pyrosulfate is sometimes recommended. Potassium hydrogen fluoride is said to be the most satisfactory flux for the dissolution of zircon.[1]

The presence of fluoride is undesirable in many analytical procedures. Its action in complexing with many metals may affect trace determinations by spectrophotometric or electrochemical techniques, and in atomic absorption the response of several elements is influenced by the presence of fluoride in the solution (p. 181).

4.1.4 Miscellaneous Decomposition Techniques

There are many decomposition techniques with specific applications to the determination of particular elements and the decomposition of particular mineral or rock types. A few of these are noted below, and more detailed information is available elsewhere.[1]

1. Decomposition of samples in an oxygen stream has been used in the determination of mercury, which distils out, and of sulfur, which is oxidized to sulfur dioxide.
2. Traces of sulfur can be oxidized to sulfur trioxide with a vanadium pentoxide catalyst at 900 to 950°C. The trioxide is subsequently reduced to sulfur dioxide for determination by one of several methods.
3. Some minerals, such as cassiterite, are satisfactorily reduced by heating in a stream of hydrogen gas.
4. Elementary chlorine and some gaseous chlorides such as hydrogen chloride and carbon tetrachloride allow the removal of elements with

volatile chlorides. Dolezal et al.[1] give details on the low-temperature volatilization of chlorides such as those of arsenic, antimony, bismuth, mercury, and tin and of the use of chlorination to liberate less common metals such as the platinum metals, hafnium, niobium, and tantalum.

5. The separation of trace amounts of tin as the volatile tetraiodide occurs when samples are heated with ammonium iodide.

6. Hydrides of elements such as arsenic, antimony, bismuth, germanium, and selenium are liberated by the action of various reducing mixtures, of which sodium borohydride solution appears to be particularly useful.

7. Several methods for the determination of fluorine in geological materials are based on the distillation of hydrofluoric acid from sulfuric or perchloric acid solution after removal of any large amounts of silica and alumina, or on steam distillation of fluorosilicic acid directly from the rock material.

8. Decomposition of some rather insoluble salts (e.g., calcium sulfate, calcium phosphate) has been achieved by equilibration with an ion exchanger (p. 44).

9. In the fire assay technique for determination of silver, gold, platinum, and other precious metals (see Chapter 6), the sample is fused with a flux such as sodium carbonate/borax, together with red lead, Pb_3O_4, and a reducing agent such as charcoal. The lead is reduced to the metallic state and forms a button containing the precious metals at the bottom of the crucible. Fusion and oxidation of the button helps to remove impurities, and the precious metals are finally obtained by cupellation, in which molten litharge formed by further oxidation of lead is absorbed by a heated cupel of magnesium oxide. The bead of precious metal that remains can then be weighed or analyzed by standard methods. Although fire assay as an analytical technique in itself has largely been superseded, a number of procedures for precious-metal analysis make use of a preliminary separation of this kind. Fire assay and other methods of isolating and separating noble metals have been reviewed by Beamish.[18]

4.1.5 Treatment of Samples Containing Organic Matter

Some geochemical samples, such as soils, contain appreciable amounts of organic matter, which should be removed prior to fusion or acid attack. When such samples are being analyzed for the total content of relatively involatile elements (e.g., aluminum, iron, calcium, magnesium), it is common practice to destroy organic matter by heating the sample in a platinum or porcelain vessel to about 800 to 900°C in a muffle furnace. In determining more volatile elements, such as lead and zinc, there is less danger of loss of the analyte if the sample is ashed at temperatures below 500°C.

Samples containing large amounts of organic matter, such as the plant material used in biogeochemical work, may be treated by dry ashing, by wet oxidation, or by a low-temperature oxidation with a reagent such as excited oxygen. Dry ashing techniques are based on atmospheric oxidation, usually at 450 to 550°C, in either open or closed systems. Small amounts of a chemical oxidant such as nitric acid or magnesium nitrate are sometimes favored as an ashing aid. Wet oxidations usually involve some combination of nitric, perchloric, and sulfuric acids, or hydrogen peroxide. The use of nitric/perchloric acid mixtures is common, although great care is essential. Nitric acid should be added first, and the proportion of perchloric acid in the mixture should be controlled so that all organic matter is destroyed before the nitric acid is completely consumed or vaporized. The temperatures used in wet oxidation are much lower than those involved in dry ashing, and there is less danger of loss of trace elements by volatilization or by reaction between the sample and its container. Even so, special techniques may be required for the most volatile elements (e.g., mercury) or for elements that can form very volatile compounds under the conditions of treatment. Ruthenium and osmium, for example, form volatile tetroxides under strongly oxidizing conditions, and elements such as arsenic and germanium can form volatile chlorides unless precautions are taken to ensure early removal of chloride during the decomposition. Some of the dangers of analyte loss during oxidation are avoided in decompositions using excited oxygen.[19,20]

A detailed account of dry ashing and wet oxidation methods, and of studies undertaken to determine the completeness of recovery of trace elements, can be found in the monograph by Gorsuch.[21]

4.2 SEPARATION AND CONCENTRATION TECHNIQUES

4.2.1 Introduction

In the determination of elements at trace levels in materials such as rocks, soils, sediments, and natural waters, the need for the traces to be concentrated or separated (or both) from the sample matrix, frequently arises. Concentration is required when the sensitivity of the analytical method is inadequate; separation serves to simplify the sample matrix, freeing traces of the wanted element from other substances that would interfere in the analysis.

It should be noted that some methods of sample decomposition provide at least a partial separation and concentration of the wanted elements. This applies particularly to the fire assay technique and to the methods of liberating from the sample such volatile substances as elemental mercury, sulfur dioxide, ruthenium and osmium tetroxides, tin(IV) iodide, and various chlorides and hydrides.

Some separation techniques, such as coprecipitation and solvent extraction, are effective in concentrating the analyte at the same time. Combinations of these techniques may be useful; coprecipitation, for example, can be used primarily as a means of concentrating traces and can be followed by a technique such as ion exchange to effect a separation.

Where a sequence of separation and/or concentration steps is required, the analysis is not only more laborious, but the danger of introducing additional errors through manipulation is greatly increased. A useful discussion of systematic errors and their minimization is given in the review by Tölg.[2]

A most valuable source of information on separation methods used for many elements that occur at trace levels in geological materials is the book by Korkisch,[22] who deals with the following elements: Li, Rb, Cs, Be, Ra, Ga, In, Tl, Ge, Se, Te, Po, Ag, Au, Ti, Zr, Hf, V, Nb, Ta, Mo, W, Tc, Re, platinum metals, lanthanides, and actinides.

4.2.2 Precipitation and Coprecipitation

In these techniques the wanted trace constituent is separated as a precipitate from the sample solution, ideally leaving macroconstituents and any interfering microconstituents in solution. Except in the case of radiochemical methods of analysis (pp. 278, 293) where traces present in the sample are supplemented by addition of carrier, direct precipitation of the trace constituent alone is often not feasible. Even when the solubility of the precipitate is very low, the separation of small quantities of material may be difficult. Losses may occur through colloid formation or supersaturation of the solution or during manipulation of the precipitate.

For these reasons the total amount of precipitate is increased by precipitating a suitable collector, and the separation of the trace constituent occurs by coprecipitation. Most frequently the collection results from ions of the trace constituent becoming adsorbed on the surface of crystals of the collector. Gelatinous precipitates, such as the hydroxides or hydrous oxides of iron, aluminum, and nickel, provide particularly large surface areas for adsorption of foreign ions. Other mechanisms of coprecipitation include the formation of solid solutions (when the foreign ion is of suitable radius and charge), and simple mechanical inclusion of foreign ions within the precipitate mass.

The widespread occurrence of the phenomenon of coprecipitation means that separations involving precipitation of undesirable *macroconstituents*, intended to leave the trace elements in solution, should be avoided as far as possible. In many cases there is considerable danger of losing at least part of the wanted trace element by coprecipitation.

In addition to the hydroxides noted above, other collector substances include various sparingly soluble sulfides, fluorides, oxalates, sulfates, and phosphates. Some metals and metalloids precipitated from solution under reducing conditions coprecipitate others. For example, tellurium prepared by reducing a tellurite solution with stannous chloride, coprecipitates a number of precious metals, and arsenic formed by reduction of As(III) with hypophosphite coprecipitates selenium, tellurium, and several other metals whose ions are readily reduced. Some metal-organic complexes carry down traces of other metals forming sparingly soluble compounds with the same reagent. The iron-cupferron precipitate collects traces of titanium, vanadium, and zirconium.

The following list gives examples of some collector precipitates, with some of the trace elements that have been collected with them shown in parentheses.

1. *Hydroxides:* ferric hydroxide (Be, Co, Ga, Ge, lanthanides, Mo, Nb, Ni, Sn, Ta, V, W), aluminum hydroxide (Be, lanthanides, Sc, Th, W, Zr).

2. *Sulfides:* copper(II) sulfide (Au, Bi, Cd, Ga, In, Pb, Pt, Ru, Sb, Sn), cadmium sulfide (Ga, Hg, In), antimony(V) sulfide (Mo), arsenic(III) sulfide (Re).

3. *Fluorides:* calcium fluoride (lanthanides, Th, Zr), lanthanum fluoride (lanthanides, Th).

4. *Oxalates:* calcium oxalate (lanthanides, Th), lanthanum oxalate (lanthanides, Th).

5. *Sulfates:* barium sulfate (Ra, Th), lead sulfate (Ra, Tl), strontium sulfate (Pb).

6. *Phosphates:* aluminum phosphate (Be, U), titanium phosphate (Be, U).

7. *Metals and metalloids:* arsenic (Se, Te), gold (Se, Te), selenium (Pd, Pt, Te), tellurium (Ag, Au, Pd, Pt).

8. *Metal-organic and organic compounds:* iron(III) + cupferron (Ti, V, Zr), lanthanum + 8-hydroxyquinoline (lanthanides), α-benzoinoxime (Mo), methyl violet + tannin (W).

Coprecipitation is particularly valuable for isolating trace constituents from large volumes of solution, and has therefore been used extensively as a first step in the determination of trace elements, including fission products such as ^{90}Sr, ^{95}Zr, ^{95}Nb, ^{99}Mo, ^{106}Ru, ^{140}Ba, ^{140}La, and ^{144}Ce, in natural waters. Table 4.1 shows some typical examples of the use of coprecipitation in the analysis of various geological materials.

4.2.3 Ion Exchange

An ion exchanger consists essentially of a stable matrix containing labile ions capable of exchanging with ions in the surrounding medium, without

TABLE 4.1 Examples of the Use of Coprecipitation Methods

Element	Collector	Material analyzed	Ref.
Au	Tellurium	Soils	23
Be	Ferric hydroxide	Seawater	24
Cd	Copper sulfide	Seawater	25
Ce	Titanium hydroxide	Rain water	26
Co, Ni	Ferric hydroxide	Natural waters	27
Eu, Y	Calcium oxalate	Synthetic solutions	28
Ga	Ferric hydroxide	Seawater	29
Lanthanides	Calcium fluoride	Synthetic solutions	30
Mo	α-Benzoinoxime	Seawater	31
Sc	Calcium oxalate	Minerals, seawater	32
Se	Ferric hydroxide	Seawater	33
Se, Te	Gold	Meteorites	34
Se, Te	Arsenic	Soils, rocks	35
Th	Barium sulfate	Rocks	36
Th	Calcium oxalate	Rocks	37
U	Aluminum phosphate	Waters	38, 39
V	Ferric hydroxide	Natural waters	40, 41
Zr	Aluminum hydroxide	Seawater	42

major structural changes taking place in the matrix itself. Early observations of the phenomenon of ion exchange, more than a century ago, dealt with the displacement of calcium ions from the clay minerals of soils by other ions, such as ammonium ions. Extensive work in soil chemistry was followed by the development of synthetic aluminosilicates, which were used in water softening, the calcium ions of hard water being replaced by sodium ions from the synthetic silicate.

From about 1935, the synthesis of resinous polymeric organic ion exchangers has provided materials useful in analytical chemistry, facilitating many inorganic separations that were time consuming by other means. More recently, extensive use has been made of synthetic inorganic ion exchangers (other than aluminosilicates), so-called liquid organic ion exchangers, and materials developed by chemical modification of a variety of naturally occurring substances such as cellulose and chitin.

Detailed discussion of the preparation, structure, and properties of ion exchangers can be found in the books by Rieman and Walton,[43] Samuelson,[44] Dorfner,[45] and Kunin.[46] Analytical applications are also covered in the reviews by Walton,[47] Kunin et al.[48] and Strelow.[49] The book by Korkisch[22] contains a wealth of information on ion-exchange separations of rarer metal ions.

Organic Ion-Exchange Resins

Organic cation-exchange resins are based on phenolic structures, or cross-linked polystyrene or methacrylic acid structures, with sulfonic acid or carboxylic acid functional groups, the latter being more weakly acidic cation exchangers. Anion-exchange resins are largely based on cross-linked polystyrene with primary, secondary, tertiary, or quaternary ammonium groups, the exchangers with quaternary ammonium groups being strongly basic.

Figure 4.1 shows portions of the structures of (*a*) a strong acid exchanger prepared by sulfonating a resin formed from the copolymerization of styrene and divinylbenzene, and (*b*) weak and strong base exchangers prepared by treating a chloromethylated polystyrene with suitable aliphatic amines.

Inorganic Ion-Exchange Resins

In addition to the synthetic aluminosilicates, many other inorganic ion exchangers have been prepared. Cation exchange properties are possessed by some complex zirconium phosphates and tungstates, for example, and the addition of ammonia to zirconyl chloride yields a material that can act as an anion exchanger. Other inorganic exchangers include zirconium arsenate and antimonate, tin(IV) phosphate, silver and molybdenum hexacyanoferrates, and hydrous oxides of chromium(III), tin(IV), and thorium(IV). The preparation and behavior of these materials has been discussed elsewhere.[50,51]

Liquid Ion Exchangers

Various amines containing long-chain substituents, dissolved in organic solvents such as chloroform or nitrobenzene, are able to extract anions from an aqueous phase by a process that can be represented by an equation such as

$$A^-(aq) + R_3NHCl(org) \rightleftharpoons Cl^-(aq) + R_3NHA(org)$$

where A^- is an anion and R is an alkyl group. Such systems have been used in the separation of metals that form anionic chloro- or sulfatocomplexes, for example.

In a similar way, cation separation is accomplished with water-insoluble acids such as dialkyl esters of phosphoric acid, $(RO)_2PO(OH)$, and monoalkyl esters of alkanephosphonic acids, $RPO(OR')(OH)$, dissolved in organic solvents. Various cations can be extracted into the organic phase, displacing an equivalent amount of hydrogen ion into the aqueous phase. It is apparent

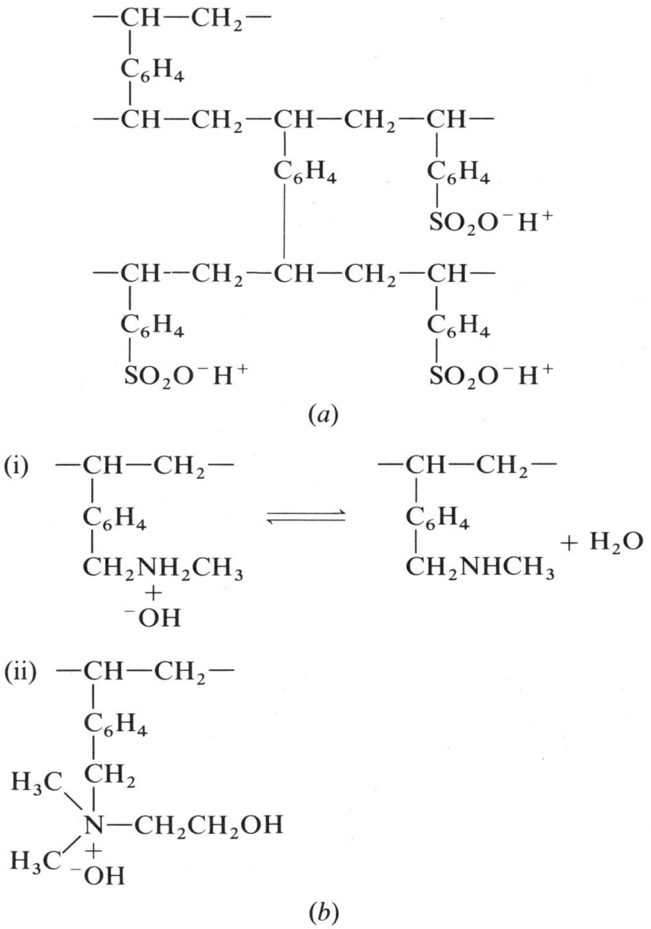

SEPARATION AND CONCENTRATION TECHNIQUES

$$\begin{array}{c}
-CH-CH_2- \\
| \\
C_6H_4 \\
| \\
-CH-CH_2-CH-CH_2-CH- \\
| | \\
C_6H_4 C_6H_4 \\
| \\
SO_2O^-H^+ \\
-CH-CH_2-CH-CH_2-CH- \\
| | \\
C_6H_4 C_6H_4 \\
| | \\
SO_2O^-H^+ SO_2O^-H^+
\end{array}$$

(a)

(i)
$$\begin{array}{c}
-CH-CH_2- \\
| \\
C_6H_4 \\
| \\
CH_2NH_2CH_3 \\
+ \\
^-OH
\end{array} \rightleftharpoons \begin{array}{c}
-CH-CH_2- \\
| \\
C_6H_4 \\
| \\
CH_2NHCH_3
\end{array} + H_2O$$

(ii)
$$\begin{array}{c}
-CH-CH_2- \\
| \\
C_6H_4 \\
| \\
H_3C CH_2 \\
\backslash | \\
N-CH_2CH_2OH \\
/ + \\
H_3C ^-OH
\end{array}$$

(b)

Fig. 4.1. Portions of structures of (a) sulfonated cross-linked polystyrene (strong acid) cation exchanger; (b) (i)—secondary amine (weak base) anion exchanger; (ii)—quaternary ammonium (strong base) anion exchanger.

that the action of so-called liquid ion exchangers should more properly be considered as a special type of liquid-liquid distribution (see Sec. 4.2.4).

Other Ion Exchangers

Ion exchangers have been produced by chemical treatment of several naturally occurring materials. Modification of cellulose leads to products such as cellulose phosphate, represented by $(C_6H_9O_4)OPO(OH)_2$, in which the phosphate protons can be exchanged for other cations, and diethylaminoethylcellulose, from which the strong-base anion exchanger

$$(C_6H_9O_4)OCH_2CH_2N^+(C_2H_5)_3I^-$$

can be prepared. Chitin, a constituent of crustacean shells, can be deacetylated to give chitosan, an exchanger with a strong affinity for heavy metals. Exchangers of this kind are the subject of a monograph by Muzzarelli.[52]

Other ion exchangers have been prepared by attaching chelating groups such as iminodiacetate, sulfoguanide, dithizonate, 8-hydroxyquinolinate and dithiocarbamate to resinous polymers, silica gel, and other inert supports.[53–57] Rieman and Walton[43] discuss other less common types of ion exchangers.

Ion-Exchange Selectivity

The equilibrium of an exchange reaction involving cations A^+ and B^+ in aqueous solution can be represented by the equation

$$R^-A^+ + B^+(aq) \rightleftharpoons R^-B^+ + A^+(aq)$$

where R^- denotes the anionic group forming part of the structure of the ion exchanger. The position of the above equilibrium can be described by an equilibrium constant, the thermodynamic exchange constant K_A^B:

$$K_A^B = \frac{a_A a_{RB}}{a_B a_{RA}} \tag{4.1}$$

where a_A, a_B are the activities of the ions in the aqueous phase and a_{RA}, a_{RB} are the activities of the ions in the exchanger phase. If each activity term is replaced by the product of a molal concentration m and an activity coefficient γ, equation 4.1 becomes

$$K_A^B = \frac{m_A m_{RB}}{m_B m_{RA}} \cdot \frac{\gamma_A \gamma_{RB}}{\gamma_B \gamma_{RA}} \tag{4.2}$$

The equilibrium distribution of *concentrations* is indicated by the selectivity coefficient E_A^B:

$$E_A^B = \frac{m_A m_{RB}}{m_B m_{RA}} \tag{4.3}$$

Data for the calculation of the selectivity coefficient are obtained by equilibrating known quantities of a salt of the cation B^+ with known quantities of the exchanger in the form R^-A^+ and determining the equilibrium concentrations of A and B in the two phases. Concentrations of A and B in the exchanger phase may be expressed as equivalent fractions, X_{RA} and X_{RB}, these being the numbers of equivalents of the ions A and B per equivalent of ions (in total) attached to the exchanger. Alternatively, these concentrations may be given as equivalents per kilogram of dry exchanger in a specified form (e.g., H^+ form). Experimental data on exchange selectivity are usually expressed in terms of the selectivity coefficient E_A^B or the equilibrium coefficient K_A^B, defined by

$$K_A^B = E_A^B \frac{\gamma_A}{\gamma_B} = \mathbf{K}_A^B \frac{\gamma_{RA}}{\gamma_{RB}} \tag{4.4}$$

The evaluation of the thermodynamic exchange constant $\mathbf{K}_A^B$ from E_A^B or K_A^B is discussed by Rieman and Walton,[43] who also indicate the way in which equation 4.3 must be modified in cases of exchange between ions of unequal charge.

Experiments show that, as is to be expected, the quantities E_A^B and K_A^B are not strictly constant, but vary with the concentrations of ions on the exchanger and in the solution.

In the case where one of the ions is present in large excess in both phases, E_A^B is relatively constant. Hence if m_A is fixed and is much greater than m_B, and $m_{RA} \gg m_{RB}$, then m_A/m_{RA} is approximately constant, and equation 4.3 becomes

$$\frac{m_{RB}}{m_B} \doteq \text{constant} \tag{4.5}$$

i.e., the ratio of the concentration of the "trace" constituent B on the exchanger to its concentration in the solution is approximately constant. The ratio of equation 4.5 is normally expressed as a distribution ratio D, defined by

$$D = \frac{\text{amount of B in unit quantity of exchanger}}{\text{amount of B in unit quantity of solution}} \tag{4.6}$$

The unit quantity of exchanger may be an equivalent (or milliequivalent) of fixed ions, or may be a kilogram (or gram) of dry exchanger. Values of D therefore are given often in milliliters per milliequivalent or milliliters per gram. These values indicate the affinity of the resin for ion B relative to ion A under specified conditions of A concentration. The distribution ratio or the selectivity coefficient can be used as a measure of the efficiency with which the exchanger is able to retain ions from a dilute solution, and of the ease with which the ions can later be removed into an eluting solution.

As the ability of an ion exchanger to retain a particular ion is related mainly to the strength of the electrical interaction between that ion in solution and the oppositely charged site on the exchanger, it is not surprising that ion-exchange selectivity depends on many factors. These include properties of the exchanging ions (charge, polarizability, extent of hydration, and effective size), the structure of the exchanger (type of functional group, degree of cross linking), and the nature of the solution (concentration of the exchanging ions, total ionic strength, presence of complexing agents and other solutes).

For a sulfonated resin cation exchanger, dilute solutions of metal chlorides exhibit a selectivity order such as $Fe^{3+} > Al^{3+} > Ba^{2+} > Pb^{2+} > Ca^{2+} > Ni^{2+} \doteq Cu^{2+} > Zn^{2+} > Mg^{2+} > Cs^+ > K^+ > Na^+ > Li^+$. However, reversion of the order occurs in concentrated solutions, with absorption of univalent ions being preferred. The lithium ion is also the most strongly held of the alkali metal cations on weak acid (carboxylic) cation-exchange resins. Strong base (quaternary ammonium) anion-exchange resins exhibit a selectivity order for univalent anions such as the following: $ClO_4^- > I^- > Br^- > Cl^- > OH^- > F^-$. Extensive generalizations on selectivity are, however, not possible because of the large number of influential variables.

The reader is referred to other sources[43–46,58] for more general discussions of the thermodynamic aspects of ion-exchange equilibria, the theory of ion-exchange chromatography, and experimental observations of ion-exchange selectivity. Practical techniques of ion exchange are also described in specialist works.[43–46]

Applications of Ion Exchange

Ion exchange has been used in several different ways in the determination of trace elements in geological materials: (*a*) the removal of interfering ions from solution; (*b*) the dissolution of insoluble salts; (*c*) the chromatographic separation of mixtures of inorganic ions; (*d*) the concentration of trace constituents from large volumes of solution.

In spectrophotometric or electroanalytical procedures the presence of certain ions may cause errors because of complexing with the analyte or reactions with added reagents. Provided the interferent is ionic it can be removed by ion exchange. Cations such as Al^{3+} and Fe^{3+} that interfere with the spectrophotometric determination of fluoride can be removed, for example.[59] Other examples are noted elsewhere.[43,44] It should be noted that some of these applications have been made unnecessary by the development of more specific and interference-free analytical methods.

Insoluble salts such as silver chloride, the sulfates of calcium, barium, and radium, and calcium phosphate can be brought into solution when the traces of ions in equilibrium with the solid salt are removed by suitable ion exchangers. Phosphate rock, for example, can be decomposed by shaking with the hydrogen form of a strong-acid cation exchanger;[60] cations can then be eluted from the resin and anions can be determined in the filtrate.

Separations of metal ions and complexes on ion-exchange columns have been used very extensively in cases where the analytical method alone does not possess sufficient specificity, or where better precision can be achieved by separating the analyte from other elements. Ions absorbed on a column of exchanger are removed selectively by passing a suitable eluting solution through the column. It is obviously desirable to achieve the best separation possible, at the same time obtaining each element in the minimum volume of eluate. For ions of similar charge type and chemical behavior, this can be done by means of a continuous or stepwise increase in the concentration of the eluting solution, such as that illustrated in Fig. 4.2. In other cases advantage can be taken of differences in behavior toward various complexing agents or of

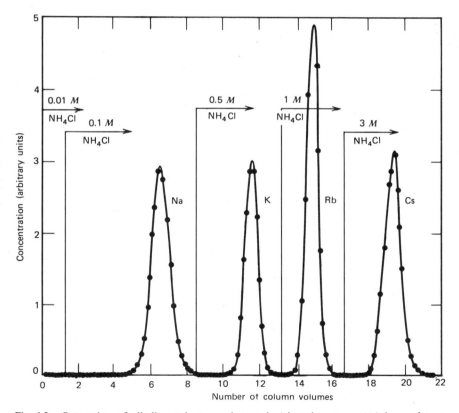

Fig. 4.2. Separation of alkali metals on an inorganic (zirconium tungstate) ion-exchange column. *Source*: Kraus et al.[61]

the sensitivity of some complex-formation equilibria to changes of pH. The use of mixed organic/aqueous eluting solutions has also been studied extensively.[62-64]

Alkali and alkaline earth metals have been separated on cation exchange resins, or, more effectively, with inorganic exchangers.[61] The lanthanides have been separated from one another on sulfonic acid resins by elution with α-hydroxyisobutyrate, citrate, lactate, EDTA, or other complexing agents.[22,65-69] Many elements have been separated on anion-exchange columns using hydrochloric acid media,[70,71] differences in behavior being related to the extent of formation and absorption of anionic chlorocomplexes. Anion-exchange separations based on complex formation with other anions, including fluoride,[72] nitrate,[73] and sulfate,[74] have also been studied extensively, as have media containing mixtures of complex-forming anions such as fluoride and chloride.[75] Separations of halide ions from one another are readily achieved on quaternary-base resins.[76] Many other separations are described and summarized elsewhere.[22,43,45,47]

In spite of recent advances in some of the most specific methods of analysis (e.g., high-resolution γ-ray spectrometry, and various atomic spectroscopic techniques), ion exchange continues to be widely used in inorganic trace analysis. Some shift in emphasis has occurred, however, from separation to collection or concentration of traces from large volumes of sample. Ion exchange is particularly valuable in the analysis of natural waters. In geochemical field work it is possible to eliminate the need to transport and store large volumes of sample by carrying out ion-exchange collection at the sampling site. It is important to note that nonionic forms of the analyte (e.g., neutral complexes, colloidal particles) may not be retained on the exchanger unless some prior chemical treatment is used. In one method for the determination of titanium in natural waters,[77] for example, colloidal matter was first taken into solution and the titanium was then converted into an anionic ascorbate complex before being absorbed on an anion exchanger.

Where unwanted ions are also present in high concentration, it is desirable to use ion exchangers with a high selectivity coefficient for the trace constituent relative to the major constituents. Cesium has been collected from seawater[78] by using a sulfonated phenolic resin with a large E_{Na}^{Cs}. Where the elements being concentrated form chelate complexes, high selectivity can be provided by the use of chelating resins,[55,79-81] cellulosic exchangers,[82] or substances such as chitin and chitosan.[83] Disks containing cation-exchange or chelating resins have been used to collect traces for direct analysis by X-ray and other methods.[84-86]

Examples illustrating the use of ion exchange in the analysis of geological materials are given in Table 4.2. Many other applications are described or summarized in works such as those of Dorfner,[45] Rieman and Walton,[43] Korkisch,[22] and Walton.[47]

TABLE 4.2 Examples of the Use of Ion-Exchange Methods

Elements determined	Exchanger type[a]	Application and notes	Ref.
Li	C, R	Determination in rock samples; elution with nitric acid/methanol	87
Na, K	C, R	Determination in silicates; separated from multivalent elements	88, 89
Rb, Cs	C, R	Determination in seawater, seaweeds, marine sediments, coals	78, 90
Rb, Cs	C, R	Determination in rocks, minerals, meteorites	91
Cs	C, I	Determination in silicates	92
Na, K, Mg, Ca	C, R	Collection from lake waters	93
Cu, Zn, Cd, Ni, Pb	Chel, R	Collection from waste waters	94
Cu, Zn, Ga	A, R	Determination in standard rocks	95
Au, Ir, Pt	A, R	Determination in meteorites	96
Au, Bi	A, R	Determination in seawater	97
B	C, R	Removal of cations interfering with spectrophotometric determination of B in rocks, minerals	98
Tl	A, R	Determination in rocks, sediments, seawater	99
Sc, Y, lanthanides	C, R	Determination in silicates	100
Y, lanthanides, Th, U	A, R	Determination in silicates	101, 102
Lanthanides	C, R	Determination in lunar fines, standard rocks	103
Ti	A, R	Determination in natural waters; complexed as ascorbate	77
Sn	A, R	Determination in rocks, sediments, soils; absorbed as chlorocomplex	104
Sn, Sb, Hg	A, R	Determination in meteorites	105
Hf, Ta	A, R	Determination in silicates; absorbed from HCl/HF medium	106
V, Mo	Chel, R	Concentration from seawater	80
Mo, W	A, R; A, Cell	Extraction from seawater from HCl/thiocyanate medium; Mo, W separated on cellulosic exchanger	82
Cr	A, R	Determination in rocks; separated as chromate	107
U	A, R	Determination in waters, rocks, soils	108, 109
U, Th, Zr	A, R	Determination in rocks	110

(*continued*)

TABLE 4.2 (*continued*)

Elements determined	Exchanger type[a]	Application and notes	Ref.
F	C, R	Removal of cations (Fe, Al, etc.) interfering in spectrophotometric determination	59
F	A, R	Collection of fluoride and separation from interfering cations	111, 112
F	A, R	Separation from interfering cations, fluoride eluted as fluoroberyllate	113
Cl	A, R	Collection from lake waters	93
I	A, R	Collection of ^{131}I from waters	114
Many elements	A, R; C, R	Determination of major and minor elements in silicates	115, 116
Many elements	Chel, R	Collection of traces of transition elements, lanthanides, etc., from seawater	79

[a] A, anionic; C, cationic; Cell, diethylaminoethylcellulose; Chel, chelating; I, inorganic; R, resin.

4.2.4 Solvent Extraction (Liquid-Liquid Distribution)

The process of solvent extraction, by which the transfer of a solute occurs from one bulk liquid phase to another, is one of the most widely used separation techniques in inorganic analysis. Extractions are most commonly carried out from aqueous solutions into organic solvents. In many cases the separation can be achieved very simply and quickly, requiring shaking of the two-phase system in a separating funnel for only a few minutes.

Solvent extraction is particularly useful for separating the analyte from interfering elements in spectrophotometry and radiochemical methods of analysis. Where the analyte is extracted in the form of a compound with strong ultraviolet or visible absorption, direct spectrophotometry of the organic phase is often possible. Even in analyses where interferences are not a major problem, solvent extraction is a valuable method for concentrating traces into a small volume of solution. The wide variety of two-phase liquid systems between which solutes can be distributed gives the method enormous versatility. The most important types of system are noted below.

Types of Extraction System

Extraction of solutes into organic solvents of low dielectric constant requires the formation of species possessing overall electrical neutrality. In a few cases *elements and simple covalent compounds* are soluble in organic solvents and can be extracted conveniently from aqueous solution in that form. The solubility of bromine, iodine, and the various allotropic forms of sulfur and selenium in solvents such as carbon disulfide, carbon tetrachloride and chloroform, and of halides of mercury(II), arsenic(III), antimony(III), germanium(IV), tin(IV), selenium(IV), and tellurium(IV) in diethyl ether, chloroform, and carbon tetrachloride, may be quoted as examples.

The formation of *chelate complexes*, in which ligands are bound to two or more sites at a central metal ion, is the basis of a large group of extraction systems. Many metal chelates are uncharged and are much more soluble in organic solvents than in water. Examples include the chelates of iron(III) with the anions of 8-hydroxyquinoline (oxine) and N-nitrosophenylhydroxylamine (cupferron), which are compounds of the type FeL_3 (L^- = ligand), soluble in chloroform. Some of the most important chelating agents, each capable of being used in the extraction of many elements, are shown in Table 4.3. Other chelating agents, used in the extraction of more limited ranges of elements, include α-dioximes such as dimethylglyoxime (Ni), toluene-3,4-dithiol [Mo(VI), W(VI), Re(VII)], hydroxyflavones such as morin (Be, Sc, Zr) and nitrosonaphthols [Co(III), Fe(II), Pd].

Many of the *heteropolyacids*, such as phosphomolybdic acid, molybdovanadophosphoric acid, tungstovanadophosphoric acid, and other similar compounds containing arsenic, titanium, and silicon, are effectively extracted into oxygen-containing organic solvents such as ethyl acetate and 1-butanol.

Where ions are poorly solvated in aqueous solution, the formation of ion pairs or *ion-association complexes* is favored. Many of these are much more soluble in organic solvents than in water, particularly if one of the ions is largely or wholly organic. For example, tetraphenylarsonium perrhenate $(C_6H_5)_4As^+ReO_4^-$, and tetraphenylstibonium fluoride $(C_6H_5)_4Sb^+F^-$ are extractable into chloroform.

Many extraction systems can be regarded as involving a *combination of complex formation and ion association*. Many metal ions react with a sufficient number of negatively charged species (either monodentate ions such as halides, or chelating ions such as oxalate and ethylenediaminetetraacetate) to form an *anionic complex*. Such an anion may then form an extractable ion-association complex with a suitable cation. For example, anionic chloro-complexes of metals such as gold(III), iron(III), and antimony(V) are readily extracted as ion pairs with hydrated hydronium ions (in a form such as $H_9O_4^+$) into oxygen-containing solvents like diethyl ether and ethyl acetate. Protonated solvent molecules, such as $(C_2H_5)_2OH^+$, may also serve as the cationic moiety. With other cations such as those of the type $(C_6H_5)_4X^+$ (X = P, As, Sb) a number of chloro- complexes (e.g. $HgCl_4^{2-}$, $SnCl_6^{2-}$) are extractable

TABLE 4.3 Some Chelating Agents Used in Solvent Extraction Systems

Reagent	Structural formula	Notes
8-Hydroxyquinoline (oxine)	(quinoline with OH at 8-position)	Over 40 elements extractable into $CHCl_3$; methyl- and dihalo-derivatives have also been used
Acetylacetone	$CH_3COCH_2COCH_3$	Over 60 elements extractable into acetylacetone (as solvent and chelating agent) or benzene
Thenoyltrifluoroacetone	(thienyl)–$COCH_2COCF_3$	Many elements, including lanthanides and actinides, extractable into benzene or toluene
Diphenylthiocarbazone (dithizone)	Ph–NH–NH–C(=S)–N=N–Ph	Many elements extractable into $CHCl_3$ or CCl_4
Sodium diethyldithio-carbamate (DEDTC, DDTC, or DDC)	$(C_2H_5)_2N$–C(=S)–S^-Na^+	Many elements extractable into $CHCl_3$, ethyl acetate, or CCl_4
Ammonium pyrrolidine-1-carbodithioate (dithiocarbamate) (APCD or APDC)	(pyrrolidinyl)–N–C(=S)–$S^-NH_4^+$	Many elements extractable into 4-methyl-2-pentanone (methyl isobutyl ketone, MIBK)
N-nitrosophenylhydroxyl-amine (ammonium salt)	Ph–N(N=O)–$O^-\overset{+}{N}H_4$	Many elements extractable into $CHCl_3$, benzene, or ethyl acetate; N-benzoyl analogue is a more stable reagent
1-(2-pyridylazo)-2-naphthol (PAN)	(pyridyl)–N=N–(2-hydroxynaphthyl)	Many elements extractable into amyl alcohol

TABLE 4.3 (continued)

Reagent	Structural formula	Notes
bis-(2-ethylhexyl)-o-phosphoric acid (HDEHP)	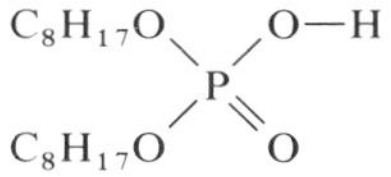 $\begin{array}{c} C_8H_{17}O \\ C_8H_{17}O \end{array} P \begin{array}{c} O-H \\ O \end{array}$	Many elements, especially lanthanides and actinides, extractable into n-heptane, toluene, or other hydrocarbons. Reagent often dimeric in organic phase, metals extracted as $M(HL_2)_n$

into hydrocarbons or chlorinated hydrocarbons. Many anionic complexes are also extracted into solvents such as chloroform or 1,2-dichloroethane in association with substituted ammonium ions of high molecular weight. Examples of suitable cations include protonated secondary amines (e.g., dodecyltrialkylmethyl amines such as Amberlite LA-2) and tertiary amines (e.g., tribenzylamine, trioctylamine, and mixtures of trioctyl- and tridecylamines such as Alamine 336), and quaternary ammonium ions (e.g., tetradecyldimethylbenzylammonium ion and mixtures of trioctylmethyl- and tridecylmethylammonium ions, as in Aliquat 336). The many anionic metal complexes that can be extracted by these reagents include complexes with monodentate ligands such as chloride and thiocyanate and with chelating agents such as ethylenediaminetetra-acetate (EDTA), cyclohexanediaminetetra-acetate (CDTA), and nitrilotriacetate (NTA).

Cations of a number of dyes (Fig. 4.3) have proved useful in extractions of anionic halocomplexes, the products giving great spectrophotometric sensitivity. Gallium has been determined following extraction of the Rhodamine

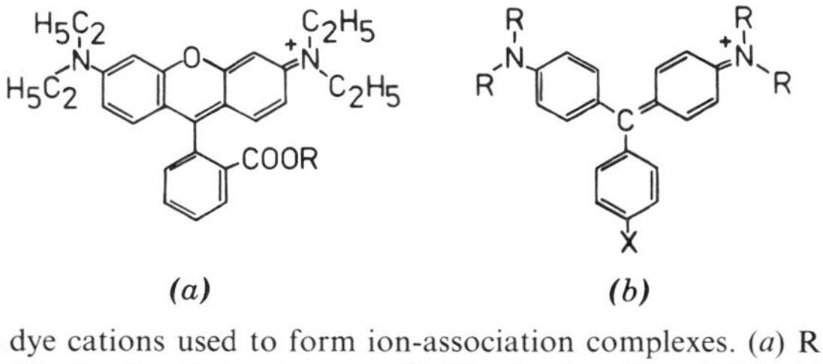

Fig. 4.3. Some dye cations used to form ion-association complexes. (a) R=H: Rhodamine B; R=n-butyl: Rhodamine B butyl ester. (b) X=H, R=CH_3: Malachite Green; X=H, R=C_2H_5: Brilliant Green; X=N$(CH_3)_2$, R=CH_3: Crystal Violet; X=NHCH_3, R=CH_3: Methyl Violet.

B-tetrachlorogallate(III) ion-association complex into a chlorobenzene/carbon tetrachloride solution, and tantalum can be separated as hexafluorotantalate(V) from sulfuric acid/hydrofluoric acid media by association with the cations of brilliant green and methylene blue.

With neutral chelating agents many metal ions yield *cationic complexes* that can pair with anions to give extractable species. An example is *tris*-(1,10-phenanthroline)iron(II) perchlorate, soluble in nitrobenzene.

In some cases where the complexing ligand does not completely take the place of coordinated water molecules, i.e., the complex is coordinatively unsaturated, the remaining water can be displaced by a second ligand or by the organic solvent itself. This can lead to increases in the extent of extraction or rate of extraction, or both. The formation of species such as $Th(TTA)_4(TBP)$, $[Fe^{III}Cl_4(Et_2O)_2]^-$, and $[Co^{II}(SCN)_3(i\text{-pentanol})_3]^-$ may be noted as examples (TTA = thenoyltrifluoroacetone; TBP = tri-n-butyl phosphate; Et_2O = diethyl ether).

Extraction Equilibria

The equilibrium distribution of a species between two liquid phases is such that the ratio of its activities in the two phases is a constant at a given temperature and pressure. In most analytically useful metal-extraction systems, however, the activities are not related in a simple way to the total measured concentrations of the metal in each phase. This is due in part to the existence of chemical reactions (in either phase) involving the extractable species. In addition, systems are often studied under conditions where many activity coefficients are very different from unity, even when the total concentration of the metal itself is low. The addition of high concentrations of salts to the aqueous phase, or the presence of high acid concentrations, for example, may cause substantial changes in the water activity and influence the positions of equilibria involving the extractable species.

Theoretical discussion of extraction is carried out by considering the equilibrium constants for all relevant equilibria, as indicated below. However, the quantity with major practical significance is the distribution ratio D of the element being distributed, defined as the equilibrium ratio of its total concentration in the organic phase (C_o) to that in the aqueous phase (C_w), i.e.,

$$D = \frac{C_o}{C_w} \tag{4.7}$$

If volumes V_o and V_w of the organic and aqueous phases, respectively, are used in a single batch extraction, the fraction extracted θ is given by

$$\theta = \frac{C_o V_o}{C_o V_o + C_w V_w} = \frac{D V_o}{D V_o + V_w} \tag{4.8}$$

A succession of n such extractions with fresh volumes of organic solvent will leave a fraction $(1 - \theta)^n$ in the aqueous phase. Hence if $D > 100$, a single extraction with equal volumes of the two phases will result in more than 99% of the element being transferred into the organic phase. Even where D is as low as 9, the same net result can be achieved by two extractions with $V_o = V_w$, or if dilution is unacceptable, by a series of four extractions with $V_o = 0.25\, V_w$.

The observed behavior of a given extraction system can be discussed with reference to schemes such as those of Fig. 4.4. Where reasonable assumptions can be made about activity coefficients, detailed quantitative explanations are possible.

1. The equilibria involved in the extraction of a neutral metal chelate, ML_m (Fig. 4.4a), formed from an ion $M(H_2O)_x^{m+}$ and a chelating agent HL, are the following (with equilibrium constants shown in brackets after each equation):

a. Distribution of the chelating agent

$$HL(aq) \rightleftharpoons HL(o)\,(K_{D,R})$$

b. Ionization of the chelating agent in the aqueous phase

$$HL(aq) + H_2O \rightleftharpoons H_3O^+ + L^-\,(K_a)$$

c. A succession of equilibria in the aqueous phase, involving displacement of coordinated water molecules by ligands L^-, represented by the overall equilibrium

$$M(H_2O)_x^{m+} + mL^- \rightleftharpoons ML_m(aq) + xH_2O\,(K_f)$$

d. Distribution of the chelate

$$ML_m(aq) \rightleftharpoons ML_m(o)\,(K_{D,C})$$

Combination of the above equilibria shows that the distribution ratio should increase as the organic-phase concentration of HL increases, and as the pH increases. For a given HL(o) concentration, as pH increases, $D \rightarrow K_{D,C}$, i.e., the distribution ratio becomes governed by the chelate-distribution equilibrium and not by its formation equilibria.

This type of scheme may be modified by the need to take other reactions into account. Such additional reactions include cases where the chelating agent contains two or more ionizable protons, or can itself be protonated, where keto-enol and similar equilibria are involved, where hydrolyzed forms of the metal aquo-complex are present, and where the extractions are carried out in the presence of other metals and other ligands.

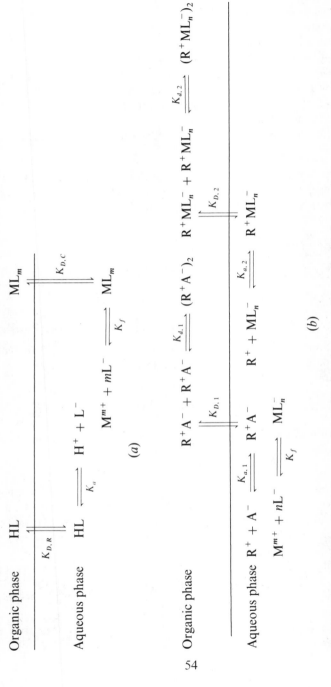

Fig. 4.4. (a) Equilibria involved in formation and distribution of a metal chelate. (b) Equilibria involved in formation and distribution of an ion-association complex $R^+ML_n^-$ from a reagent R^+A^- and an anionic complex ML_n^-. After Freiser.[117]

2. The equilibria involved in the extraction of an ion-association complex $R^+ML_n^-$ (Fig. 4.4b), in which an anionic complex becomes associated with a cation added to the system as a salt R^+A^-, include the following:

a. Formation of the anionic complex by a series of equilibria with overall form

$$M(H_2O)_x^{m+} + nL^- \rightleftharpoons ML_n^-(aq) + xH_2O\,(K_f)$$

b. Association of the salt in the aqueous phase

$$R^+(aq) + A^-(aq) \rightleftharpoons R^+A^-(aq)(K_{a,1})$$

c. Distribution of the ion-pair R^+A^-

$$R^+A^-(aq) \rightleftharpoons R^+A^-(o)(K_{D,1})$$

d. Formation in the aqueous phase of the metal-containing ion-association complex

$$R^+(aq) + ML_n^-(aq) \rightleftharpoons R^+ML_n^-(aq)(K_{a,2})$$

e. Distribution of the ion-association complex

$$R^+ML_n^-(aq) \rightleftharpoons R^+ML_n^-(o)(K_{D,2})$$

This scheme may be further complicated by the formation of dimers and higher polymeric species in the organic phase, as shown in Fig. 4.4b, and by the participation of organic solvent molecules or hydrated hydronium ions in the formation of extractable species. As before, it is possible to discuss the behavior of the system by expressing D as a function of the equilibrium constants and the total concentrations of L^-, R^+, and M.

Quantitative treatment of solvent extraction equilibria, including detailed discussion of some specific systems (e.g., metal oxinates and dithizonates, iron(III) chlorocomplexes, uranyl nitrate complexes), can be found in specialized texts on solvent extraction[118-120] and in some works dealing with spectrophotometry (e.g., Sandell[121]).

Selectivity

Many organic reagents of the type shown in Table 4.3 are relatively nonspecific in their action, and form chelates with 20 to 60 or more elements. Where solvent extraction is being used to concentrate a number of elements from a large volume of an aqueous phase, prior to analysis by a highly specific technique, this lack of selectivity may be no disadvantage. However, many analyses by spectrophotometric or radiometric methods, for example, must gain their specificity from the separation stage of the procedure.

Separation of two species, A and B, is possible if there is a significant difference between the distribution ratios D_A and D_B, optimum separation being achieved when $D_A D_B = 1$. A number of means exist for adjusting distribution ratios to give satisfactory separation. These include the choice of reagent and solvent, the control of pH of the aqueous phase, and the addition of masking agents to form stable, nonextractable complexes with elements that are to be retained in the aqueous phase. For separation of many elements from one another in a complex multicomponent system, use can be made of discontinuous countercurrent distribution with an apparatus such as that developed by Craig.[122] Alternatively, extraction systems may be used in which the organic solvent is caused to pass continuously through the aqueous medium; in continuous countercurrent distribution the two phases are made to flow in opposite directions.

Kinetic Factors

The rate of approach to equilibrium in a solvent extraction step depends on the rate of formation of the extractable species and on the rate of transfer of this species across the phase boundary. For the latter, equilibrium is usually achieved after no more than a few minutes' gentle shaking of the system. The rate of formation of extractable species, however, is in some cases a measurably slow process, particularly for the formation of some metal chelates from solutions containing only trace concentrations of the metal concerned. The influence of time on the completeness of extraction should therefore be checked when trace element extraction techniques are being developed. In favorable cases, kinetic differences can be used as a means of improving selectivity.

Applications to Trace Analysis of Geological Materials

The widespread uses of solvent extraction in trace analysis of geological materials include (a) the separation of groups of elements (e.g., transition elements, lanthanides, actinides) from other groups (e.g., alkali metals, alkaline earths), (b) the separation of closely related elements from one another (e.g., zirconium and hafnium, niobium and tantalum), (c) the separation of analytes from interfering elements, and (d) the concentration of traces into a smaller volume of solution.

Comprehensive periodic reviews[123,124] on both fundamental and applied aspects of solvent extraction appeared up to 1968. More recent work has been covered in several publications.[125–131] A great deal of information on the behavior of specific extraction systems is given in standard texts,[118–120] in which conditions have been summarized for the extraction of many metals with reagents such as sodium diethyldithiocarbamate, 2-thenoyltrifluoroacetone, and 8-hydroxyquinoline. Data relevant to the use of dithizone,[121] N-benzoyl-N-phenylhydroxylamine,[132] and ammonium pyrrolidine-1-carbodithioate[133] have also been given for many elements. Applications of extraction procedures using so-called liquid anion exchangers have been reviewed.[134] Extraction systems used in separations of most of the rarer metals, including the actinides, lanthanides, and platinum metals, have been covered in detail by Korkisch;[22] extractive separations of the actinides are also discussed by Bagnall.[135]

Solvent extraction is an integral part of many colorimetric procedures, being used chiefly to separate the wanted element from potentially interfering elements. In some cases spectrophotometric measurements are made directly on the organic extract; in other cases, back-extraction (stripping) of the analyte into an aqueous medium is used, followed by reaction with the color-forming reagent. Many procedures useful in silicate rock analysis are described in detail by Sandell,[121] Jeffery,[136] and Stanton.[137] Stary[127] has reviewed concisely the status of extraction methods for all elements of the periodic table, with particular reference to situations where a spectrophotometric finish to the analysis is possible.

Notable applications to analysis by atomic absorption spectrometry include the extraction of many elements by ammonium pyrrolidine-1-carbodithioate (APDC) into methyl isobutyl ketone,[133,138–141] the extraction of 1,10-phenanthroline complexes into nitrobenzene,[142] and the extraction of heteropolyacids[143–145] into solvents such as isobutyl acetate, butanol, and methyl isobutyl ketone. These solvents are suitable for direct aspiration into air/acetylene or nitrous oxide/acetylene flames. Diethyldithiocarbamate extraction has been used in an automated procedure for determining several transition elements by atomic fluorescence.[146]

Brooks[147] has demonstrated the use of discontinuous countercurrent extraction prior to emission spectrography for a number of trace elements in geological materials. The development of atomic spectroscopic methods less prone to interferences and matrix effects than emission spectrography has, however, lessened the need for complete separation of individual elements.

A variation of the usual procedure can be found in the extraction of transition elements with molten 8-hydroxyquinoline (m.p. 75°C), the organic phase after solidification being powdered and used for X-ray fluorescence analysis.[148]

In radiochemical work, solvent extraction has become a standard separation method, along with precipitation and coprecipitation procedures and ion exchange. Typical examples of extractions following radioactivation include those of gallium from a hydrochloric acid medium into diethyl ether,[149,150] gold from hydrochloric acid into ethyl acetate,[151–153] vanadium with cupferron into chloroform,[154] zirconium and hafnium with 2-thenoyltrifluoroacetone (TTA) into xylene,[155] and the separation of zirconium from hafnium by extraction of the zirconium with TTA into benzene.[156] Substoichiometric extractions[157,158] (pp. 280, 294) have been developed and used as part of various analyses by neutron activation or isotope dilution.[159–162] The use of solvent extraction or solvent extraction/ion exchange combinations in multielement activation analysis has been demonstrated,[163–165] and several workers[166–169] have devised comprehensive trace-element separation schemes based on a sequence of solvent extraction steps, prior to a finish by absorptiometric, fluorimetric, or electroanalytical methods.

Table 4.4 summarizes some representative applications of solvent extraction in the analysis of materials such as rocks, minerals and natural waters.

TABLE 4.4 Examples of Solvent Extraction Applications

Element	Extraction system	Analytical method, notes	Ref.
Al	8-Hydroxyquinoline/$CHCl_3$	Spectrophotometry	170
Au	HBr/isopropyl ether	Spectrophotometry	171
Be	Acetylacetone/benzene	Spectrophotometry; extraction of other elements masked with EDTA	172
Cd	Dithizone/CCl_4	Polarography	173
Co	Dithizone/CCl_4	Spectrophotometry of thiocyanate/tri-n-butylamine complex in benzene/amyl alcohol	173
Cu	2,2'-Diquinolyl/n-hexanol (neocuproine)	Spectrophotometry	174
F	Triphenylstibine dichloride/CCl_4	Spectrophotometry; CDTA liberates F^- from complexes; F^- back-extracted into aq. solution for analysis	175
Ga	Rhodamine B/chlorobenzene-CCl_4	Spectrophotometry	176
Ge	HCl/CCl_4	Spectrophotometry of phenylfluorone complex	177
Hg	NaI/benzene	Gas chromatography of organomercury halides	178
I	CCl_4	Neutron activation analysis of Te, using activity of ^{131}I daughter; I_2 back-extracted and converted into AgI	179
In	HBr/diethyl ether	Neutron activation analysis	180
Mo	Toluene-3,4-dithiol/amyl acetate	Spectrophotometry	181
Nb	Thiocyanate/ethyl acetate	Spectrophotometry	182, 183
Ni	Dimethylglyoxime/$CHCl_3$	Spectrophotometry	184
Os	$CHCl_3$	Neutron activation analysis; extracted as tetroxide	185

SEPARATION AND CONCENTRATION TECHNIQUES 59

TABLE 4.4 (*continued*)

Element	Extraction system	Analytical method, notes	Ref.
P	Molybdate/isoamyl alcohol	Radiometric determination of ^{32}P as phosphomolybdate	186
Pb	DEDTC/pentanol-toluene	Spectrophotometry with dithizone	187
Pd	Dimethylglyoxime/CHCl$_3$	Neutron activation analysis	188
Pt	Dithizone/MIBK	Atomic absorption spectroscopy	189
Re	H$_2$SO$_4$/benzyl alcohol; NaOH/methyl ethyl ketone	Neutron activation analysis	190
Ru	CCl$_4$	Neutron activation analysis; extracted as tetroxide	185
Sb	HCl/brilliant green/toluene	Spectrophotometry	191
Se	Toluene-3,4-dithiol/CCl$_4$-dichloroethane	Fluorimetry with 3,3'-diaminobenzidine	192
Sn	TTA/MIBK	Neutron activation analysis	193
Ta	HF-H$_2$SO$_4$/MIBK	Neutron activation analysis	194
Te	HCl/MIBK	Atomic absorption spectroscopy	195
Th	TTA/xylene	Radiotracer method	196
Tl	HBr/diethyl ether	Spectrophotometry of complex with brilliant green	136, 197
U	HCl/TBP-MIBK	Spectrophotometry with 8-hydroxyquinoline	198
U	HDEHP/CCl$_4$	Pulse polarography	199
V	PAR/Zephiramine-CHCl$_3$	Spectrophotometry	200
W	Toluene-3,4-dithiol/amyl acetate	Spectrophotometry	201
Zr	TTA/benzene	Neutron activation analysis	156, 202

Abbreviations: CDTA, cyclohexanediaminetetra-acetic acid; DEDTC, diethyldithiocarbamate; EDTA, ethylenediaminetetra-acetic acid; HDEHP, *bis*-(2-ethylhexyl)-*o*-phosphoric acid; MIBK, 4-methyl-2-pentanone (methyl isobutyl ketone); PAR, 4-(2-pyridylazo)resorcinol; TBP, tri-*n*-butyl phosphate; TTA, 2-thenoyltrifluoroacetone; Zephiramine, tetradecyldimethylbenzyl ammonium chloride.

4.2.5 Other Separation Methods

Several other separation methods are less widely used in the analysis of geological materials than those discussed in the previous sections, or are applicable only to smaller groups of elements. The most important of these methods are noted briefly below.

Volatilization

Several useful separations are based on transfer of the analyte into the vapor phase. Some analytical methods deal with the vapor directly, whereas others involve condensation and further chemical treatment.

Elemental mercury is readily separated from solid samples by heating and can also be liberated from solution after reduction of its compounds with a variety of reducing agents. This separation has been extensively adopted in the determination of mercury by atomic absorption and atomic fluorescence spectroscopy.[203]

Compounds that can be volatilized readily from suitable solid or solution media include hydrides (e.g., HF, AsH_3, SbH_3, BiH_3, SeH_2, TeH_2, GeH_4), oxides (e.g., OsO_4, RuO_4, Re_2O_7) and halides (e.g., arsenic and antimony bromides, $AsCl_3$, $GeCl_4$, $SeBr_4$, SnI_4). Notes on some of these separations are given in Table 4.5.

Work on the volatilization of various halides from media containing perchloric, sulfuric, and phosphoric acids, together with hydrobromic or hydrochloric acids, has been tabulated by Sandell (Ref. 121, p. 73). Vaporization of organometallic compounds has been used in gas chromatographic separations (see p. 61).

TABLE 4.5 Separations by Volatilization

Elements	Volatile compounds	Notes	Ref.
As	AsH_3	Reduced by zinc and sulfuric or hydrochloric acid and other reducing agents.	121, 204–207
As, Sb	Bromides	Distilled from sulfuric/hydrochloric acid medium containing bromide and hydrazine sulfate.	136, 208
As, Bi, Ge, Pb, Sb, Se, Sn, Te	Hydrides	Liberated by reaction with sodium borohydride from acid solution.	209–211
F	HF	Distilled from sulfuric or perchloric acid. Can be collected on anion-exchange resin.	112, 136, 212
Ge	$GeCl_4$	Distilled from hydrochloric acid.	213
Os, Ru	OsO_4, RuO_4	Distilled from solution containing strong oxidizing agent, such as bismuthate or periodate. Usually collected in reducing or alkaline solutions, or hydrogen peroxide.	22, 185, 214, 215
Re	Re_2O_7	Distilled from sulfuric acid.	22, 151
Se	$SeBr_4$	Distilled from bromine/hydrobromic acid, or from hydrobromic/hydrochloric/sulfuric acid mixture.	22, 216
Sn	SnI_4	Distilled from mixture of solid with ammonium iodide.	136, 217, 218

Electrodeposition

Electrolysis of solutions with careful control of voltage or current provides selectivity in a number of electroanalytical methods, dealt with in more detail in Chapter 14. However, as a form of solution pretreatment for other analytical methods, electrochemical separation is not as widely used in dealing with geological samples as it is in some other fields, such as metallurgical analysis. Nevertheless, in some cases it is possible to use electrodeposition on an inert electrode to concentrate traces of easily reduced metals (e.g., Ag, Cd, Cu, Zn) from a large volume of solution. Alternatively, it may be possible to remove an interfering element by electrolytic means. By controlled-potential electrolysis from a sulfuric acid solution, using a mercury cathode, many elements (e.g., Ag, Bi, Cd, Co, Cr, Cu, Fe, Mn, Ni, Pb, Zn) can be deposited, leaving alkali metals, alkaline earths, Al, Be, La, Ti, U, V, and Zr in solution. In general, the ease of removal depends on the valence state of the metal and on the nature of the electrolyte (acid concentration, presence of complex-forming ligands, etc.).

Several workers[219-221] have studied extensively the deposition of metals on to a mercury cathode from sulfuric acid solutions and uses of separations involving the deposition of mercury analgams have been reviewed.[222] Electrolytic removal of interfering ions has been used in the colorimetric determination of vanadium, titanium, and chromium,[223] and in the determination of beryllium, prior to extraction of its complex with acetylacetone.[224] Applications of electrodeposition on platinum[225] and pyrolytic graphite[226] in neutron activation analysis have been described. Deposition of hydrous oxides of elements such as uranium and thorium is used in the preparation of samples for measurements of α-particle radioactivity (p. 270).

The development of electrothermal methods of atomization for atomic absorption and atomic fluorescence spectroscopy has led to renewed interest in electrochemical processes for concentrating traces on to metal collectors. Traces of mercury have been concentrated by spontaneous deposition on copper[227] or silver.[228] Electroreduction of metals on to a hanging mercury drop[229,230] or tungsten filament[231] can be followed by electrothermal atomization.

Gas-Liquid Chromatography

This technique, in which volatile compounds in a moving gas stream are partitioned between the gas phase and a stationary liquid phase on a solid support, has been used for a limited range of inorganic separations. Complexes of aluminum, beryllium, gallium, and indium with acetylacetone or its trifluoro- and hexafluoro- derivatives have been separated in this way. The separation may be combined with an element-specific detection system (e.g., atomic emission or atomic absorption) to complete a quantitative analysis.

Extraction of the organometallic compound into an organic solvent such as benzene or toluene can provide very high sensitivity, as exhibited by the method developed for beryllium using its complex with trifluoroacetylacetone.[232] Other recent work in this field includes separations of β-diketone complexes of Al, Be, Cr, Cu, Fe, Ga, and In,[233,234] the determination of aluminum in seawater by separation of its trifluoroacetylacetonate,[235] and the development of a method for determining microgram amounts of selenium based on extraction and chromatography of an organoselenium compound with acetophenone.[236]

Gas-liquid chromatography is also useful in the separation and determination of alkylmercury compounds[237,238] and in the determination of alkyls of other elements, such as lead[239,240] and selenium.[241] Increasing use of systems combining gas chromatography and atomic spectroscopy is likely as interest develops in the chemical form of trace elements, especially in connection with the occurrence of pollutants in waters and sediments, and with their movement through the biosphere.

Partition Chromatography

This involves the distribution of solutes between two immiscible solvents, a stationary aqueous phase adsorbed on a column of material such as silica gel or cellulose, and a mobile phase (mainly organic) that passes through the column of adsorbent. Separation of solutes with different distribution ratios occurs as a result of continuous partitioning between the two phases.

High-speed (or high-pressure) liquid chromatography is a recent development, which so far has found few applications in inorganic separations. However, there are several other important modifications of the basic partition technique.

In *paper chromatography* the aqueous phase is supported on a paper sheet, and the largely organic mobile phase moves vertically (ascending or descending) through the paper by capillary action. This is particularly useful in separations where only small volumes of sample (e.g., 10–100 μl) are available. The solutes in the developed chromatogram are located by spraying the paper with a reagent solution that forms colored or fluorescent products with the metal ions concerned. Alternatively, radiometric methods with tracers may be used to follow the solutes. The position of each solute in the developed chromatogram is indicated by its R_f value, this being the ratio of the distance traveled by the solute to the distance traveled by the solvent front.

Separations on paper can also be carried out by developing the chromatogram outwards from a central spot. In the ring-oven technique[242] a drop of solution is applied to the center of a filter-paper disk. Some components of the mixture are immobilized in the paper by a precipitant, whereas soluble components are carried outward in concentric rings by a developing solvent that evaporates near the outer edge of the paper.

In *electrochromatography* the separation is based on the differential migration of ions under an applied potential difference. The aqueous phase is stabilized by adsorption on a material such as paper, gelatin, or silica gel. Separations can be achieved in two dimensions simultaneously by allowing the developing solvent to descend through the adsorbent while differential electromigration is made to occur at right angles.

In *thin-layer chromatography* the stationary phase takes the form of a thin layer of adsorbent on a glass plate, the mobile phase being allowed to ascend through the layer. The stationary phase often consists of water adsorbed on silica gel containing binders and other additives. Among the advantages of thin-layer chromatography is the possibility of using corrosive solutions that would destroy paper and similar materials.

In *reversed-phase partition chromatography* (extraction chromatography) the stationary phase consists of an organic liquid attached to a porous support, and the mobile phase is aqueous. In inorganic applications it is common to incorporate a selective extractant in the stationary phase. Widely used reagents include tri-*n*-butyl phosphate (TBP), *bis*-(2-ethylhexyl)-*o*-phosphoric acid (HDEHP), tri-*n*-octylphosphine oxide (TOPO), 2-thenoyltrifluoroacetone (TTA), tertiary amines such as tri-*n*-octylamine (TOA), and quaternary ammonium salts such as Aliquat 336.

For detailed discussion of the theory and practice of the various forms of liquid-liquid partition chromatography, the reader is referred to specialized texts.[243–248] The book by Pollard and McOmie[249] deals with early work on inorganic chromatography; later work on inorganic paper chromatography is described by Kakáč.[250] Short sections on applications to inorganic systems are contained in the biennial reviews on chromatography.[251,252] Reviews devoted to inorganic systems include a wide-ranging account by Nickless,[253] and articles on applications of chromatography to geology,[254] paper chromatography of halide and thiocyanate complexes with liquid anion exchangers as the mobile phase,[255] chromatography on columns of naturally occurring and substituted celluloses,[256] partition chromatography of the lanthanides and actinides,[257] thin-layer chromatography,[258] and reversed-phase liquid chromatography.[259] A large amount of data on thin-layer chromatography of inorganic systems has been collected by Brinkman et al.[260]

The following examples are noted as an illustration of the range of inorganic separations achieved by using various liquid-liquid chromatographic methods. Uranium has been separated from other elements by classical partition chromatography on cellulose columns,[261-264] with diethyl ether/nitric acid as the mobile phase; methods for finishing the analysis include polarography, colorimetry, and fluorimetry. Niobium and tantalum have been separated from each other on cellulose with a mobile phase consisting of methyl ethyl ketone and hydrofluoric acid,[265] and the extraction of gold on cellulose has been described.[266] High-speed liquid chromatography has been used to separate a number of organometallic and metal chelate compounds.[267,268]

Paper chromatographic separations form the basis of semiquantitative methods for determining vanadium, molybdenum, and gold in rocks, minerals, and soils,[269] and for determining thorium and uranium in monazite.[270] Applications of paper chromatography in mineral prospecting have been described.[271] Majumdar and Chakrabartty[272] have investigated the separation of a large number of inorganic ions on paper. The use of paper impregnated with a complexing agent (TTA) for separating mixtures of transition-metal ions, with a ketone/hydrochloric acid developing solvent, has been demonstrated.[273] A ring-oven method for the separation of beryllium has been reported,[274] the fluorescence of the complex with morin being used for the determination of 10–200 ng beryllium.

Thin-layer chromatography with a colorimetric finish has been used to separate and determine tellurium at levels of about 10 μg/g in cinnabar.[275] It has also found applications in qualitative and semiquantitative analysis of trace elements in minerals[276] and waters.[277]

The electrochromatographic behavior of many cations in various electrolyte media has been reported by Gross.[278] High-voltage electrophoresis has been used in the determination of lanthanides in meteorites.[279]

Among the wide range of separations by reverse-phase (extraction) liquid chromatography, particular note may be made of separations involving the lanthanides, actinides, thorium, and uranium. Lanthanides have been separated on silica columns treated with HDEHP,[280] (see Fig. 4.5), and on Kieselguhr columns treated with TBP,[281,282] the columns being eluted with nitric acid or hydrochloric acid. TBP on Kel-F (powdered polytrifluorochloroethylene) has been used in the separation of uranium from processing wastes.[283] Lanthanides, thorium, uranium, and zirconium have been separated with a liquid anion exchanger on silica gel,[284] using a sulfuric acid/ammonium sulfate aqueous phase. Both paper and Kel-F treated with TOPO have been used in studies of the separation of a large number of metal ions.[285] The incorporation of substituted hydroquinones as reducing agents with TOPO on a microporous polyethylene column has allowed traces of

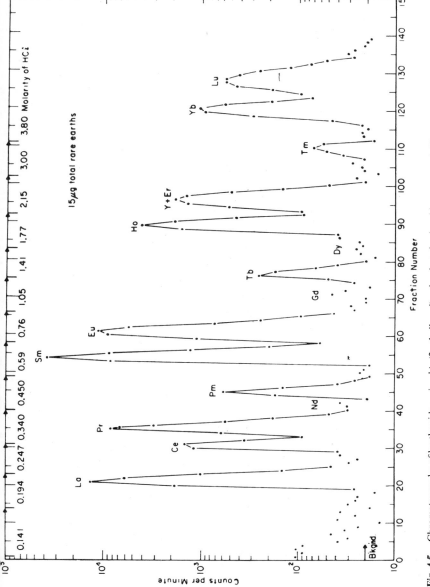

Fig. 4.5. Chromatography of lanthanides using *bis*-(2-ethylhexyl)orthophosphoric acid on a treated silica column. Separation monitored by measuring radioactivity induced by 1-hr neutron activation. *Source*: Winchester.[280]

actinides to be separated by red-ox extraction chromatography,[286] the separation being based on properties of lower valence states [e.g., Pu(III), Np(IV)] than exist in the initial aqueous phase.

Although increasing use is being made of more specific and interference-free analytical techniques that make fewer demands on separation stages, chromatographic and other separation methods are likely to continue to be of importance in the analysis of geological materials. In the first place such methods permit the extraction of selected groups of trace elements from solution and their concentration in a small volume. Second, chromatographic techniques allow the separation of different compounds of the same element and are becoming important adjuncts to analytical methods that simply yield the total concentration of any given element in the sample, regardless of its chemical form.

References

1. J. Dolezal, P. Povondra, and Z. Sulcek, *Decomposition Techniques in Inorganic Analysis*, Elsevier, New York, 1968; Iliffe, London, 1969.
2. G. Tölg, *Talanta*, **19**, 1489 (1972).
3. F. J. Langmyhr and S. Sveen, *Anal. Chim. Acta*, **32**, 1 (1965).
4. J. Ito, *Bull. Chem. Soc. Jap.*, **35**, 225 (1962).
5. I. May and J. J. Rowe, *Anal. Chim. Acta*, **33**, 648 (1965).
6. F. J. Langmyhr and P. E. Paus, *Anal. Chim. Acta*, **43**, 397 (1968).
7. B. Bernas, *Anal. Chem.*, **40**, 1682 (1968).
8. F. J. Langmyhr, *Anal. Chim. Acta*, **39**, 516 (1967).
9. G. K. Hoops, *Geochim. Cosmochim. Acta*, **28**, 405 (1964).
10. F. J. Langmyhr and P. E. Paus, *Anal. Chim. Acta*, **50**, 515 (1970).
11. T. A. Rafter and F. T. Seelye, *Nature*, **165**, 317 (1950).
12. C. B. Belcher, *Talanta*, **10**, 75 (1963).
13. M. S. Wang, *Appl. Spectrosc.*, **16**, 141 (1962).
14. V. S. Biskupsky, *Anal. Chim. Acta*, **33**, 333 (1965).
15. C. O. Ingamells, *Talanta*, **11**, 665 (1964).
16. N. H. Suhr and C. O. Ingamells, *Anal. Chem.*, **38**, 730 (1966).
17. C. O. Ingamells, *Anal. Chem.*, **38**, 1228 (1966).
18. F. E. Beamish, *Talanta*, **5**, 1 (1960); **14**, 991, 1133 (1967).
19. C. E. Gleit and W. D. Holland, *Anal. Chem.*, **34**, 1454 (1962).
20. G. Kaiser, P. Tschöpel, and G. Tölg, *Z. Anal. Chem.*, **253**, 177 (1971).
21. T. T. Gorsuch, *The Destruction of Organic Matter*, Pergamon, Oxford, 1970.

REFERENCES

22. J. Korkisch, *Modern Methods for the Separation of Rarer Metal Ions*, Pergamon, Oxford, 1969.
23. R. E. Stanton and A. J. McDonald, *Analyst*, **89**, 767 (1964).
24. T. Shigematsu, M. Tabushi, and F. Isojima, *Jap. Anal.*, **11**, 752 (1962).
25. M. Ishibashi, T. Shigematsu, M. Tabushi, Y. Nishikawa, and S. Goda, *J. Chem. Soc. Jap., Pure Chem. Sect.*, **83**, 295 (1962).
26. B. L. Hampson, *Analyst*, **89**, 651 (1964).
27. D. C. Burrell, *Perkin-Elmer At. Absorpt. Newsl.*, **4**, 309 (1965).
28. B. C. Purkayastha and S. N. Bhattacharyya, *J. Inorg. Nucl. Chem.*, **10**, 103 (1959).
29. F. Culkin and J. P. Riley, *Nature*, **181**, 180 (1958).
30. H. Onishi and C. V. Banks, *Talanta*, **10**, 399 (1963).
31. H. V. Weiss and M. G. Lai, *Talanta*, **8**, 72 (1961).
32. T. Shigematsu, M. Tabushi, Y. Nishikawa, K. Hiraki, S. Goda, and R. Inoue, *J. Chem. Soc. Jap. Pure Chem. Sect.*, **84**, 336 (1963).
33. Y. K. Chau and J. P. Riley, *Anal. Chim. Acta*, **33**, 36 (1965).
34. A. DuFresne, *Geochim. Cosmochim. Acta*, **20**, 141 (1960).
35. B. C. Severne and R. R. Brooks, *Talanta*, **19**, 1467 (1972).
36. C. W. Sill and C. P. Willis, *Anal. Chem.*, **36**, 622 (1964).
37. S. Abbey, *Anal. Chim. Acta*, **30**, 176 (1964).
38. A. P. Smith and F. S. Grimaldi, *U.S. Geol. Surv. Bull.*, **1006**, 125 (1954).
39. W. L. Kehl and R. G. Russell, *Anal. Chem.*, **28**, 1350 (1956).
40. K. Sugawara, M. Tanaka, and H. Naito, *Bull. Chem. Soc. Jap.*, **26**, 417 (1953).
41. H. Naito and K. Sugawara, *Bull. Chem. Soc. Jap.*, **30**, 799 (1957).
42. T. Shigematsu, Y. Nishikawa, K. Hiraki, and H. Nakagawa, *J. Chem. Soc. Jap. Pure Chem. Sect.*, **85**, 490 (1964).
43. W. Rieman and H. F. Walton, *Ion Exchange in Analytical Chemistry*, Pergamon, New York, 1970.
44. O. Samuelson, *Ion Exchange Separations in Analytical Chemistry*, Wiley, New York, 1963.
45. K. Dorfner, *Ion Exchangers. Properties and Applications*, 3rd ed., Ann Arbor Science Publishers, Ann Arbor, Mich., 1972.
46. R. Kunin, *Ion Exchange Resins*, 2nd ed., Wiley, New York, 1958.
47. H. F. Walton, *Anal. Chem.*, **36**, 51R (1964); **38**, 79R (1966); **40**, 51R (1968); **42**, 86R (1970); **44**, 256R (1972); **46**, 398R (1974); **48**, 52R (1976).
48. R. Kunin and F. X. McGarvey, *Anal. Chem.*, **36**, 142R (1964); R. Kunin, *Anal. Chem.*, **38**, 176R (1966); R. Kunin and R. L. Gustafson, *Anal. Chem.*, **40**, 136R (1968).
49. F. W. E. Strelow, in *Ion Exchange and Solvent Extraction*, Vol. 5 (eds. J. A. Marinsky and Y. Marcus), Marcel Dekker, New York, 1973.
50. C. B. Amphlett, *Inorganic Ion Exchangers*, Elsevier, New York, 1964.

51. V. Vesely and V. Pekarek, *Talanta*, **19**, 219, 1245 (1972).
52. R. A. A. Muzzarelli, *Natural Chelating Polymers*, Pergamon, Oxford, 1973.
53. F. Vernon and H. Eccles, *Anal. Chim. Acta*, **63**, 403 (1973).
54. G. Koster and G. Schmuckler, *Anal. Chim. Acta*, **38**, 179 (1967).
55. A. J. Bauman, H. H. Weetall, and N. Weliky, *Anal. Chem.*, **39**, 932 (1967).
56. J. M. Hill, *J. Chromatogr.*, **76**, 455 (1973).
57. D. M. Hercules, L. E. Cox, S. Onisick, G. D. Nichols, and J. C. Carver, *Anal. Chem.*, **45**, 1973 (1973).
58. H. L. Rothbart, in *An Introduction to Separation Science* (eds. B. L. Karger, L. R. Snyder, and C. Horvath), Wiley-Interscience, New York, 1973.
59. F. I. Brownley and C. W. Howle, *Anal. Chem.*, **32**, 1330 (1960).
60. H. N. S. Schafer, *Anal. Chem.*, **35**, 53 (1963).
61. K. A. Kraus, T. A. Carlson, and J. S. Johnson, *Nature*, **177**, 1128 (1956).
62. J. Korkisch, F. Feik, and S. S. Ahluwalia, *Talanta*, **14**, 1069 (1967).
63. T. Cummings and J. Korkisch, *Talanta*, **14**, 1185 (1967).
64. W. Koch and J. Korkisch, *Mikrochim. Acta*, 101, 117, 225, 245, 263, 877 (1973).
65. H. L. Smith and D. C. Hoffman, *J. Inorg. Nucl. Chem.*, **3**, 243 (1956).
66. K. Rengan and W. W. Meinke, *Anal. Chem.*, **36**, 157 (1964).
67. B. H. Ketelle and G. E. Boyd, *J. Am. Chem. Soc.*, **69**, 2800 (1947).
68. A. W. Mosen, R. A. Schmitt, and J. Vasilevskis, *Anal. Chim. Acta*, **25**, 10 (1961).
69. G. Duyckaerts and R. Lejeune, *J. Chromatogr.*, **3**, 58 (1960).
70. K. A. Kraus and G. E. Moore, *J. Am. Chem. Soc.*, **75**, 1460 (1953).
71. K. A. Kraus and F. Nelson, *Proceedings of the 1st International Conference on the Peaceful Uses of Atomic Energy*, **7**, 113 (1955).
72. J. P. Faris, *Anal. Chem.*, **32**, 520 (1960).
73. J. P. Faris and R. F. Buchanan, *Anal. Chem.*, **36**, 1158 (1964).
74. F. W. E. Strelow and C. J. C. Bothma, *Anal. Chem.*, **39**, 595 (1967).
75. K. A. Kraus, F. Nelson, and G. E. Moore, *J. Am. Chem. Soc.*, **77**, 3972 (1955).
76. D. C. DeGeiso, W. Rieman, and S. Lindenbaum, *Anal. Chem.*, **26**, 1840 (1954).
77. J. Korkisch, *Z. Anal. Chem.*, **178**, 39 (1960).
78. A. A. Smales and L. Salmon, *Analyst*, **80**, 37 (1955).
79. J. P. Riley and D. Taylor, *Anal. Chim. Acta*, **40**, 479 (1968).
80. J. P. Riley and D. Taylor, *Anal. Chim. Acta*, **41**, 175 (1968).
81. S. L. Law, *Science*, **174**, 285 (1971).
82. K. Kawabuchi and R. Kuroda, *Anal. Chim. Acta*, **46**, 23 (1969).
83. R. A. A. Muzzarelli and O. Tubertini, *Talanta*, **16**, 1571 (1969).
84. C. L. Luke, *Anal. Chem.*, **36**, 318 (1964).
85. J. G. Bergmann, C. H. Ehrhardt, L. Granatelli, and J. L. Janik, *Anal. Chem.*, **39**, 1258 (1967).

REFERENCES

86. T. E. Green, S. L. Law, and W. J. Campbell, *Anal. Chem.*, **42**, 1749 (1970).
87. F. W. E. Strelow, C. H. S. W. Weinert, and T. N. van der Walt, *Anal. Chim. Acta*, **71**, 123 (1974).
88. L. E. Reichen, *Anal. Chem.*, **30**, 1948 (1958).
89. F. W. E. Strelow, F. S. Toerien, and C. H. S. W. Weinert, *Anal. Chim. Acta*, **50**, 399 (1970).
90. E. Bolter, K. K. Turekian, and D. F. Schutz, *Geochim. Cosmochim. Acta*, **28**, 1459 (1964).
91. M. J. Cabell and A. A. Smales, *Analyst*, **82**, 390 (1957).
92. O. Osterreid, *Z. Anal. Chem.*, **199**, 260 (1963).
93. F. Nydahl, *Proc. Int. Assoc. Theor. Appl. Limnol.*, **11**, 276 (1951).
94. D. G. Biechler, *Anal. Chem.*, **37**, 1054 (1965).
95. A. O. Brunfelt, O. Johansen, and E. Steinnes, *Anal. Chim. Acta*, **37**, 172 (1967).
96. J. H. Crocket, *Geochim. Cosmochim. Acta*, **36**, 517 (1972).
97. R. R. Brooks, *Analyst*, **85**, 745 (1960).
98. M. E. Fleet, *Anal. Chem.*, **39**, 253 (1967).
99. A. D. Matthews and J. P. Riley, *Anal. Chim. Acta*, **48**, 25 (1969).
100. R. A. Edge and L. H. Ahrens, *Anal. Chim. Acta*, **26**, 355 (1962).
101. J. Korkisch and G. Arrhenius, *Anal. Chem.*, **36**, 850 (1964).
102. N. E. Cohen, R. D. Reeves, and R. R. Brooks, *Talanta*, **15**, 1449 (1968).
103. U. Krähenbühl, H. P. Rolli, and H. R. von Gunten, *Helv. Chim. Acta*, **55**, 697 (1972).
104. J. D. Smith, *Anal. Chim. Acta*, **57**, 371 (1971).
105. W. Kiesl, *Z. Anal. Chem.*, **227**, 13 (1967).
106. L. P. Greenland, *Anal. Chim. Acta*, **42**, 365 (1968).
107. A. O. Brunfelt and E. Steinnes, *Anal. Chem.*, **39**, 833 (1967).
108. J. Korkisch and W. Koch, *Mikrochim. Acta*, 157, 865 (1973).
109. J. Korkisch and L. Gödl, *Anal. Chim. Acta*, **71**, 113 (1974).
110. T. Kiriyama and R. Kuroda, *Anal. Chim. Acta*, **71**, 375 (1974).
111. H. M. Nielsen, *Anal. Chem.*, **30**, 1009 (1958).
112. W. H. Evans and G. A. Sergeant, *Analyst*, **92**, 690 (1967).
113. F. S. Kelso, J. M. Matthews, and H. P. Kramer, *Anal. Chem.*, **36**, 577 (1964).
114. R. E. Bently, R. P. Parker, D. M. Taylor, and M. P. Taylor, *Nature*, **194**, 736 (1962).
115. L. H. Ahrens, R. A. Edge, and R. R. Brooks, *Anal. Chim. Acta*, **28**, 551 (1963).
116. F. W. E. Strelow, C. J. Liebenberg, and F. S. Toerien, *Anal. Chim. Acta*, **47**, 251 (1969).
117. H. Freiser, in *An Introduction to Separation Science* (eds. B. L. Karger, L. R. Snyder, and C. Horvath), Wiley-Interscience, New York, 1973.

118. G. H. Morrison and H. Freiser, *Solvent Extraction in Analytical Chemistry*, Wiley, New York, 1957.
119. J. Stary, *The Solvent Extraction of Metal Chelates*, Pergamon, Oxford, 1964.
120. Y. Marcus and A. S. Kertes, *Ion Exchange and Solvent Extraction of Metal Complexes*, Wiley, New York, 1969.
121. E. B. Sandell, *Colorimetric Determination of Traces of Metals*, 3rd ed., Interscience, New York, 1959.
122. L. C. Craig, W. Hausmann, E. H. Ahrens, and E. J. Harfenist, *Anal. Chem.*, **23**, 1236 (1951).
123. G. H. Morrison and H. Freiser, *Anal. Chem.*, **36**, 93R (1964).
124. H. Freiser, *Anal. Chem.*, **38**, 131R (1966); **40**, 552R (1968).
125. A. S. Kertes and Y. Marcus, *Solvent Extraction Research*, Wiley, New York, 1970.
126. C. Hanson (ed.), *Recent Advances in Liquid—Liquid Extraction*, Pergamon, Oxford, 1971.
127. J. Stary, in *M.T.P. International Review of Science*, Physical Chemistry Series One, Vol. 12: Analytical Chemistry, Part I (ed. T. S. West), Butterworths, London, 1973.
128. J. A. Marinsky and Y. Marcus (eds.), *Ion Exchange and Solvent Extraction*, Vol. V, Marcel Dekker, New York, 1973.
129. H. M. N. H. Irving, *Chem. Ind.*, 660 (1974).
130. N. Suzuki, T. Kato, K. Saitoh, and H. Watarai, *Jap. Anal.*, 13R (1972).
131. T. Sekine, Y. Zeniya, H. Honda, N. Masui, and Y. Hasegawa, *Jap. Anal.*, 11R (1974).
132. A. Riedel, *J. Radioanal. Chem.*, **13**, 125 (1973).
133. W. J. Price, *Analytical Atomic Absorption Spectrometry*, Heyden, London, 1972.
134. H. Green, *Talanta*, **20**, 139 (1973).
135. K. W. Bagnall, *The Actinide Elements*, Elsevier, Amsterdam, 1972.
136. P. G. Jeffery, *Chemical Methods of Rock Analysis*, 2nd ed., Pergamon, Elmsford, N.Y., 1975.
137. R. E. Stanton, *Rapid Methods of Trace Analysis*, Arnold, London, 1966.
138. J. E. Allan, *Spectrochim. Acta*, **17**, 459 (1961); *Analyst*, **86**, 530 (1961).
139. R. E. Mansell, *Perkin-Elmer At. Absorpt. Newsl.*, **4**, 276 (1965).
140. J. Ramirez-Munoz, *Beckman Flame Notes*, **1**, 8 (1966).
141. K. Kremling and H. Petersen, *Anal. Chim. Acta*, **70**, 35 (1974).
142. A. A. Schilt, R. L. Abraham, and J. E. Martin, *Anal. Chem.*, **45**, 1808 (1973).
143. G. F. Kirkbright, A. M. Smith, and T. S. West, *Analyst*, **92**, 411 (1967).
144. T. V. Ramakrishna, J. W. Robinson, and P. W. West, *Anal. Chim. Acta*, **45**, 43 (1969).
145. G. F. Kirkbright and H. N. Johnson, *Talanta*, **20**, 433 (1973).
146. M. Jones, G. F. Kirkbright, L. Ranson, and T. S. West, *Anal. Chim. Acta*, **63**, 210 (1973).

147. R. R. Brooks, *Talanta*, **12**, 505, 511 (1965).
148. B. Magyar and F. I. Lobanov, *Talanta*, **20**, 55 (1973).
149. H. Brown and E. D. Goldberg, *Science*, **109**, 347 (1949).
150. D. F. C. Morris and M. E. Chambers, *Talanta*, **5**, 147 (1960).
151. E. D. Goldberg and H. Brown, *Anal. Chem.*, **22**, 308 (1950).
152. E. A. Vincent and A. A. Smales, *Geochim. Cosmochim. Acta*, **9**, 154 (1956).
153. A. R. DeGrazia and L. Haskin, *Geochim. Cosmochim. Acta*, **28**, 559 (1964).
154. D. M. Kemp and A. A. Smales, *Anal. Chim. Acta*, **23**, 397 (1960).
155. E. Merz, *Geochim. Cosmochim. Acta*, **26**, 347 (1962).
156. J. L. Setser and W. D. Ehmann, *Geochim. Cosmochim. Acta*, **28**, 769 (1964).
157. J. Ruzicka and J. Stary, *Talanta*, **8**, 228 (1961).
158. J. Ruzicka and J. Stary, *Substoichiometry in Radiochemical Analysis*, Pergamon, Oxford, 1968.
159. T. B. Pierce and P. F. Peck, *Analyst*, **88**, 603 (1963).
160. J. Stary and J. Ruzicka, *Talanta*, **18**, 1 (1971).
161. E. E. Rakovskii, T. D. Krylova, M. I. Starozhitskaya, and A. Y. Frolova, *Z. Anal. Khim.*, **28**, 1554 (1973).
162. J. W. Mitchell and R. Ganges, *Talanta*, **21**, 735 (1974).
163. A. Alian and R. Shabana, *Microchem. J.*, **12**, 427 (1967).
164. J. Korkisch, *Nature*, **210**, 626 (1966).
165. G. H. Morrison, J. T. Gerard, A. Travesi, R. L. Currie, S. F. Peterson and N. M. Potter, *Anal. Chem.*, **41**, 1633 (1969).
166. P. C. van Erkelens, *Anal. Chim. Acta*, **25**, 129 (1961).
167. R. A. Chalmers and D. M. Dick, *Anal. Chim. Acta*, **32**, 117 (1965).
168. H. Förster and K. Schwabe, *Anal. Chim. Acta*, **45**, 511 (1969).
169. B. Morsches and G. Tölg, *Z. Anal. Chem.*, **250**, 81 (1970).
170. J. P. Riley, *Anal. Chim. Acta*, **19**, 413 (1958).
171. W. A. E. McBryde and J. H. Yoe, *Anal. Chem.*, **20**, 1094 (1948).
172. J. R. Merrill, M. Honda, and J. R. Arnold, *Anal. Chem.*, **32**, 1420 (1960).
173. R. E. Stanton, A. J. McDonald, and I. Carmichael, *Analyst*, **87**, 134 (1962).
174. J. P. Riley and P. Sinhaseni, *Analyst*, **83**, 299 (1958).
175. H. Chermette, M. Perrousset, and J. Ratelade, *Anal. Chim. Acta*, **70**, 217 (1974).
176. F. Culkin and J. P. Riley, *Analyst*, **83**, 208 (1958).
177. W. A. Schneider and E. B. Sandell, *Mikrochim. Acta*, 263 (1954).
178. J. A. Ealy, W. D. Shults, and J. A. Dean, *Anal. Chim. Acta*, **64**, 235 (1973).
179. N. Lavi, *Anal. Chim. Acta*, **70**, 199 (1974).
180. A. A. Smales, J. van R. Smit and H. Irving, *Analyst*, **82**, 539 (1957).
181. L. J. Clark and J. H. Axley, *Anal. Chem.*, **27**, 2000 (1955).
182. F. S. Grimaldi, *Anal. Chem.*, **32**, 119 (1960).

183. J. Esson, *Analyst*, **90**, 488 (1965).
184. L. F. Rader and F. S. Grimaldi, U.S. Geol. Surv. Prof. Paper 391-A (1961).
185. G. L. Bate and J. R. Huizenga, *Geochim. Gosmochim. Acta*, **27**, 345 (1963).
186. W. W. Flynn and W. R. Meehan, *Anal. Chim. Acta*, **63**, 483 (1973).
187. J. C. Gage, *Analyst*, **80**, 789 (1955); **82**, 453 (1957).
188. D. F. C. Morris, N. Hill, and B. A. Smith, *Mikrochim. Acta*, 962 (1963).
189. A. Simonsen, *Anal. Chim. Acta*, **49**, 368 (1970).
190. D. F. C. Morris and F. W. Fifield, *Talanta*, **8**, 612 (1961).
191. R. E. Stanton and A. J. McDonald, *Analyst*, **87**, 299 (1962).
192. J. H. Watkinson, *Anal. Chem.*, **32**, 981 (1960).
193. D. Schmidt, *Isotopenpraxis*, **7**, 416 (1971).
194. D. H. F. Atkins and A. A. Smales, *Anal. Chim. Acta*, **22**, 462 (1960).
195. R. D. Beaty, *Anal. Chem.*, **45**, 234 (1973).
196. S. A. Reynolds, *Talanta*, **10**, 611 (1963).
197. N. T. Voskresenskaya, *Zavod. Lab.*, **24**, 395 (1958).
198. A. R. Eberle and M. W. Lerner, *Anal. Chem.*, **29**, 1134 (1957).
199. G. W. C. Milner, J. D. Wilson, G. A. Barnett, and A. A. Smales, *J. Electroanal. Chem.*, **2**, 25 (1961).
200. M. Nishimura, K. Matsunaga, T. Kudo, and F. Obara, *Anal. Chim. Acta*, **65**, 466 (1973).
201. B. F. Quin and R. R. Brooks, *Anal. Chim. Acta*, **58**, 301 (1972).
202. W. D. Ehmann and J. L. Setser, *Science*, **139**, 594 (1963).
203. A. M. Ure, *Anal. Chim. Acta*, **76**, 1 (1975).
204. H. Onishi and E. B. Sandell, *Geochim. Cosmochim. Acta*, **7**, 1 (1955).
205. D. Liederman, J. E. Bowen, and O. I. Milner, *Anal. Chem.*, **31**, 2052 (1959).
206. E. W. Fowler, *Analyst*, **88**, 380 (1963).
207. E. F. Dalton and A. J. Malanoski, *Perkin-Elmer At. Absorpt. Newsl.*, **10**, 92 (1971).
208. Y. V. Morachevskii and A. I. Kalinin, *Zavod. Lab.*, **27**, 272 (1961).
209. F. J. Schmidt and J. L. Royer, *Anal. Lett.*, **6**, 17 (1973).
210. F. J. Fernandez, *Perkin-Elmer At. Absorpt. Newsl.*, **12**, 93 (1973).
211. K. C. Thompson and D. R. Thomerson, *Analyst*, **99**, 595 (1974).
212. H. H. Willard and O. B. Winter, *Ind. Eng. Chem., Anal. Ed.*, **5**, 7 (1933).
213. H. J. Cluley, *Analyst*, **76**, 523, 530 (1951).
214. A. D. Westland and F. E. Beamish, *Mikrochim. Acta*, 625 (1957).
215. K. S. Chung and F. E. Beamish, *Talanta*, **15**, 823 (1968).
216. A. O. Brunfelt and E. Steinnes, *Geochim. Cosmochim. Acta*, **31**, 283 (1967).
217. B. Martinet, *Chim. Anal.*, **43**, 483 (1961).

REFERENCES

218. B. J. Heffernan, R. O. Archbold, and T. J. Vickers, *Proc. Australas. Inst. Min. Met.*, **223**, 65 (1967).
219. R. Bock and K.-G. Hackstein, *Z. Anal. Chem.*, **138**, 339 (1953).
220. H. O. Johnson, J. R. Weaver, and L. Lykken, *Anal. Chem.*, **19**, 481 (1947).
221. T. D. Parks, H. O. Johnson, and L. Lykken, *Anal. Chem.*, **20**, 148 (1948).
222. J. A. Page, J. A. Maxwell, and R. P. Graham, *Analyst*, **87**, 245 (1962).
223. R. M. Sherwood and F. W. Chapman, *Anal. Chem.*, **27**, 88 (1955).
224. T. Y. Toribara and P. S. Chen, *Anal. Chem.*, **24**, 539 (1952).
225. F. Lux, *Radiochim. Acta*, **1**, 20 (1962).
226. H. B. Mark and F. J. Berlandi, *Anal. Chem.*, **36**, 2062 (1964).
227. H. Brandenburger and H. Bader, *Helv. Chim. Acta*, **50**, 1409 (1967).
228. M. J. Fishman, *Anal. Chem.*, **42**, 1462 (1970).
229. C. Fairless and A. J. Bard, *Anal. Chem.*, **45**, 2289 (1973).
230. F. O. Jensen, J. Dolezal, and F. J. Langmyhr, *Anal. Chim. Acta*, **72**, 245 (1974).
231. W. Lund and B. V. Larsen, *Anal. Chim. Acta*, **70**, 299 (1974).
232. W. D. Ross and R. E. Sievers, *Talanta*, **15**, 87 (1968).
233. H. Kawaguchi, T. Sakamoto, and A. Mizuike, *Talanta*, **20**, 321 (1973).
234. H. Kawaguchi, T. Sakamoto, Y. Yoshida, and A. Mizuike, *Jap. Anal.*, **22**, 1434 (1973).
235. M.-L. Lee and D. C. Burrell, *Anal. Chim. Acta*, **66**, 245 (1973).
236. N. Iordanov, K. Daskalova, and N. Rizov, *Talanta*, **21**, 1217 (1974).
237. P. Zarnegar and P. Mushak, *Anal. Chim. Acta*, **69**, 389 (1974).
238. J. E. Longbottom, R. C. Dressman, and J. J. Lichtenberg, *J. Assoc. Off. Anal. Chem.*, **56**, 1297 (1973).
239. D. T. Coker, *Anal. Chem.*, **47**, 386 (1975).
240. D. A. Segar, *Anal. Lett.*, **7**, 89 (1974).
241. Y. K. Chau, P. T. S. Wong, and P. D. Goulden, *Anal. Chem.*, **47**, 2279 (1975).
242. H. Weisz, *Microanalysis by the Ring Oven Technique*, Pergamon, London, 1970; *Analyst*, **101**, 152 (1976).
243. G. Zweig and J. R. Whitaker, *Paper Chromatography and Electrophoresis, Vol. 1: Electrophoresis in Stabilizing Media*, Academic Press, New York, 1967.
244. J. Sherma and G. Zweig, *Paper Chromatography and Electrophoresis, Vol. 2: Paper Chromatograpy*, Academic Press, New York, 1971.
245. J. J. Kirkland (ed.), *Modern Practice of Liquid Chromatography*, Interscience, New York, 1971.
246. S. G. Perry, R. Amos, and P. I. Brewer, *Practical Liquid Chromatography*, Plenum, New York, 1972.
247. E. Stahl (ed.), *Thin-layer Chromatography; A Laboratory Manual*, 2nd ed (transl. M. R. F. Ashworth), Allen and Unwin, London, 1969.

248. B. L. Karger, L. R. Snyder, and C. Horvath, *An Introduction to Separation Science*, Wiley-Interscience, New York, 1973.
249. F. H. Pollard and J. F. W. McOmie, *Chromatographic Methods of Inorganic Analysis*, Butterworths, London, 1953.
250. B. Kakáč, in *Paper Chromatography* (eds. I. M. Hais and K. Macek), Academic Press, New York, 1963; also Publishing House of the Czechoslovak Academy of Sciences, Prague, 1963.
251. G. Zweig, R. B. Moore, and J. Sherma, *Anal. Chem.*, **42**, 349R (1970).
252. G. Zweig and J. Sherma, *Anal. Chem.*, **44**, 42R (1972); **46**, 73R (1974).
253. G. Nickless, *Adv. Chromatogr.*, **7**, 121 (1968).
254. A. S. Ritchie, *Adv. Chromatogr.*, **2**, 169 (1966).
255. S. Przeszlakowski, *Chromatogr. Rev.*, **15**, 29 (1971).
256. R. A. A. Muzzarelli, *Adv. Chromatogr.*, **5**, 127 (1968).
257. I. Stronski, *Isotopenpraxis*, **9**, 273 (1973).
258. M. Lederer, *Chromatogr. Rev.*, **9**, 115 (1967).
259. E. Cerrai and G. Ghersini, *Adv. Chromatogr.*, **9**, 3 (1970).
260. U. A. T. Brinkman, G. de Vries, and R. Kuroda, *J. Chromatogr.*, **85**, 187 (1973).
261. F. H. Burstall and R. A. Wells, *Analyst*, **76**, 396 (1951).
262. N. F. Kember, *Analyst*, **77**, 78 (1952).
263. D. I. Legge, *Anal. Chem.*, **26**, 1617 (1954).
264. J. A. S. Adams and W. J. Maeck, *Anal. Chem.*, **26**, 1635 (1954).
265. R. A. Mercer and R. A. Wells, *Analyst*, **79**, 339 (1954).
266. N. F. Kember and R. A. Wells, *Analyst*, **76**, 579 (1951).
267. J. F. K. Huber, J. C. Kraak, and H. Veening, *Anal. Chem.*, **44**, 1554 (1972).
268. D. R. Jones, H. C. Tung, and S. E. Manahan, *Anal. Chem.*, **48**, 7 (1976).
269. H. Agrinier, *Compt. Rend.*, **246**, 2761 (1958); **249**, 2365 (1959); **255**, 2801 (1962).
270. I. I. M. Elbeih and M. A. Abou-Elnaga, *Anal. Chim. Acta*, **19**, 123 (1958).
271. E. C. Hunt, A. A. North, and R. A. Wells, *Analyst*, **80**, 172 (1955).
272. A. K. Majumdar and M. M. Chakrabartty, *Anal. Chim. Acta*, **17**, 415 (1957); **19**, 129, 132 (1958).
273. A. K. De and C. R. Bhattacharyya, *Separ. Sci.*, **6**, 621 (1971).
274. P. W. West and E. Jungreis, *Anal. Chim. Acta*, **45**, 188 (1969).
275. I. I. Nazarenko and M. P. Volynets, *Z. Anal. Khim.*, **24**, 1209 (1969).
276. W. S. Sijperda and G. de Vries, *Geol. Mijnbouw*, **45**, 315 (1966); **47**, 197 (1968).
277. R. M. Cassidy, V. Miketukova, and R. W. Frei, *Anal. Lett.*, **5**, 115 (1972).
278. D. Gross, *J. Chromatogr.*, **10**, 221 (1963).
279. F. Grass and R. Kittl, *Mikrochim. Acta*, 371 (1971).
280. J. W. Winchester, *J. Chromatogr.*, **10**, 502 (1963).
281. S. Siekierski and I. Fidelis, *J. Chromatogr.*, **4**, 60 (1960).

282. I. Fidelis and S. Siekierski, *J. Chromatogr.*, **5**, 161 (1961).
283. A. G. Hamlin, B. J. Roberts, W. Loughlin, and S. G. Walker, *Anal. Chem.*, **33**, 1547 (1961).
284. T. Shimizu and K. Ikeda, *J. Chromatogr.*, **85**, 123 (1973).
285. E. Cerrai and C. Testa, *J. Chromatogr.*, **7**, 112 (1962); **9**, 216 (1962).
286. A. Delle Site and C. Testa, *Anal. Chim. Acta*, **72**, 155 (1974).

CHAPTER

5

ABUNDANCE DATA AND STANDARD ROCKS

5.1 INTRODUCTION

One of the major purposes of trace analysis of geological materials is to assist the geochemist in his study of the chemistry of the earth and of the cosmos. Many advances in geochemistry depend on the gathering and utilization of a pool of reliable abundance data and therefore are linked closely with the progress of analytical chemistry.

Although analytical chemistry had its beginnings as far back as Greek and Roman times, it was not until the middle of the 19th Century that the discipline had progressed far enough to allow geologists to have any real concept of the chemical composition of the earth's crust. By the end of the 19th Century the classical methods of rock analysis had been developed. Therefore, it is surprising that it was not until 1924 that the first attempt was made to assess the overall composition of the common rocks and thereby to arrive at an approximate composition of the earth's crust.

In 1924 Clarke and Washington[1] compiled a list of elemental abundances in rocks, based on 5159 "superior" analyses. Their estimates for the major constituents at least, were fairly close to those accepted today, but most of the data for trace elements were both sketchy and unreliable. Although it is surprising that the world had to wait 25 years until already developed analytical techniques were used to obtain comprehensive abundance data, it is even more surprising that a further quarter century was to pass before a really serious attempt was made to assess the reliability of the data upon which so much theoretical work had been based.

In the late 1940s Fairbairn and his co-workers of the U. S. Geological Survey prepared about 50-kg samples of each of two crushed rock samples (granite G-1 and diabase W-1). These samples were sent to a large number of laboratories and were initially analyzed by 34 reputable analysts for a number of elements. The data were published in 1951[2] and showed quite appreciable variability for each constituent determined by the various analysts. The rocks were later analyzed by many more workers for nearly all the elements of the periodic table.[3-7] In 1957 Ahrens[8] showed that if the logarithm of the relative standard deviation (rsd) of replicate analyses for individual elements

INTRODUCTION

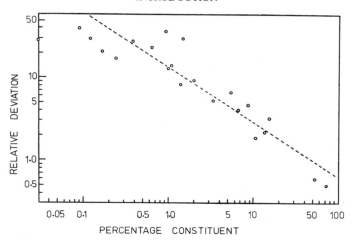

Fig. 5.1. The relationship between the relative standard deviation (reproducibility) and the concentration of each major constituent of standard rocks G-1 and W-1. The rsd is based on replicate analysis by various laboratories using classical methods of chemical analysis. *Source*: Ahrens.[8] Reprinted with permission from Pergamon Press Ltd. Copyright 1957.

in the standard rocks was plotted as a function of the logarithm of the mean concentration of that constituent in the rock, an approximately linear inverse relationship became apparent. In other words, the lower the concentration of an element in a rock, the less precise (reproducible) the analytical data. Ahrens' data for G-1 and W-1 are shown in Fig. 5.1. This relationship would be suspected intuitively by most workers, but it was left to Ahrens to show it quantitatively.

The relationship derived by Ahrens referred entirely to data obtained by the so-called classical "wet-chemical" techniques, such as gravimetry, volumetric analysis, and colorimetry. In 1957 when the paper appeared, the only widely used instrumental technique for the analysis of rocks was emission spectrography, which had been developed as a quantitative tool in the early 1930s by Goldschmidt and his co-workers at Göttingen. In the late 1950s a profound change occurred in the field of analytical chemistry. This was due to the development of a whole range of new instrumental techniques that revolutionized chemical analysis and stimulated those disciplines (such as geochemistry) that were dependent upon it. These new techniques such as atomic absorption, X-ray fluorescence, and neutron activation have shown a spectacular growth during the past 25 years.[9] This growth is shown in Fig. 5.2 and probably arose because the new methods have afforded many new opportunities for studying elements previously difficult to determine and have also extended the accessible concentration ranges of many elements.

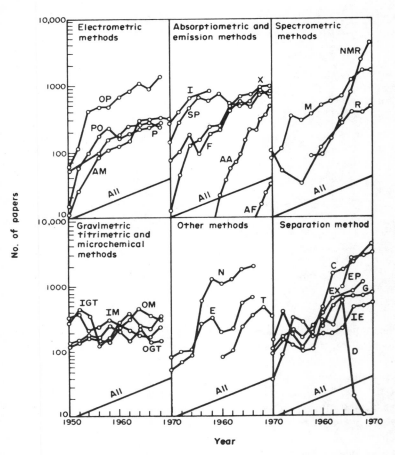

Fig. 5.2. Numbers of papers published in various fields of analytical chemistry, 1950–1970. Line labeled 'All' shows overall relative rate of growth of total published papers in analytical chemistry. *Electrometric methods*—AM, amperometry; OP, organic polarography; P, polarography; PO, potentiometry. *Absorptiometric and emission methods*—AA, atomic absorption; AF, atomic fluorescence; F, fluorimetry; I, infrared spectroscopy; SP, spectrophotometry; X, X-ray fluorescence. *Spectrometric methods*—R, Raman spectrometry; M, mass spectrometry; NMR, nuclear magnetic resonance. *Gravimetric, titrimetric and microchemical methods*—IGT, inorganic gravimetry and titrimetry; IM, inorganic microchemistry; OGT, organic gravimetry and titrimetry; OM, organic microchemistry. *Other methods*—E, electron microscopy; N, nucleonics; T, thermal analysis. *Separation methods*—C, chromatography; D, distillation; EP, electrophoresis; EX, extraction; G, gas chromatography; IE, ion-exchange. *Source*: Brooks and Smythe.[9]

The benefits of these instrumental methods can be summarized under three general headings. First, analyses with a relative standard deviation in the range 0.5–10% (acceptable for most trace element work) can now be made for many elements in geological samples at concentrations down to 1 μg/g (or even in some cases down to 1 ng/g), by one or more instrumental methods. Second, the new methods usually involve less operator skill than the original classical "wet" methods. Finally, analysis times have been greatly reduced. The result of the first factor is that analyses previously considered impossible or at least extremely difficult, have now become commonplace. An example of this is the determination of mercury, which is such an important element in environmental studies and can now easily be measured at the microgram per gram level with relatively simple equipment. The decreased demands on operator skill and analysis time have made chemical analysis relatively much cheaper for many elements. This has been of particular benefit in the fields of economic geology and exploration geochemistry, where success at finding new ore deposits depends largely on the rapid and inexpensive analysis of a large number of geological samples.

5.2 ABUNDANCE DATA

The chemical composition of the earth's crust is essentially that of the igneous rocks of the lithosphere since the hydrosphere and metamorphic and sedimentary rocks are a small proportion ($<5\%$) of the overall mass. An approximation to the chemical composition of the crust can be obtained by averaging elemental abundances in acid and basic rocks such as granites and basalts. For this purpose the mean of abundance data for the typical rocks G-1 and W-1 may be used. Although this means extrapolation from two samples only, these standard rocks have now been analyzed very extensively and useful data are available for most of the chemical elements. It would also seem that G-1 and W-1 are very "typical" rocks because the mean values for the major constituents agree very well with abundance data derived from large numbers of other rocks.

Table 5.1 lists the crustal abundances of most of the chemical elements. Concentrations in granites, basalts, shales and seawater are also included. Values for the major elements in rocks are derived mainly from Taylor.[10] Trace element data are usually those for the standard rocks G-1 and W-1.[5–7] The shale data are from Vinogradov[11] and the seawater values are mainly from Goldberg.[12] It will readily be seen that if the hydrosphere is excluded, 11 elements have crustal abundances of 1000 μg/g (0.1%) or greater. The other 68 elements listed could be regarded as trace elements in terms of their abundances in the crust as a whole. However, even some major constituents (e.g., magnesium) have low abundances in some geological materials.

TABLE 5.1 Abundances (µg/g) of the Chemical Elements (Listed in Decreasing Crustal Abundance)

Element	Crust	Granite	Basalt	Shale	Seawater
O	464,000	490,000	459,700	395,200	857,000
Si	282,000	323,000	240,000	238,000	3.0
Al	82,000	77,000	88,000	80,000	0.01
Fe	56,000	27,000	86,000	47,000	0.01
Ca	41,000	16,000	67,000	25,000	400
Na	24,000	28,000	19,000	6,600	10,500
Mg	23,000	1,600	45,000	13,400	1,350
K	21,000	33,000	8,300	23,000	380
Ti	5,700	2,300	9,000	4,500	0.001
H	1,400	400	600	—	108,000
P	1,050	700	1,400	770	0.07
Mn	1,000	400	1,500	850	0.002
Ba	600	1,200	180	580	0.03
F	500	700	250	500	1.3
S	260	270	250	220	885
Sr	215	250	180	450	8.0
C	200	300	100	1,000	28
V	180	16	240	130	0.002
Zr	155	210	100	200	—
Cl	150	100	200	160	19,000
Rb	125	210	40	270	0.12
La	100	100	112	40	0.000012
Ce	96	170	23	50	0.0000052
Ni	75	1	100	95	0.002
Cr	71	22	120	100	0.00005
Zn	70	45	82	80	0.01
Cu	60	13	110	57	0.003
Nd	36	55	17	23	0.0000092
Co	26	2	50	20	0.0001
Sc	20	3	34	10	0.00004
Y	19	13	25	23	0.0003
Li	18	24	12	60	0.17
Ga	17	18	16	19	0.00003
Nb	15	20	10	20	0.00001
N	15	10	20	60	0.5
Pb	14	20	5	20	0.00001
Th	10	17	2	11	0.00005
B	10	15	5	100	4.6
Pr	10	17	4	5	0.0000026
Sm	7	9	4	7	0.0000017
Gd	5	5	4	7	0.0000024
Hf	4	6	2	6	—
Sn	4	4	3	6	0.0008
Dy	3	2	4	5	0.0000029
Er	3	2	3	3	0.0000024

TABLE 5.1 (*continued*)

Element	Crust	Granite	Basalt	Shale	Seawater
Mo	2	3	0.5	2	0.01
Yb	2	1.0	3	3	0.000002
U	2	3	0.50	3	0.003
Be	1.9	3	0.8	3	0.0000007
As	1.6	0.80	2	6	0.003
Ge	1.4	1.0	1.7	2	0.00006
Cs	1.3	1.5	1.0	5	0.0005
Eu	1.2	1.3	1.1	1.1	0.00000046
Ta	1.2	1.6	0.70	—	—
Ho	0.70	0.50	1.0	1.0	0.00000088
Sb	0.70	0.40	1.1	1.5	0.0005
Tb	0.70	0.60	0.80	0.90	—
Tl	0.70	1.3	0.13	1.0	<0.00001
W	0.40	0.40	0.45	2	0.0001
Tm	0.30	0.20	0.30	0.25	0.00000052
Lu	0.27	0.20	0.35	0.7	0.00000048
Br	0.26	0.13	0.40	6	65
Bi	0.20	0.10	0.25	0.01	0.00002
Cd	0.18	0.06	0.30	0.30	0.00011
Hg	0.17	0.24	0.11	0.40	0.00003
Se	0.10	—	0.11	0.60	0.0004
Ag	0.05	0.04	0.05	0.10	0.00004
I	0.05	0.05	0.05	1.00	0.06
In	0.05	0.025	0.07	0.05	<0.02
Pd	0.01	0.001	0.02	<0.05	<0.00001
Pt	0.01	0.001	0.02	<0.05	<0.00001
Te	0.01	0.01	0.01	—	—
Au	0.005	0.005	0.005	—	0.00001
Rh	0.005	0.005	0.005	<0.05	<0.00001
Ir	0.001	0.006	0.00005	<0.05	<0.00001
Re	0.0005	0.0006	0.0004	—	—
Os	0.0002	0.0001	0.0004	<0.05	<0.00001
Ru	0.0001	—	—	<0.05	<0.00001

Sources: Various, including Taylor[10] and Vinogradov.[11]

5.3 STANDARD ROCKS

It is generally true that the more accurately an analyst knows the composition of a sample, the more accurately he can analyze it. Accurate trace analysis, in particular, requires some prior knowledge of the gross composition of the sample. Such information can be useful at all stages of the analytical procedure. It can help the analyst to decide on the best methods

TABLE 5.2 Reference Standards

Name of standard	Code number	Source[a]	Ref.
Adularia	PSU-Or-1	12	23
Albite	PSU-Ab-1	12	23
Andesite	USGS-AGV-1	11	24
Anhydrite	ZGI-AN	6	25
Basalt	CRPG-BR	5	26
	ZGI-BM	6	25, 27
	USGS-BCR-1	11	24
	USGS-BHVO-1	11	24
	GSJ-JB-1	13	28
Bauxite	ANRT-BX-N	5	26
Biotite	CRPG-Mica-Fe	5	26
Calcsilicate	QMC-M-3	4	
Carnotite	NBL-4	16	
Carnotite ore	IAEA-S-3	25	
Clay			
Attapulgus	NBS-GM-2007	10	
Flint	NBS-97a	10	
Plastic	NBS-98a	10	
Clay minerals			
Attapulgite	CMS-PFl-1	21	
Bentonite	CMS-SAz-1	21	
	CMS-STx-1	21	
	CMS-SWy-1	21	
Cookeite	CMS-CAr-1	21	
Hectorite	CMS-SHCa-1	21	
Kaolin	CMS-KGa-1	21	
	CMS-KGa-2	21	
Montmorillonite	CMS-Sny-1	21	
Rectorite	CMS-RAr-1	21	
Smectite	CMS-SWa-1	21	
Diabase (dolerite)	QMC-I-3	4	
	USGS-W-1	11	24
Diorite	ANRT-DR-N	5	26
Disthene (kyanite)	ANRT-DT-N	5	26
Dolomite	BCS-368	3	
	GFS-400	9	
Dunite	USGS-DTS-1	11	24
Dunite (chrysolite)	NIM-D	14	29
Feldspar, potash	BCS-376	3	
	NBS-70a	10	
Feldspar, soda	BCS-375	3	
	NBS-99a	10	
Fluorspar	NBS-79a	10	
	NBS-180	10	
	NBS-1621	10	
Gabbro	USGS-GSM-1	11	24

TABLE 5.2 (*continued*)

Name of standard	Code number	Source[a]	Ref.
Galena	FF-1	28	30
Gold-quartz	USGS-GQS-1	11	31
Granite	GIB-G-B	2	32
	QMC-I-1	4	
	ANRT-GS-N	5	
	CRPG-GA	5	26
	ZGI-GM	6	25, 27
	USGS-G-1	11	6
	USGS-G-2	11	24
	NIM-G	14	29
Granodiorite	USGS-GSP-1	11	24
	GSJ-JG-1	13	28
Hematite	GFS-453	9	
Hornblende	Basel 1-H	17	
	USGS-Hbl-1	11	
Jasperoid	USGS-GX-1	22	
Kaolin	UNS-KK	29	33
Larvikite	ASK-1	27	
Lepidolite	NBS-183	10	
Limestone	ZGI-KH	6	25, 27
	GFS-401	9	
Argillaceous	NBS-1b	10	
Dolomite	NBS-88a	10	
Lujavrite	NIM-L	14	29
Magnesite	UNS-MK	29	33
Magnetite	GFS-450	9	
Meteorite, Allende	USNM-3529	30	
Monazite sand	NBL-7-A	16	
Mud, marine	USGS-MAG-1	11	
Muscovite	Bern 4M	18	
	USGS-P-207	11	
Nepheline syenite	Len-X	7	34
	USGS-STM-1	11	24
Norite	NIM-N	14	29
Ores			
Cr, Grecian	BCS-308	3	
Cu, mill heads	NBS-330	10	
	USGS-GX-4	22	
Cu, mill tails	NBS-331	10	
Cu, concentrate	NBS-332	10	
Cu-Mo	CSRM-HV-1	8	35
Fe, Lincolnshire	BCS 301/1	3	
Fe, Nimba	BCS 175/2	3	
Fe, Northamptonshire	BCS-302	3	
Fe, Sibley	NBS-27e	10	

TABLE 5.2 (*continued*)

Name of standard	Code number	Source[a]	Ref.
Fe, sinter	BCS-303	3	
Mn	BCS-176/1	3	
	NBS-25c	10	
Mo	CSRM-PR-1	8	
Mo, concentrate	NBS-333	10	
Pt, blacksand	CSRM-PTA-1	8	36
Pt, min. dressing prod.	CSRM-PTC-1	8	36, 37
Pt, Ni-Cu matte	CSRM-PTM-1	8	36, 38
Pt, Merensky reef	NIM-PTO-1	14	
Sn, sulfide	CSRM-MP-1	8	35
Zn	NBS-113	10	
	NBS-329	10	
Peridotite	USGS-PCC-1	11	24
Petalite	NBS-182	10	
Phlogopite	CRPG-Mica-Mg	5	26
Phosphate	NBS-120a	10	
	NBL-1	16	
Pitchblende	NBL-3-A	16	
Pitchblende ore	NBL-6	16	
	IAEA-S-12	25	
Pyrite	PS-1	28	30
Pyroxene	PSU-Px-1	12	23
Pyroxenite	NIM-P	14	29
Quartz, Brazilian	GEM-430	19	
Quartz, latite	USGS-QLO-1	11	
Rhyolite	USGS-RMG-1	11	24
Sand, glass	NBS-81a	10	
	UNS-SpS	29	33
Sand, glass, low Fe	NBS-165	10	
Schist	QMC-M-2	4	
	ASK-2	27	
Schist, mica	USGS-SDC-1	11	24
Sediment, lake	IAEA-SL-1	25	
Sediment, marine	IAEA-SD-B-1	26	
Serpentine	ANRT-UB-N	5	26
Shale	KnC-Shp-1	15	
	JOS-1	20	
	ZGI-TS	6	25, 27
	USGS-SCo-1	11	24
	USGS-SGR-1	11	24
Slate	ZGI-TB	6	25, 27
Soils (in preparation)	CSRM	8	
	CSRM	23	
	CSRM	24	
	CSRM	22	
(now available)	IAEA-Soil-5	25	

TABLE 5.2 (continued)

Name of standard	Code number	Source[a]	Ref.
Sphalerite	SF-1	28	30
Spodumene	NBS-181	10	
Sulphide	CSRM-SU-1	8	17, 39
Syenite	NIM-S	14	29
	CSRM-SY-2 & 3	8	36
Tonalite	MRT-T-1	1	40, 41
Torbernite ore	IAEA-S-1	25	
Ultramafic rock	CSRM-UM-1	8	39, 42
Uraninite ore	IAEA-S-4	25	
Zircon	BCS-388	3	

Source: After Flanagan.[21]
[a] See Table 5.3.

of chemical attack; it also can indicate what separations may be needed or what complexing agents or masking agents should be used and may even influence the choice of instrumental method.

Most instrumental methods are susceptible to "matrix effects" in some degree. In other words, the signal produced by a given concentration of analyte is influenced by the bulk composition of the sample. Examples of these effects can be seen in emission spectrography where the matrix may influence the intensity of spectral lines by modifying the volatilization and excitation processes, in X-ray fluorescence where the matrix may influence X-ray penetration and the efficiency of conversion of incident into fluorescent radiation, and in atomic absorption spectroscopy where the efficiency of trace-element atom production in a flame may depend on other constituents of the aspirated solution. Since it is often impossible to prepare standards with exactly the same bulk composition as that of the sample, techniques such as the method of standard additions (see, for example, pp. 187, 220, 306), or the use of standard rocks, have to be employed to minimize matrix effects.

Before 1949, when Fairbairn prepared the now famous G-1 and W-1 rock standards, very few geological standards were available. By 1965 these standards had been analyzed for 74 elements by 105 laboratories in 22 countries and the data had been published in 256 different papers. The data have been summarized by various workers.[2,5-7,13]

The problem with the earlier standards was that insufficient material had been prepared to allow for sufficient analyses to obtain reliable estimates of elemental abundances and at the same time leave sufficient material for general laboratory use as reference standards for calibration of instruments.

TABLE 5.3 Sources for Reference Standards Listed in Table 5.2

No.	Source
1	Mineral Resources Division, Box 903, Dodoma, Tanzania.
2	E. Aleksiev, Geological Institute, Sofia, Bulgaria.
3	Bureau of Analysed Samples, Newham Hall, Middlesborough, U. K.
4	A. B. Poole, Department of Geology, Queen Mary College, London.
5	H. de la Roche, Centre de Recherches Petrographiques et Géochimiques, 15 Rue N.-D. des Pauvres, Nancy, France.
6	K. Schmidt, Zentrales Geologisches Institut, Invalidenstrasse 44, 104 Berlin, Germany (D. D. R.).
7	A. A. Kukharenko, Department of Mineralogy, Leningrad State University.
8	A. H. Gillieson, Mines Branch, 555 Booth St, Ottawa.
9	G. F. Smith Chemical Co., Box 23344, Columbus, Ohio 43223.
10	National Bureau of Standards, Washington D.C. 20234.
11	F. J. Flanagan, U. S. Geological Survey, Reston, Virginia 22092.
12	N. H. Suhr, Mineral Constitution Laboratories, Pennsylvania State University, University Park, Pennsylvania 16802.
13	Atsushi Ando, Geological Survey of Japan, Misamoto-cho, Kawasaki, Japan.
14	H. P. Beyers, South African Bureau of Standards, Pvt. Bag 191, Pretoria, South Africa.
15	D. M. Moore, Department of Geology, Knox College, Galesburg, Illinois 61401.
16	New Brunswick Laboratory, U. S. Atomic Energy Commission, Box 150, New Brunswick, New Jersey 08903.
17	H. Schwander, Mineralogisches-Petrographisches Institut der Universität, CH-4000, Basel, Switzerland.
18	E. Jaeger, Mineralogisches-Petrographisches Institut der Universität, CH-3000 Bern, Switzerland.
19	Engineered Materials, Box 363, Church St Station, New York 10008.
20	S. Y. Taha, Laboratories Division, Natural Resources Authority, Box 2220, Amman, Jordan.
21	W. D. Johns, Source Clay Repository, Department of Geology, University of Missouri, Columbia, Missouri 65201.
22	Branch of Exploration Research, U. S. Geological Survey, Denver, Colorado 80225.
23	R. I. Barnhisel, Department of Agronomy, University of Kentucky, Lexington, Kentucky 40506.
24	G. Holmgren, Soil Conservation Service, 1325 North St., Lincoln, Nebraska 68508.
25	International Atomic Energy Agency, Analytical Quality Control Services, Laboratory Seibersdorf, Box 590, A-1011 Vienna, Austria.
26	Laboratory of Marine Radioactivity, Monaco.
27	O. H. J. Christie, Mass Spectrometric Laboratory, University of Oslo, Box 1048, Blindern, Oslo 3, Norway.
28	H. J. Rösler, Bergakademie Freiberg, Sektion Geowissenschaften, 92 Freiberg, Germany (D. D. R.).
29	Václav Zýka, Institute of Mineral Raw Materials, 284 03, Kutná Hora, Czechoslovakia.
30	E. Jarosewich, Department of Mineral Sciences, Smithsonian Institute, Washington, D. C. 20560.

Source: Flanagan.[21]

The G-1 and W-1 standards were quickly followed by a number of others such as U.S.G.S. standards: granite G-2, granodiorite GSP-1, andesite AGV-1, peridotite PCC-1, dunite DTS-1, and basalt BCR-1.[14,15] Two other extensively analyzed standards are the syenite and sulfide ore prepared by the Canadian Association for Applied Spectroscopy (now Spectroscopy Society of Canada).[16-18]

There had been widespread criticism of the possible inhomogeneity of G-1 and W-1 (particularly G-1) because they had been ground only to -80 mesh size.[4,19,20] Because of this criticism, special care was taken with the U.S.G.S. successors of G-1 and W-1. They were ground more finely, were passed through a sample splitter, and special precautions were taken to avoid contamination from grinding and pulverizing equipment.[14] The standard G-2 was in fact derived from the same original material (Westerley granite) as G-1, although sampled some distance away, and it was therefore interesting to see whether better precision and accuracy were obtained with the new standard. Compared with G-1, the multilaboratory analyses of replicates showed better reproducibility for some elements and worse for others. It does not seem that the special precautions taken for G-2 have resulted in any appreciable improvement of the reproducibility of the data.

Over 200 geological reference standards are now available.[15,21,22] Some of these are listed in Table 5.2, and the sources are presented in Table 5.3. In some cases as much as 1000 kg of material have been prepared.

To date there are no standard seawater samples for determining trace constituents, and it will obviously be difficult to form a stock of such samples because of the problems inherent in storing natural waters (see Chapter 2).

The great increase in the number of geological standards presents its own problems. One of these is the problem of communication because of the vastness of the relevant literature. Flanagan[21] has suggested that a yearly bound volume of up-to-date abundance data be issued by some competent authority and this would seem to be the most sensible approach.

Despite some problems, there is no doubt that the ready availability of reference standards is doing much to improve the accuracy of many analytical techniques and the time cannot be far off when standards will be available for every conceivable geological material.

References

1. F. W. Clarke and H. S. Washington, U. S. Geol. Surv. Prof. Pap. **127**, 1 (1924).
2. H. W. Fairbairn et al., U. S. Geol. Surv. Prof. Pap. **980**, 1 (1951).
3. H. W. Fairbairn, *Geochim. Cosmochim. Acta*, **4**, 143 (1953).
4. F. J. Flanagan, *U. S. Geol. Surv. Bull.*, **1113**, 113 (1960).

5. M. Fleischer, *Geochim. Cosmochim. Acta*, **29**, 1263 (1965).
6. M. Fleischer, *Geochim. Cosmochim. Acta*, **33**, 65 (1969).
7. M. Fleischer and R. E. Stevens, *Geochim. Cosmochim. Acta*, **26**, 525 (1962).
8. L. H. Ahrens, in *Progress in Physics and Chemistry of the Earth*, Vol. 2, Pergamon, New York, 1957, p. 30.
9. R. R. Brooks and L. E. Smythe, *Talanta*, **22**, 495 (1975).
10. S. R. Taylor, *Geochim. Cosmochim. Acta*, **28**, 1280 (1964).
11. A. P. Vinogradov, *Geokhimiya*, 560 (1962).
12. E. D. Goldberg, in *Chemical Oceanography*, Vol. 1 (eds. J. P. Riley and G. Skirrow), Academic Press, New York, 1965, p. 164.
13. R. E. Stevens et al., *U. S. Geol. Surv. Bull.*, **1113**, 1 (1960).
14. F. J. Flanagan, *Geochim. Cosmochim. Acta*, **33**, 81 (1969).
15. F. J. Flanagan, *Geochim. Cosmochim. Acta*, **34**, 121 (1970).
16. Canadian Association of Applied Spectroscopy, *Appl. Spectrosc.*, **15**, 159 (1961).
17. N. M. Sine, W. O. Taylor, G. R. Webber, and C. L. Lewis, *Geochim. Cosmochim. Acta*, **33**, 121 (1969).
18. G. R. Webber, *Geochim. Cosmochim. Acta*, **29**, 229 (1965).
19. D. H. F. Atkins and A. A. Smales, *Anal. Chim. Acta*, **22**, 462 (1960).
20. I. Carmichael and A. J. McDonald, *Geochim. Cosmochim. Acta*, **22**, 87 (1961).
21. F. J. Flanagan, *Geochim. Cosmochim. Acta*, **38**, 1731 (1974).
22. F. J. Flanagan and M. E. Gwyn, *Geochim. Cosmochim. Acta*, **31**, 1211 (1967).
23. S. S. Goldich, C. O. Ingamells, N. H. Suhr, and D. H. Anderson, *Can. J. Earth Sci.*, **4**, 747 (1967).
24. F. J. Flanagan, U. S. Geol. Surv. Prof. Pap. **840**, 1 (1974).
25. H. Grassman, *Z. Angew. Geol.*, **18**, 278 (1972).
26. H. De la Roche and K. Govindaraju, *Rev. GAMS.*, **7**, 314 (1971).
27. R. Schindler, *Z. Angew. Geol.*, **18**, 221 (1972).
28. A. Ando, H. Kurasawa, T. Ohmori, and E. Takeda, *Geochem. J. Jap.*, **5**, 151 (1971).
29. B. G. Russell, R. G. Goodvis, G. Domel, and J. Levin, S. Afr. Natl. Inst. Materials Job Rep. **1351**, 1 (1972).
30. H. J. Rösler, unpublished compilation of data, see Ref. 21.
31. H. T. Millard, J. Marinenko, and J. E. McLane, *U. S. Geol. Surv. Circ.*, **598**, 1 (1969).
32. E. Aleksiev and R. Boyadjieva, *Geochim. Cosmochim. Acta*, **30**, 511 (1966).
33. V. Zýka, unpublished compilation of data, see Ref. 21.
34. A. A. Kukharenko et al., *Vses. Mineral. Obschchest. Zap.*, **97**, 133 (1968).
35. G. H. Faye, Can. Dept. Energy Mines Resour., Mines Branch, Tech. Bull., **155**, 1 (1972).
36. G. H. Faye, Can. Dept. Energy Mines Resour., Mines Branch, Inf. Circ. **294**, 1 (1972).

REFERENCES

37. R. C. McAdam, Sutarno and P. E. Moloughney, Can. Dept. Energy Mines Resour., Mines Branch, Tech. Bull. **176**, 1 (1973).
38. R. C. McAdam, Sutarno, and P. E. Moloughney, Can. Dept. Energy Mines Resour., Mines Branch, Tech. Bull. **182**, 1 (1973).
39. G. H. Faye, W. S. Bowman, and Sutarno, Can. Dept. Energy Mines Resour., Mines Branch, Tech. Bull. **177**, 1 (1973).
40. S. Abbey, Geol. Surv. Can. Pap. 72-30, 1 (1972).
41. S. Abbey, Geol. Surv. Can. Pap. 73-36, 1 (1973).
42. E. M. Cameron, Geol. Surv. Can. Pap, 71-35, 1 (1972).

CHAPTER

6

GRAVIMETRIC AND TITRIMETRIC METHODS OF ANALYSIS

6.1 INTRODUCTION

Gravimetric, titrimetric, and colorimetric methods of analysis were the mainstay of the analyst until the 1950s. Unfortunately, the classical techniques of gravimetry and titrimetry usually demand a lengthy separation of interfering elements by a series of chemical operations, each of which increases the risk of obtaining an imprecise result. In addition, these techniques became less applicable as analysts became interested in precise determination of elements in submilligram, and eventually submicrogram, amounts.

The heavy dependence of geologists upon gravimetry, titrimetry, and colorimetry is illustrated by Vincent's summary[1] of methods for the major constituents of silicate rocks. Calcium, magnesium, silica, sodium, and potassium were determined gravimetrically; iron (in each oxidation state) titrimetrically; manganese, phosphorus, and titanium colorimetrically; and aluminum was determined by difference after obtaining a gravimetric value for the sum of all R_2O_3 oxides. Rapid methods of analysis were developed[2] in which the alkali metals were determined by flame photometry; aluminum, manganese, phosphorus, silica, titanium, and total iron were found colorimetrically, and calcium, magnesium, and ferrous iron were determined by complexometric titrations.

More recently, instrumental methods such as emission spectrography, X-ray fluorescence, and atomic absorption, have been used for major constituents, but wet chemical techniques are still extensively employed. In the 1969 summary,[3] 24 new sets of data for major constituents of W-1 were reported, of which 17 were obtained by wet chemical methods including gravimetry and titrimetry, three were obtained by emission spectrography, and four by X-ray fluorescence.

The limited uses of gravimetry and titrimetry in trace analysis of geological materials include the fire assay procedures for gold, silver, and other precious metals,[4,5] and complexometric titration methods.[6] Both gravimetry and titrimetry have benefited from the development of amplification procedures,[7-10] discussed in Section 6.2.3 and 6.3.3.

6.2 GRAVIMETRIC TECHNIQUES

6.2.1 Classical Gravimetry

Classical gravimetric analysis is seldom useful where submilligram amounts of an element are to be determined. Factors limiting the value of trace gravimetry include difficulties of manipulating small masses of precipitates without loss, difficulties of weighing them precisely, and solubility-product limitations. Near the upper part of the trace-element range (0.01–0.1%), gravimetry can be used where the precipitate:analyte mass ratio is particularly favorable (at least 10:1). Table 6.1 lists a number of substances for which the precipitate mass is at least 10 times that of the ion being determined.

6.2.2 Fire Assay Analysis

Fire assay procedures[4,5] are used for the determination of the noble metals (gold, silver, and the platinum metals) in geological specimens. The sequence of operations has been described briefly in Section 4.1.4. Fire assay gravimetry is still sometimes used for the noble metals, partly as a result of improvements in sensitivity of microbalances and partly because the

TABLE 6.1 Gravimetric Conversion Factors (Mass of Analyte Ion/Mass of Precipitate)

Analyte	Precipitant	Compound weighed	Conversion factor
Al^{3+}	8-Hydroxyquinoline	$Al(C_9H_6ON)_3$	0.05871
Co^{3+}	1-Nitroso-2-naphthol	$Co(C_{10}H_6O_2N)_3 \cdot 2H_2O$	0.09640
	1-Nitro-2-naphthol	$Co(C_{10}H_6O_3N)_3$	0.09454
Cu^{2+}	Ethylene diamine, tetra-iodomercurate	$Cu(C_2H_8N_2)_2HgI_4$	0.07126
F^-	Lead nitrate	PbClF	0.07261
	Triphenyltin chloride	$(C_6H_5)_3SnF$	0.05149
Mg^{2+}	8-Hydroxyquinoline	$Mg(C_9H_6ON)_2 \cdot 2H_2O$	0.06976
		$Mg(C_9H_6ON)_2$	0.07780
	Diammonium hydrogen phosphate	$Mg(NH_4)PO_4 \cdot 6H_2O$	0.09909
Na^+	Zinc uranyl acetate	$NaZn(UO_2)_3(C_2H_3O_2)_9 \cdot 6H_2O$	0.01495
	Magnesium uranyl acetate	$NaMg(UO_2)_3(C_2H_3O_2)_9 \cdot 6.5H_2O$	0.01527
NH_4^+	Chloroplatinic acid	$(NH_4)_2PtCl_6$	0.08125
Ni^{2+}	α-Benzildioxime	$Ni(C_{14}H_{11}O_2N_2)_2$	0.10926
PO_4^{3-}	Ammonium molybdate	$(NH_4)_3PMo_{12}O_{40}$	0.05061
		$P_2O_5 \cdot 24MoO_3$	0.05281

method is suitable for the large samples that are often taken for noble metal analysis because of the inhomogeneity of their distribution in rocks. Fire assay may also be used as a means of separation, and the noble metal bead can then be analyzed by instrumental techniques such as emission spectrography.

6.2.3 Amplification and Cascade Processes

In some cases it is possible to alter the normal stoichiometry of gravimetric and volumetric processes to make a more favorable measurement. Because amplification reactions are used more often in volumetric work, examples are given in Section 6.3.3.

A modification of amplification procedures is seen in the development of cascade methods,[10] usually with a gravimetric finish. These involve repetition of a cycle of operations until there is sufficient material available for precise measurement, as in the following example. Silver can be precipitated with chromate and the precipitate treated with barium chloride solution:

Step 1:

$$2Ag^+ + CrO_4^{2-} \longrightarrow Ag_2CrO_4$$
$$Ag_2CrO_4 + BaCl_2 \longrightarrow 2AgCl + BaCrO_4$$

More silver chromate is produced by adding silver nitrate, and this is then converted into silver chloride as before:

Step 2:

$$2AgCl + BaCrO_4 \xrightarrow{AgNO_3} 2AgCl + Ag_2CrO_4$$
$$2AgCl + Ag_2CrO_4 \xrightarrow{BaCl_2} 4AgCl + BaCrO_4$$

This step, involving recycling of the chromate, can be repeated n times, to give $2(n + 1)$ moles of silver chloride in place of the two moles formed in Step 1. The cascade therefore results in an amplification of the mass of silver chloride by the factor $n + 1$. Other examples are given by Belcher.[7]

6.3 TITRIMETRIC TECHNIQUES

6.3.1 Introduction

Recent developments in titrimetry have been included in the reviews by West and West,[11] and applications to the analysis of geological materials have been noted by Dinnin.[12] Woodward and Redman[13] discuss high-precision titrimetry, with emphasis on methods giving precision better than 0.1%. The book by Wagner and Hull[14] is a comprehensive work on contemporary titrimetry, and recent developments have been considered by

Curran.[15] Much of the recent work in classical titrimetry has centered on improving the methods of end-point determination. Bishop[16] has edited a collection of articles on visual indicators under the following headings: acid-base, organic dye, metallochromic, metal adsorption, red-ox, fluorescent, and chemiluminescent indicators, indicators for titrations with nonchelating liquids and for nonaqueous acid-base titrations. Other significant recent publications include discussions by Johansson[17] on the choice of indicators for photometric titrations, and by Berka et al.[18] on the factors influencing the usefulness of titrimetric methods.

Because of space limitations, only a few titrimetric techniques of particular value for measuring low concentrations are mentioned here.

6.3.2 Complexometric Titrations

Titrations involving formation of a soluble complex of known stoichiometry between a metal ion and a suitable complexing agent, have increased in importance during the last 30 years. The availability of aminopolycarboxylic acids, in particular, has led to the development of many methods suitable for determining metal ions in solution at concentrations of approximately 10^{-2}–10^{-5} mol/l.

The most important of these reagents is ethylenediamine-N,N,N',N'-tetraacetic acid (EDTA), which is sometimes represented as a tetraprotic acid, H_4Y. The following properties make EDTA particularly useful: (a) it forms soluble 1:1 complexes with many metal ions, (b) selectivity can be controlled by appropriate choice of pH, (c) the disodium salt $Na_2H_2Y \cdot 2H_2O$ can be used as a primary standard, (d) all the metal complexes are soluble, (e) the end-point is readily determined visually or instrumentally, (f) the titration is suitable for the semimicro or even the micro scale. This last point is significant for trace analysis.

Complexometric techniques for determining metal ions have several advantages over other titrimetric methods: they are more sensitive, and require less "wet chemistry," since the effect of interfering ions can be eliminated or reduced by suitable buffering and by masking with other complexing agents. End-points can be determined visually or instrumentally. Visual indicators include dyestuffs, such as Eriochrome Black T, that form colored complexes with many metals. Better precision or sensitivity can often be obtained by spectrophotometric or potentiometric methods of end-point detection.

The acid dissociation constants of EDTA are such that the species H_2Y^{2-} predominates at pH 2.7 to 6.2, and HY^{3-} predominates between pH 6.2 and 10.2. The reaction with a metal ion can be represented as follows:

$$M^{n+} + H_2Y^{2-} \rightleftharpoons MY^{(4-n)-} + 2H^+$$

TABLE 6.2 Stability Constants for Metal-EDTA Complexes

Cation	K_s (1 mol^{-1})	log K_s
Ag$^+$	2.0×10^7	7.30
Ba^{2+}	5.8×10^7	7.76
Sr^{2+}	4.3×10^8	8.63
Mg^{2+}	4.9×10^8	8.69
Ca^{2+}	5.0×10^{10}	10.70
Mn^{2+}	6.2×10^{13}	13.79
Fe^{2+}	2.1×10^{14}	14.33
Al^{3+}	1.3×10^{16}	16.13
Co^{2+}	2.0×10^{16}	16.31
Cd^{2+}	2.9×10^{16}	16.46
Zn^{2+}	3.2×10^{16}	16.50
Pb^{2+}	1.1×10^{18}	18.04
Ni^{2+}	4.2×10^{18}	18.62
Cu^{2+}	6.3×10^{18}	18.80
Hg^{2+}	6.3×10^{21}	21.80
Th^{4+}	1.6×10^{23}	23.20
Fe^{3+}	1.3×10^{25}	25.10
V^{3+}	8.0×10^{25}	25.90

Source: Schwarzenbach and Flaschka.[6]

The stability constant, K_s, is usually given for the reaction with Y^{4-}:

$$K_s = \frac{[MY^{(4-n)-}]}{[M^{n+}][Y^{4-}]}$$

Table 6.2 gives values of K_s for various metal-EDTA complexes. The fraction of the metal existing in the form of the complex is clearly dependent on the pH of the solution, leading to restrictions on the pH range in which titrations can successfully be performed. The minimum pH for successful EDTA titration of a number of metal ions is shown in Fig. 6.1. Detailed calculations of the course of a given titration may have to take into account other equilibria, such as metal-ion hydrolysis and complexing with other ligands present in the buffer solution used.

The dyestuffs used as visual indicators are characterized by their intense color, enabling the transition between the metal-indicator complex and the free indicator to be detected when only low formal concentrations ($<10^{-5}$ mol/l) of the indicator are present. The stability constant K_{MIn} of the metal-indicator complex should be one to two orders of magnitude lower than that

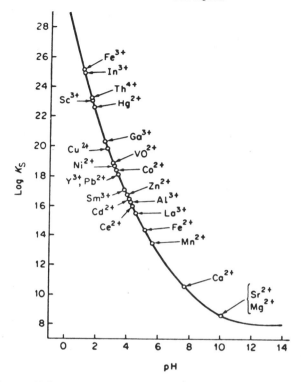

Fig. 6.1 Minimum pH for successful EDTA titrations of several metal ions. *Source*: Reilley and Schmid.[19] Reprinted with permission from *Analytical Chemistry*. Copyright by the American Chemical Society.

of the corresponding metal-EDTA complex. Such a condition is fulfilled, for example, by the indicator Eriochrome Black T used in EDTA titration of magnesium ($K_{MgE^-} = 10^7$). If K_{MIn} is too large, the end-point occurs later than the true equivalence point, and if the complex is too unstable a premature end-point is found. The metal-indicator complex must also be more stable than complexes which might be formed between the metal and the components of the buffer (e.g., ammine complexes, if an ammonia/ammonium chloride buffer is used).

Table 6.3 lists some of the more important titrants and indicators for complexometric titrations. Before the advent of atomic absorption spectroscopy, complexometric titrations with EDTA and Eriochrome Black T were widely used for the determination of magnesium and calcium in natural waters. Many ingenious titration procedures for low concentrations of

TABLE 6.3 Some Titrants and Indicators Used in Complexometric Titrations

Name	Applications
Titrants	
Ethylenediaminetetra-acetic acid (EDTA)	Most metals
Nitrilotriacetic acid (NTA)	Some advantages for alkali metals
Triethylenetetramine (trien)	Cu^{2+}, Hg^{2+}, Ni^{2+} in alkaline solution
trans-1,2-diaminocyclohexanetetra-acetic acid (DCTA, CDTA)	Forms more stable complexes than EDTA
Diethylenetriaminepenta-acetic acid (DTPA)	Large cations, e.g., lanthanides
Ethyleneglycol bis(2-aminoethylether)-tetra-acetic acid (EGTA)	Mg^{2+} in presence of other alkaline earths
Indicators	
Eriochrome Black T, Calmagite	EDTA titration of Mg^{2+}, Ca^{2+}, Sr^{2+}, Mn^{2+} in alkaline solution
Naphthyl azoxine, NAS	EDTA titration of Cu^{2+}, Co^{2+}, Ni^{2+}, Cd^{2+}, Zn^{2+}, Al^{3+}
Calcein	EDTA titration of Ca^{2+}, Ba^{2+}, Sr^{2+}
Arsenazo I	EDTA titration of lanthanides, Th^{4+}
Xylenol orange	EDTA titration of Bi^{3+}, Th^{4+}, Zr^{4+}, Fe^{3+} in acid solution

other elements were also devised. Flaschka and Carley,[20] for example, were able to determine as little as 30 μg cadmium and 10 μg zinc by EGTA titration. Some of the most sensitive methods make use of spectrophotometric[20] or potentiometric[19] end-point detection. The reader is referred to the comprehensive work by Schwarzenbach and Flaschka[6] for further details of all aspects of complexometric titrations.

6.3.3 Amplification Techniques in Titrimetry

The sensitivity of a number of conventional titrimetric procedures can be increased by amplification reactions, pioneered by Leipert.[21] In a standard iodometric titration, for example, iodide is oxidized to iodine, which is titrated with a standard thiosulfate solution:

$$2I^- + 2Ce^{4+} \longrightarrow I_2 + 2Ce^{3+}$$
$$I_2 + 2S_2O_3^{2-} \longrightarrow 2I^- + S_4O_6^{2-}$$

The normal stoichiometry is therefore two moles of thiosulfate for two moles of iodide consumed originally. In Leipert's method the iodide ion is oxidized to iodate with bromine water, the excess bromine is destroyed, iodide is added in excess, and the liberated iodine can then be titrated with thiosulfate as follows:

$$2I^- + 6Br_2 + 6H_2O \longrightarrow 2IO_3^- + 12Br^- + 12H^+$$
$$2IO_3^- + 10I^- + 12H^+ \longrightarrow 6I_2 + 6H_2O$$
$$6I_2 + 12S_2O_3^{2-} \longrightarrow 12I^- + 6S_4O_6^{2-}$$

The overall reaction now involves 12 moles of thiosulfate ion and there has therefore been a sixfold amplification of the original reaction.

Unfortunately, many of the most useful amplification reactions involve the halogens, which are of limited interest to geologists and geochemists. However, certain elements may be determined by indirect amplification procedures. For example, many metals can be precipitated with 8-hydroxyquinoline and the precipitate titrated with bromine (from bromide and bromate) as follows:

$$M^{2+} + 2C_9H_7NO \longrightarrow (C_9H_6NO)_2M + 2H^+$$
$$(C_9H_6NO)_2M + 4Br_2 + 2H^+ \longrightarrow 2C_9H_5NOBr_2 + 4HBr + M^{2+}$$

Hence one mole of metal ion requires four moles of bromine.

Phosphate at very low levels can be determined by a modification of the procedure of Birnbaum and Walden.[22] Kirkbright et al.[9] formed the phosphomolybdate complex, extracted it into isobutyl acetate and after back-extraction into water, passed the solution through a silver reductor to reduce Mo^{6+} to Mo^{5+}. The solution was then titrated with 10^{-3} M ceric sulfate solution. A twelve-fold amplification was achieved because the ceric salt reacts with each of the twelve molybdate ions associated with each phosphorus atom.

Many other amplification schemes are discussed in the review by Belcher.[7]

References

1. E. A. Vincent, in *Methods in Geochemistry* (eds. A. A. Smales and L. R. Wager), Interscience, New York, 1960, p. 23.
2. L. Shapiro and W. W. Brannock, *U. S. Geol. Surv. Bull.*, **1036-C**, 1 (1956).
3. M. Fleischer, *Geochim. Cosmochim. Acta*, **33**, 65 (1969).
4. E. A. Smith, *The Sampling and Assay of the Precious Metals*, 2nd ed., Griffin, London, 1947.
5. F. E. Beamish, *Talanta*, **5**, 1 (1960); **14**, 1133 (1967).
6. G. Schwarzenbach and H. Flaschka, *Complexometric Titrations*, Methuen, London, 1969.
7. R. Belcher, *Talanta*, **15**, 357 (1968).

8. R. Belcher, J. W. Hamya, and A. Townshend, *Anal. Chim. Acta*, **47**, 149 (1969).
9. G. F. Kirkbright, A. M. Smith, and T. S. West, *Analyst*, **93**, 224 (1968).
10. H. Weisz, *Mikrochim. Ichnoanal. Acta*, 258 (1965).
11. P. W. West and F. K. West, *Anal. Chem.*, **40**, 138R (1968); **42**, 99R (1970); **44**, 251R (1972).
12. J. I. Dinnin, *Anal. Chem.*, **47**, 97R (1975).
13. C. Woodward and H. N. Redman, *High Precision Titrimetry*, Soc. Anal. Chem., London, 1973.
14. W. Wagner and C. J. Hull, *Inorganic Titrimetric Analysis*, Vol. 1, Dekker, New York, 1971.
15. D. J. Curran, in *New Developments in Titrimetry* (ed. J. Jordan), Dekker, New York, 1974.
16. E. Bishop (ed.), *Indicators*, Pergamon, New York, 1972.
17. A. Johansson, *Anal. Chim. Acta*, **61**, 285 (1972).
18. A. Berka, J. Sevcik, and R. A. Chalmers, *Talanta*, **19**, 747 (1972).
19. C. N. Reilley and R. W. Schmid, *Anal. Chem.*, **30**, 947 (1958).
20. H. Flaschka and F. B. Carley, *Talanta*, **11**, 423 (1963).
21. T. Leipert, *Mikrochem. Pregl Festschr.*, 266 (1929).
22. N. Birnbaum and G. H. Walden, *J. Am. Chem. Soc.*, **60**, 64 (1938).

CHAPTER

7

SOLUTION ABSORPTIOMETRY

7.1 INTRODUCTION

7.1.1 Ultraviolet and Visible Absorption

The absorption of visible or ultraviolet light by solutions has been the basis of methods of trace element determination for many years. Such methods involve the measurement of the extent to which radiation in a particular wavelength range is absorbed by an absorbing species in the solution.

The term *colorimetry* is used where the wavelengths concerned are part of the visible spectrum (approximately 400–760 nm). *Visual* colorimetric methods usually employ natural or artificial white light and rely on comparisons of color intensity made by the eye of the observer. Better control of the radiant-energy source is provided in instruments with an incandescent light source. In *photoelectric colorimetry* the use of a photoelectric cell as a detector eliminates the major source of subjective error inherent in the use of the human eye. More precise control of the wavelengths at which absorption is studied can be achieved by using filters (in filter photometers) or monochromators (in spectrophotometers). Most modern analytical work is carried out with spectrophotometers, although some visual techniques are still used, for example, in hydrogeochemical field work.

The absorption of radiant energy in the visible and ultraviolet parts of the electromagnetic spectrum is usually the result of electronic excitation of an absorbing molecule or ion in solution. Absorption therefore occurs at wavelengths determined by the electronic structure and electronic energy levels of the absorbing species. Because changes of vibrational and rotational energy accompany electronic transitions and because the energy levels of the absorbing molecules are modified by a variety of interactions with neighboring molecules in the solution, the absorption occurs not at a single wavelength but over a wavelength range that may cover 50 to 100 nm or more.

The absorbing species used in trace element determination may be a relatively simple ion containing the element in question (e.g., MnO_4^-, TeI_6^{2-}),

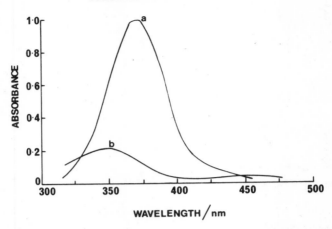

Fig. 7.1. Absorption spectra of (*a*) chromate ion and (*b*) dichromate ion in aqueous solutions each containing 10.7 µg chromium/ml.

but is more often the product of a reaction between ions of the element and a complex organic molecule. Selective absorption in the visible and ultraviolet is characteristic of organic compounds containing double bonds, and is particularly marked in compounds containing conjugated double-bond systems. In principle, if such a compound absorbs at one wavelength, and forms with a metal ion a soluble complex that absorbs at a different wavelength, this can be made the basis of a spectrophotometric method of analysis of the metal ion. Typical visible absorption spectra are shown in Figs. 7.1 and 7.2. Various aspects of solution absorptiometry in the visible and ultraviolet are well covered in a number of standard works,[1-14] some of which emphasize the general practice of spectrophotometry and others deal mainly with specific chemical analyses. Methods suitable for use in dealing with geological materials are described in detail by Jeffery,[15] Stanton,[16] Sandell,[3] Snell and Snell,[5] Allport and Brocksopp,[7] and Charlot,[8] among others. Recent work has been summarized, with comprehensive bibliographies, in the reviews by Boltz and Mellon.[17]

7.1.2 Spectrophotometric Terminology

A confusing variety of names, symbols, and units has been used in spectrophotometry during the last 40 years or so. It is to be hoped that, in the future, increasing use will be made of the International System of units (SI),[18] and of the names of the quantities approved for use in spectrophotometry by the

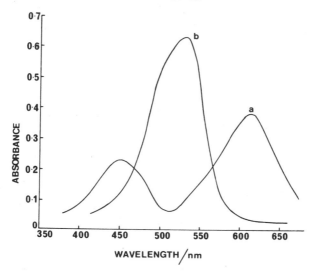

Fig. 7.2. Absorption spectra of (*a*) diphenylthiocarbazone ("dithizone") and (*b*) zinc dithizonate in carbon tetrachloride solution.

Commission Internationale d'Éclairage (CIE) and the International Union of Pure and Applied Chemistry (IUPAC).

SI quantities relevant to the study of solution spectrophotometry are discussed below and set out in Table 7.1a. Because alternative names, symbols, and units, especially those recommended for use by periodicals such as *Analytical Chemistry*,[19] are frequently found in the literature, these are also noted in Table 7.1b. Other names and symbols used in earlier work are noted in Table 7.1c.

At any wavelength, the extent of the absorption, which depends on the optical path length *l* in the absorbing medium, and on the concentration of the absorbing species in the solution, is obtained from measurements of the radiant power *P* reaching the detector. The effects of radiant-energy losses from reflection at the container walls and from absorption by the container and other constituents of the solution, are eliminated by comparing the power *P* of a beam that has passed through the absorbing solution with the power P_0 of the same beam (or of a second identical beam), which has passed through a suitable "blank" solution.

Radiant intensity *I* is defined by $I = dP/d\omega$, and can be regarded as radiant power per unit solid angle in a specified direction. Under the usual conditions of spectrophotometric measurements, the ratio of powers P/P_0 is identical to the ratio of intensities I/I_0, and is known as the *internal transmittance*,

TABLE 7.1 Quantities and Units Used in Spectrophotometry

(a) *SI Names and Units*

Name of quantity: description	Recommended symbol	Coherent SI Unit	Practical SI unit
Radiant power	Φ^a	W	
Radiant intensity $I = d\Phi/d\omega$	I	W	
Transmittance $T = I/I_0$	T	1	
Internal transmittance	T_i, T	1	
Absorbance $A = \log_{10}(1/T)$	A	1	
Pathlength	l	m	cm
(Linear) absorption coefficient $a = A/l$	a	m^{-1}	cm^{-1}
Concentration	c	$mol\ m^{-3}$	$mol\ dm^{-3\,b}$
Molar (linear) absorption coefficient $\varepsilon = A/lc$	ε	$m^2 mol^{-1}$	$cm^{-1} mol^{-1} dm^3$

(b) *Other Names, Symbols and Units Widely Used*[c]

Name of quantity: description	Symbol	Unit
Pathlength	b	cm
Concentration	c	$g\ l^{-1}$
Absorptivity $a = A/bc$	a	$cm^{-1} g^{-1}\ l$
Molar absorptivity $\varepsilon = a\mathrm{M}$	ε	$cm^{-1} mol^{-1}\ l$

(c) *Obsolete Names and Symbols*

Obsolete name	Obsolete symbol
Absorbance: optical density, extinction, absorbancy	D, E
Pathlength	d
Absorptivity: extinction coefficient, specific extinction, absorbancy index	k
Molar absorptivity: molar extinction coefficient, molar absorbancy index	a_m

[a] The symbol P is used in this text to prevent confusion with the fluorescence quantum efficiency.
[b] The liter (symbol l) is frequently used as a special name for the cubic decimeter.
[c] See, for example, Ref. 19.

T (or T_i). The term "internal" is used to indicate that this is the transmittance of the absorbing solution, disregarding the influence of container and boundaries. (The term *transmittance* has been widely used with this implication.)

Absorbance A is defined by

$$A = \log_{10} \frac{1}{T}$$

INTRODUCTION

This quantity has previously been known as optical density, extinction, and absorbancy. The (linear) absorption coefficient a is defined by

$$a = \frac{A}{l}$$

It is important to note that the name absorptivity and symbol a have been widely used for the quantity A/bc where b is the pathlength and c is the concentration of the absorbing species. When b is expressed in centimeters and c in g l^{-1}, the units of absorptivity are cm^{-1} l g^{-1}. When the molecular weight M, of the absorbing species is known, the use of the molar absorptivity ε, being the product of absorptivity and molecular weight, has often been preferred. The term molar (linear) absorption coefficient, defined by $\varepsilon = A/lc$ has been adopted for SI use.

The unit cm^{-1}mol^{-1}l has been used almost exclusively for the molar absorption coefficient; this arises from the practice of expressing concentrations in mol l^{-1} and pathlengths in centimeters. The SI form of this unit is cm^{-1}mol^{-1}dm^3, which is permissible, but contains two different multiples of the unit of length. The coherent SI unit is therefore m^{2}mol^{-1} (Table 7.1a), and 1 cm^{-1}mol^{-1}dm^3 = 10^{-1}m^{2}mol^{-1}. An ε value in the coherent unit m^{2}mol^{-1} is therefore 0.1 times the value in cm^{-1}mol^{-1}l. Particular care must be taken to check the units of ε values quoted in the literature; too often, no units are given, since the unit cm^{-1} mol^{-1} l (or cm^{-1} mol^{-1} dm^3) is "understood."

7.1.3 Beer's Law

It has been noted that the extent of absorption at any wavelength depends on the length of the optical path and on the concentration of the absorbing species. The dependence on pathlength in a homogeneous absorbing medium was studied by Bouguer (1729) and Lambert (1768), among others, and the statement sometimes known as the Bouguer-Lambert law was developed: the radiant power decreases by the same fraction for each successive increment of distance dl traversed through the medium.

The work of Beer (1852) emphasized the dependence of absorption on the number of absorbing centers, and led to a statement of the following form: *successive increments*, dn, *in the number of absorbing centers traversed by a beam of monochromatic radiation absorb equal fractions of the radiant energy incident upon them, that is,*

$$-\frac{dP}{P} = k\, dn$$

or
$$\ln\frac{P_0}{P} = kN \qquad (7.1)$$

where N is the total number of absorbing centers in the beam. For absorption by a species in solution, the number of absorbing centers in the beam is readily shown to be proportional to the product of pathlength l and concentration c. Equation 7.1 then takes the form

$$\log\frac{P_0}{P} = \varepsilon l c$$

or
$$A = \varepsilon l c \qquad (7.2)$$

Because the dependence of A on both length and concentration is considered to be implicit in Beer's work,[20,21] and for the sake of convenience, equation 7.2 is generally known as Beer's law.

Beer's law can be expected to hold where monochromatic radiation is absorbed in a volume of uniform cross-section, provided the absorbing species act independently of one another. In practice, deviations from the law arise from instrumental imperfections (e.g., presence of stray light, use of nonmonochromatic radiation) and from various concentration-dependent phenomena in the absorbing solution (e.g., solute-solvent reactions, polymerization, ionization, and other chemical equilibria), as noted in more detail in Section 7.3.1.

7.2 INSTRUMENTATION

Many instruments, with widely varying degrees of sophistication, have been developed for solution absorptiometry. Designs have generally been dictated by the state of development of optical and electronic components and by the particular purpose for which the instrument was intended. Components that have found widespread use at various times and for various purposes may be summarized briefly as follows.

1. *Light source.* Daylight, incandescent lamp (e.g., tungsten filament), hydrogen- or deuterium-discharge lamp.
2. *Intensity control.* Iris diaphragm, variable slit, lamp rheostat.
3. *Wavelength selection.* Color filter, monochromator (prism, diffraction grating).
4. *Sample cell.* Test tube, parallel-sided cuvet.
5. *Detector.* Human eye, photocell, vacuum phototube, photomultiplier.
6. *Indicator.* Galvanometer, potentiometer, chart recorder.

INSTRUMENTATION 105

Before the advent of photometric instruments, absorption in the visible spectrum was studied by visual comparison of color intensity. Precision of about $\pm 3\%$ was achieved in favorable cases. The Nessler tube technique involved inspection of intensity of light passed from a uniform diffuse source through identical long tubes containing the sample and standards. The sample was eventually bracketed between two standards, one slightly more concentrated, and one slightly less concentrated, than the sample. The inconvenience of preparing many closely spaced standards was avoided in the Duboscq comparator, which allowed variation of the pathlength in the standard and sample solutions. The pathlengths were adjusted to give equal color intensities, and the sample concentration was obtained from the equality of the product of concentration and pathlength for the two solutions.

The period between approximately 1930 and 1945 saw the development of *filter photometers* using incandescent light sources, filters, and photocell detectors, and of *photoelectric spectrophotometers*, incorporating monochromators for wavelength selection and vacuum phototubes or photomultiplier tubes as detectors. Many spectrophotometers have now been equipped with devices for automatic wavelength scanning and the recording of spectra. The principal optical components are discussed briefly below.

7.2.1 Light Sources

For the study of absorption in the visible region (extending into the near infrared) an incandescent tungsten filament lamp is used. The output of this source is satisfactory down to wavelengths of about 350 nm. For work at shorter wavelengths in the ultraviolet, it is necessary to use the continuous spectrum of a gas-discharge lamp, such as those of hydrogen or deuterium. A useful range from approximately 190 to 400 nm can be obtained, the output from a deuterium lamp being two or three times as great as that from a similar hydrogen lamp.

7.2.2 Wavelength Selectors

The simplest photoelectric photometers use filters of colored glass for wavelength selection. Absorption can be studied over a number of wavelength ranges with spectral bandwidths of 30 to 60 nm. Combinations of filters can be used to isolate narrower wavelength bands, but the radiant power transmitted through the filter system becomes so small that the photocell detectors are insufficiently sensitive.

Although systems employing filters are satisfactory for many types of simple analysis, there are important disadvantages. The molar absorption coefficient ε of the absorbing species may vary considerably over the wavelength range being used, making it impossible to obtain true absorption spectra. Some sensitivity is lost by not examining the solution at the wavelength where ε is a maximum, and deviations occur from Beer's law if ε is not uniform throughout the wavelength range used.

In spectrophotometers a prism or a diffraction grating is used to provide wavelength dispersion of the source radiation. By incorporating a device for rotating the dispersing element, the band of wavelengths that passes through the exit slit can be varied continuously. Prisms can be made of glass if only the visible region is of interest, otherwise quartz may be used (down to about 210 nm). The use of vitreous silica for optical components allows absorption to be studied below 200 nm. The dispersion of a prism monochromator, as indicated by the effective bandwidth in nanometers per millimeter of slit width, varies markedly with wavelength. The same slit width that gives an effective bandwidth of 1 nm at 300 nm may only isolate a band of 20 nm at 800 nm. Adjustable slit widths are therefore essential. A grating monochromator produces a *normal* spectrum, that is, uniformly dispersed on a wavelength scale; a given slit width therefore isolates a wavelength band of constant width throughout the spectrum. The effective bandwidth of grating monochromators in general-purpose spectrophotometers may range from about 20 nm in less expensive instruments, down to less than 0.5 nm.

7.2.3 Sample Cells

These may be of glass for work in the visible spectrum; cylindrical test tubes are used in a number of colorimeters. The nonuniform pathlength through the absorbing solution in a cylindrical tube is, however, a contributor to deviations from Beer's law, and cells with plane parallel faces are used in the most precise work. Cells of Vycor, quartz, or fused silica are required for work in the ultraviolet.

7.2.4 Detectors

The simplest photoelectric instruments use a barrier-layer photocell as detector and a microammeter as indicator. With a low resistance in the external circuit, the photocell current is proportional to the radiant flux reaching the photosensitive surface. The ratio of the two meter readings obtained with the sample and blank solutions, respectively, gives the transmittance directly.

Most spectrophotometers use a vacuum phototube or photomultiplier tube, as photocells tend to suffer fatigue if they are continuously subjected to illumination. In addition, the narrower bandwidth of spectrophotometers makes amplification of the photoinduced current desirable; this is more readily carried out with a phototube as detector. Two or more phototubes with different photoemissive surfaces can be used to obtain optimum sensitivity over the whole ultraviolet-visible range. The additional amplification provided by a photomultiplier tube is an advantage when very low levels of radiant power reach the detector, for example, when very narrow slits are used, or when solutions of high absorbance are examined.

7.2.4 Complete Instruments

Two instruments representative of many that are in common use are shown schematically in Figs. 7.3 and 7.4. In single-beam instruments the reference and sample solutions are introduced sequentially into the light path. The meter is set to read zero transmittance with the photodetector in darkness (to compensate for the thermal emission, or dark current, of the detector), and unit transmittance with the reference solution in place. The reference solution is then replaced by the sample solution and its transmittance is measured directly. The sequential nature of the operation makes source stability very important.

Fluctuations in source power are less important in double-beam spectrophotometers. The beam is split by some technique such as the insertion of a rotating sector mirror or vibrating mirror that alternately passes the beam

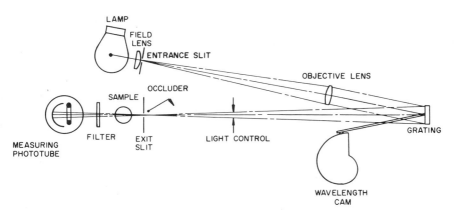

Fig. 7.3. Simplified optical diagram of a single-beam colorimeter (Bausch and Lomb Spectronic 20). Dispersion by diffraction grating, 600 lines/mm; effective bandwidth 20 nm. Tungsten filament lamp for wavelength range 350–650 nm; extension to 950 nm by adding red filter and changing phototube. Vacuum phototube detector. Readout of transmittance or absorbance. *Source*: courtesy of Bausch and Lomb, Inc.

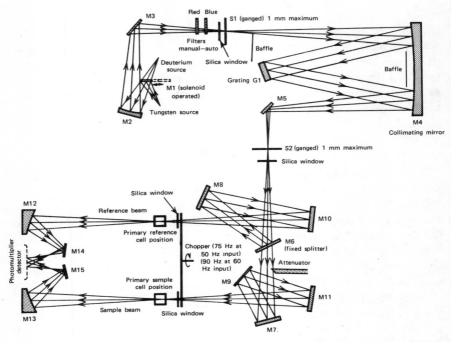

Fig. 7.4. Simplified optical diagram of a double-beam UV-visible spectrophotometer (Unicam SP1800). Dispersion by diffraction grating; nominal bandwidth 3 nm/mm slit width; variable slit width, 0.01–1 mm; limiting resolution at small slit widths, 0.1 nm. Tungsten filament lamp for 330–700 nm; extension to 850 nm with red-sensitive photomultiplier; deuterium discharge lamp for 190–330 nm. Photomultiplier detector. Readout of absorbance, recorded as a function of wavelength or as a function of time at fixed wavelength; direct concentration readout; single-beam operation gives energy vs wavelength or time. *Source*: reproduced by permission of Pye Unicam Ltd., Cambridge, England.

along one path and reflects it along the second path. In *optical null* readout systems the two chopped beams fall alternately on the detector. With equal intensities a d.c. output results; otherwise, there is an a.c. signal at the chopping frequency. The unbalance signal can be amplified and used to drive an optical attenuator into or out of the reference beam until the two beam intensities become identical. The absorption is indicated by the attenuator position. Better precision, especially at high absorbances, is obtained by *ratio-recording* readout systems that make use of potentiometric electrical nulling of the two amplified signals from the phototube. Further details on various automatically controlled operations of recording spectrophotometers are beyond the scope of this book and can be found in the manuals of instrument manufacturers.

7.3 ASPECTS OF ABSORPTIOMETRIC ANALYSIS

7.3.1 Selection of Analytical Methods

Absorptiometric analyses in the visible and ultraviolet can be carried out directly on the analyte in some cases, for example, aquocomplexes of many transition metals and lanthanides, and oxyanions such as chromate and permanganate. In other cases relatively simple complex ions, such as thiocyanatoiron(III) and hexaiodotellurite, can be formed. However, the greatest number of methods for the determination of traces of metal ions involves reaction with organic reagents to form products with high molar absorption coefficients.

The choice of chromogenic reagent may be governed by many factors, which include simplicity and rapidity of the chemical procedure, chemical stability of the absorbing species, sensitivity, and specificity. Because the reactions used are seldom completely specific, some form of chemical pretreatment is often necessary for geological materials, apart from the initial process of bringing the sample into solution. Commonly used separation methods, including precipitation, solvent extraction, and ion exchange, are discussed in Chapter 4.

The need for separations can often be averted by the use of masking agents. The interfering ion can be converted (e.g., by complex formation) into a species that no longer gives reactions characteristic of the original ion. The use of masking agents in spectrophotometry and other analytical procedures has been well covered in the book by Perrin.[22]

It is also desirable that any absorption by excess of the chromogenic reagent be well separated from that of the species being used for the analysis and that the molar absorption coefficient of the latter be as large as possible.

The molar absorption coefficients of most species used in trace analysis range from approximately 10^3 to 10^5 cm^{-1}mol^{-1}dm^3. If the smallest detectable value of absorbance is 0.001 and the pathlength is 1 cm, the smallest detectable concentration of a substance with $\varepsilon = 10^5$ cm^{-1}mol^{-1}dm^3 is therefore 10^{-8} mol dm^{-3}. Detection limits are usually rather poorer than this, first because ε is seldom as large as 10^5 cm^{-1}mol^{-1}dm^3, and second because instrumental and chemical factors may prevent absorbances as small as 0.001 from being detected or attributed to the species being sought. In spectrophotometry the term sensitivity is used by many authors to represent the mass of element which, in the form of the absorbing species in a column of solution of 1 cm^2 cross section, has an absorbance of 0.001. The sensitivity in μg cm^{-2} is therefore numerically equal to nM/ε, where M is the atomic mass of the element, n atoms of which are contained in the molecule of the absorbing species with molar absorption coefficient ε cm^{-1}mol^{-1}dm^3. The sensitivities of most absorptiometric procedures used in trace analysis lie in the range 0.001 to 0.05 μg cm^{-2}, enabling many elements to be determined satisfactorily in geological materials at concentrations above about 0.1 μg/g, provided adequate separation can be achieved. The best sensitivity and adherence to Beer's law are obtained at the wavelength, designated λ_{max}, where ε has its greatest value. However, it is sometimes desirable to work at a wavelength displaced from λ_{max} to minimize absorption by an interfering species or by excess of the chromogenic reagent.

Many hundreds of organic reagents have been investigated for their suitability in trace analysis, but only a relatively small number have found regular use; some of these are noted in Section 7.4. The book by Perrin[23] on organic complexing agents discusses such aspects as thermodynamics and kinetics of metal complex formation, solubility in aqueous and organic solvents, and other topics of analytical significance. More recent developments have been covered in reviews such as those of Boltz and Mellon[17] and Ueno.[24]

Some of the chromogenic reagents most commonly used are listed in Table 7.2. Many other reagents (e.g., substituted phenanthrolines and hydroxyquinolines, α-furildioxime, 1-(2-pyridylazo)-2-naphthol, and various phthaleins and dithiols) are structurally related to those tabulated.

7.3.2 Deviations from Beer's Law

Deviations from Beer's law are described as positive or negative, according to whether the absorbance-concentration curves bend toward the absorbance or concentration axis, respectively. A plot of A/c versus c to investigate the apparent constancy of ε is a more sensitive test for Beer's law deviations than the plot of A versus c. Conformity to Beer's law is not essential in analytical work, but poorer precision results from working on the curved part of an absorbance-concentration curve. In addition, there is the danger that the curvature is caused by some chemical factor that has not been properly controlled in the standard solutions and that may vary from sample to sample.

Instrumental limitations on the applicability of Beer's law have been reviewed by Lothian.[25] Deviations arise from such factors as the nonuniform pathlength when cylindrical cells are employed, the use of polychromatic radiation, and the presence of stray light. If the pathlength l varies significantly over the width of the light beam or if ε varies significantly over the wavelength band being used, the curves show a negative deviation from Beer's law. The presence of stray light, scattered or reflected to the detector from optical components, also leads to a negative deviation. This can be serious in dealing with solutions of high absorbance, when the stray light may be a

TABLE 7.2 Some Commonly Used Reagents for Absorptiometry

1. 8-Hydroxyquinoline ("oxine")

2. Sodium diethyldithiocarbamate (DTC, DEDTC)

3. Dimethylglyoxime

4. Toluene-3,4-dithiol ("dithiol")

5. 1,10-Phenanthroline

6. Diphenylthiocarbazone ("dithizone")

7. 1,5-Diphenylcarbohydrazide ("diphenylcarbazide")

8. 4-(2-Pyridylazo)resorcinol (PAR)

9. Pyrocatechol violet (catechol violet, catecholsulphonphthalein II)

10. Rhodamine B (cation)

significant part of the light reaching the detector. For a solution with a true absorbance of 1.00, if the stray light is equivalent to 0.25% of the incident power, the measured absorbance is 0.99 (1% error). Modern instruments are designed to keep stray light levels below about 0.1%.

Many types of chemical reaction involving the absorbing species can lead to the appearance of deviations from Beer's law. The following examples may be noted.

1. *Acid-base equilibria.* Where an absorbing species (e.g., A^-) is protonated to give its conjugate acid,

$$A^- + H_3O^+ \rightleftharpoons HA + H_2O$$

the absorbance is not a linear function of the total concentration of A (A^- and HA) unless the solution is buffered to constant pH, or studied at a wavelength where ε is the same for both A^- and HA.

2. *Polymerization equilibria.* The chromate-dichromate equilibrium is a good example

$$2CrO_4^{2-} + 2H_3O^+ \rightleftharpoons Cr_2O_7^{2-} + 3H_2O$$

In this case, buffering at constant pH is not sufficient to maintain a linear relation between absorbance and chromium concentration. The pH must be high enough to keep all the chromium as chromate or low enough to keep it all as dichromate, or the solution could be studied at a wavelength where the absorbance is independent of the chemical form of the chromium (e.g., 446 nm, see Fig. 7.1).

3. *Multiple metal-ligand equilibria.* Many metals form a series of complexes with ligand:metal ratios such as 1:1, 2:1, 3:1. These complexes in general have different values of ε at any given wavelength. The concentration of free ligand may be pH-dependent, and the relative concentrations of the various metal complexes change with ligand concentration. Beer's law can only be expected to hold if the same complex predominates throughout the concentration range being studied, and it may be necessary to ensure that enough ligand is present to keep the metal in the form of the complex with the highest ligand:metal ratio.

7.3.3 Sources of Error

Photometric Errors

These are due to factors such as nonlinear response of photodetectors and amplifiers and to inaccuracies of other electrical and mechanical parts of the recording instrument. In instruments that give a readout linear in transmittance, the smallest detectable change in power ΔP is often approximately constant, regardless of the actual power. The absolute error ΔT in the transmittance is therefore nearly constant over the whole range.

Where Beer's law is followed and ΔT is constant,

$$\varepsilon lc = -\log T = -\frac{\ln T}{\ln 10} \tag{7.3}$$

$$\varepsilon l\, dc = -d \log T = -\frac{dT}{T \ln 10} \tag{7.4}$$

Hence the relative error in the concentration determined is

$$\frac{dc}{c} = \frac{dT}{T \ln T}$$

or, replacing the differentials by finite increments,

$$\frac{\Delta c}{c} = \frac{\Delta T}{T \ln T} \tag{7.5}$$

Differentiation of equation 7.5 with respect to T shows that $\Delta c/c$ is a minimum when $\ln T = -1$, (i.e., when $T = 0.368$). It follows from equation 7.5 that a transmittance error of 0.01 gives rise to concentration errors of about 2.7% at $T = 0.368$, 3.5% at $T = 0.15$ and $T = 0.64$, and 7% at $T = 0.047$ and $T = 0.84$. The U-shaped curve of equation 7.5 can be found in most texts on analytical chemistry, and is the basis of recommendations that measurements be made with solutions giving transmittances of 0.8–0.1 (absorbances of 0.1–1.0), if possible. This simple analysis has been modified to take account of setting errors at each end of the transmittance scale.[26]

In most recording spectrophotometers the phototube or photomultiplier response is transformed into a logarithmic function before amplification, providing an output linear in absorbance, and the photometric error depends in a complex way on the errors in a series of electrical and mechanical processes. The preceding simple analysis is not applicable, and the conclusions concerning the optimum absorbance range depend on some assumption or observation of a relationship between dT and T.[27] In general, the qualitative result remains—photometric precision is poor for solutions of very high or very low absorbance. The range of absorbances within which good precision of photometric measurement can be obtained with any given instrument is probably best decided experimentally by determining the repeatability of measurements on solutions in different parts of the absorbance range.

Other Sources of Error

Apart from the photometric measurement itself, any of the factors that can lead to deviations from Beer's law (Section 7.3.2) must be regarded as capable of influencing the precision of the analysis. Instrumental errors such as loss of wavelength calibration and the use of nonmatched absorption cells, and errors of technique (e.g., errors in preparing solutions, presence of foreign matter in the solutions or on cell windows, variation of sample temperature) also contribute. Both precision and accuracy may be affected by shortcomings, sometimes unavoidable, in the chemical part of the experimental procedure. These include chemical instability of the absorbing species, reactions with dissolved or atmospheric oxygen, slowness of the color-forming reaction to reach equilibrium (especially at very low analyte concentrations), and the need to correct absorbances for the effects of interfering substances.

More extensive discussion of absorptiometric errors can be found in the literature[25-30] and in standard texts.

7.3.4 Precision (Differential) Absorptiometry

In dealing with solutions of very high or very low absorbance, techniques are available for increasing the precision of photometric measurement. In the case of a strongly absorbing solution, for example, the blank or reference solution can be replaced by one that absorbs nearly as strongly as the sample. With a suitable increase of slit width or amplifier gain, the full-scale deflection of the meter then corresponds to only a fraction of the transmittance range (e.g., 0–0.1 instead of 0–1.0). The same error in photocurrent now corresponds to a much lower error in transmittance than before, and better precision is achieved at the expense of limiting the range of transmittances that can be measured. This is the principle of precision, or differential, absorptiometry.

Other portions of the transmittance scale can be "expanded" in a similar way. The various differential techniques are summarized in Table 7.3.

TABLE 7.3 Techniques of Precision (Differential) Absorptiometry[a]

	Transmittance scale settings	
Application	$T = 0$	$T = 1$
High absorbance solutions	No light ("dark current")	Standard solution, concentration c_1
Low absorbance solutions (trace determination)	Standard solution concentration c_2	Solvent blank
Maximum precision	Standard solution, concentration c_2	Standard solution, concentration c_1

[a] For these applications the concentrations c_1 and c_2 are chosen such that c_1 is rather less than the sample concentration c, and c_2 is rather greater than c.

In the second and third of these applications ("trace determination" and "maximum precision") the values of absorbance obtained by conversion from the transmittance scale are not accurately linear functions of concentration, and preparation of a working curve with several standards is recommended. These techniques are discussed extensively in the literature.[27,28,31–35]

7.3.5 Multicomponent Analysis and Derivative Spectrophotometry

If two absorbing species do not interact when mixed, the spectrum of the mixture is the sum of the spectra of the two components. Quantitative analysis from overlapping spectra is possible if absorbance measurements are made on the mixture at two different wavelengths, λ_1 and λ_2. Denoting properties of individual components by the subscripts 1 and 2, and those of the mixture by the subscript m, we have

$$A_m(\lambda_1) = A_1(\lambda_1) + A_2(\lambda_1) = [\varepsilon_1(\lambda_1)c_1 + \varepsilon_2(\lambda_1)c_2]l \tag{7.6}$$

$$A_m(\lambda_2) = A_1(\lambda_2) + A_2(\lambda_2) = [\varepsilon_1(\lambda_2)c_1 + \varepsilon_2(\lambda_2)c_2]l \tag{7.7}$$

If the molar absorption coefficients of each component at the two wavelengths are known, these equations become simultaneous equations in the unknowns, c_1 and c_2. This technique is most satisfactory when much of the absorbance at λ_1 is due to component 1, whereas much of the absorbance at λ_2 is due to component 2. In more complex cases of multicomponent absorption, it may be desirable to carry out additional chemical separations or to seek alternative chromogenic reagents.

Where serious overlapping of absorption bands occurs, better resolution can be obtained by the use of derivative spectrophotometry. The instrumental methods by which direct recording of first-derivative absorption spectra can be performed are discussed elsewhere.[14] General aspects of derivative spectrometry have been reviewed,[36] and some applications to inorganic analysis have been demonstrated by Shibata et al.[37] This technique has not yet been widely used in trace-metal analysis, since most workers prefer to avoid the problem of overlapping spectra by using separation or masking techniques, or alternative reagents.

7.3.6 Photometric Titrations

In some titrimetric procedures, the color change at the equivalence point of the titration is not sufficiently sharp for visual detection to be satisfactory. Good precision may, however, be obtained by measuring the absorbance of the solution at a suitable wavelength during the course of the titration.

A photometric titration curve is a plot of absorbance versus volume of titrant. Absorbances are corrected for the effect of dilution during the titration. Several forms of titration curve are possible, depending on whether the analyte, the titrant, or the reaction product is strongly absorbing. If Beer's law is obeyed by the absorbing species, the corrected absorbance curve takes the form of two straight lines, the intersection of which indicates the equivalence point.

In the field of metal analysis, photometric titration with complex-forming reagents such as EDTA (see also Chapter 6) has been extensively investigated.[38–40] The technique has been used more often for determining metals at concentrations of 10^{-4} to 10^{-2} mol/liter than at lower concentrations. Detailed discussion and applications are given in Headridge,[41] Underwood,[42] and Schwarzenbach and Flaschka.[43]

7.4 APPLICATIONS TO THE DETERMINATION OF VARIOUS ELEMENTS

The size of the literature on absorptiometric methods—many thousands of papers covering most of the elements in the periodic table—precludes any detailed discussion in this volume. Comprehensive discussions of individual methods, and experimental procedures for dealing with a variety of materials, including water, ores, rocks, and soils, are given in standard texts such as those of Sandell[3] and Snell and Snell.[5] The book by Jeffery[15] is a valuable source of procedures dealing particularly with silicate rock analysis, whereas that of Stanton[16] emphasizes rapid colorimetric methods suitable for soil and sediment samples in geochemical exploration.

The following examples serve to illustrate a range of operations and techniques used in trace inorganic analysis of silicates and related materials by solution absorptiometry.

Chromium, Diphenylcarbazide Method

Alkali fusion of the sample (with sodium carbonate/magnesium oxide) ensures that only traces of iron are extracted from the solidified melt into aqueous solution. The alkaline solution containing chromium(VI) is acidified and diphenylcarbazide solution in acetone is added. After dilution to the appropriate volume, the absorbance is measured at 540 nm. If necessary, interference by large amounts of vanadium is avoided by extraction of the vanadium as the 8-hydroxyquinoline complex. For more information on these methods, see Refs. 3, 44, and 45.

APPLICATIONS TO THE DETERMINATION OF VARIOUS ELEMENTS 117

Boron, Carmine Method

The sample is fused with potassium carbonate. The solidified melt is digested with water, and mannitol is added to complex the borate. An acidified cation-exchange resin is then added. The borate remains in solution after the resin and precipitated silica are removed by filtration. Concentrated hydrochloric and sulfuric acids are added, followed by carmine in sulfuric acid. The absorbance of the solution is measured at 585 nm. This method is described in Refs. 46 and 47.

Cobalt, Tri-n-butylamine/Thiocyanate Method

The sample is attacked by a hydrofluoric/perchloric acid mixture, and the residue is taken up in hydrochloric acid. The solution is buffered, and cobalt is extracted into carbon tetrachloride with diphenylthiocarbazone. The extracted cobalt in buffered acidic aqueous solution is treated with sodium hexametaphosphate (to mask traces of iron) and thiocyanate, and tri-n-butylamine in a benzene-based solvent is added. Tri-n-butylammonium tetrathiocyanatocobaltate(III) is extracted into the benzene solution and the absorbance is measured at 625 nm. (See Refs. 48 and 49.)

Germanium, Phenylfluorone Method

The sample is decomposed with a mixture of hydrofluoric, nitric, and sulfuric acids. Concentrated hydrochloric acid is added, and germanium tetrachloride is extracted into carbon tetrachloride. The germanium is back-extracted into water, the solution is acidified with sulfuric acid, and ethanolic phenylfluorone solution added, together with gum arabic or polyvinyl alcohol as a protective colloid to prevent gradual flocculation of the product. The absorbance of the germanium-phenylfluorone complex is measured at 504 to 510 nm. For more details, see Refs. 50–53.

Fluorine, by Bleaching of Zirconium/Eriochrome cyanine R Complex

The sample is fused with sodium carbonate and zinc oxide. The solidified melt is digested with water and washed, and the residue is discarded. Concentrated nitric acid is added, and the solution diluted to an appropriate volume. Sodium hydroxide is added to an aliquot of solution which is then added to a solution containing the colored complex formed between zirconyl chloride and eriochrome cyanine R. The extent of the bleaching of this complex (examined at 525 nm) is a measure of the amount of fluoride added. (See Refs. 54–56.)

Some of the more widely used methods for many elements are listed, together with references, in Table 7.4.

TABLE 7.4 Selected Spectrophotometric Methods for Individual Elements

Element	Reagent or method	Ref.
Ag	Dithizone	3, 57
	p-Dimethylaminobenzylidenerhodanine	58, 59
Al	8-Hydroxyquinoline	60–62
	Pyrocatechol violet	63
As	Molybdate ("molybdenum blue")	64
	Silver diethyldithiocarbamate	65–67
Au	p-Diethylaminobenzylidenerhodanine	68
	Rhodamine B	69
	2,2′-Dipyridylketoxime	70
B	Curcumin	71, 72
	Carmine, carminic acid	46, 47
	Dianthrimide	73
Be	Beryllon II	15, 74
	p-Nitrobenzeneazo-orcinol	75, 76
	Fast sulfon black F	77
Bi	Dithizone	78
	APDC	79
Cl	Mercuric thiocyanate (SCN⁻ displacement)	56, 80
Co	Thiocyanate	81–83
	Thiocyanate/tri-n-butylamine	48, 49
	Nitroso-R salt	3, 84–86
	APDC	79
	2-Nitroso-1-naphthol	87, 88
	1-Nitroso-2-naphthol	83
Cr	Chromate	89
	Diphenylcarbazide	3, 44, 45, 90
Cu	Dithizone	3, 85
	Diethyldithiocarbamate	91
	2,2′-Biquinolyl (cuproine)	92
	2,9-Dimethyl-1,10-phenanthroline (neocuproine)	93, 94
F	Zr/eriochrome cyanine R (bleached by F^-)	54–56
	Alizarin complexan	95, 96
Fe	1,10-Phenanthroline	3, 97, 98
	2,2′-Bipyridyl	99
	Thiocyanate	3, 100
Ga	Rhodamine B	101, 102
	Rhodamine B n-butyl ester	103, 104
Ge	Phenylfluorone	50–53
I	Iodide-iodate-iodine conversion	105
Mg	Titan yellow	3, 106, 107
	Magon	108–111
Mn	Permanganate	3, 112, 113
Mo	Thiocyanate	3, 114, 115
	Toluene-3,4-dithiol	15, 116–118

TABLE 7.4 (*continued*)

Element	Reagent or method	Ref.
Nb	Thiocyanate	15, 119–121
	4-(2-Pyridylazo)resorcinol	15, 122
Ni	Dimethylglyoxime	3, 85, 123–126
	α-Furildioxime	127
P	Vanadomolybdate	128
	Molybdenum blue	60, 129, 130
Pb	Dithizone	131, 132
	Diethyldithiocarbamate	133
	4-(2-Pyridylazo)resorcinol	134
Pd	*p*-Nitrosodiphenylamine	135
	p-Nitrosodimethylaniline	136
Pt	*p*-Nitrosodimethylaniline	137
Sb	Rhodamine B	138, 139
	Brilliant green	140, 141
Sc	Sulfonazo	142
Se	3,3′-Diaminobenzidine	143–145
Si	Silicomolybdate	146, 147
	Molybdenum blue (reduced silicomolybdate)	129, 148–151
Sn	Toluene-3,4-dithiol	3, 152, 153
	Salicylideneamino-2-thiophenol	154
	Gallein	155
Ta	Rhodanine 6J	156
Te	Hexaiodotellurite	157
Th	Arsenazo III	15, 158
Ti	Hydrogen peroxide	3, 159
	Tiron	160, 161
	Diantipyrylmethane	162
Tl	Dithizone	163
	Brilliant green	15, 164
U	Thiocyanate	165, 166
	Hydrogen peroxide	3, 167
	1-(2-Pyridylazo)-2-naphthol	168
	4-(2-Pyridylazo)resorcinol	169
	Arsenazo III	170
V	*N*-Benzoyl-*o*-tolylhydroxylamine	171, 172
	Diaminobenzidine	173, 174
	Hydrogen peroxide	175
	Phosphomolybdate	176
	Phosphotungstate	3, 177, 178
W	Thiocyanate	179–182
	Toluene-3,4-dithiol	118, 183, 184
Zn	Dithizone	185–188
	Zincon	189, 190
Zr	Xylenol orange	191, 192
	Quinalizarin sulfonic acid	193

References

1. M. G. Mellon (ed.), *Analytical Absorption Spectroscopy*, Wiley, New York, 1950.
2. G. F. Lothian, *Absorption Spectrophotometry*, 2nd ed., Hilger and Watts, London, 1958.
3. E. B. Sandell, *Colorimetric Determination of Traces of Metals*, 3rd ed., Interscience, New York, 1959.
4. D. F. Boltz (ed.), *Colorimetric Determination of Nonmetals*, Interscience, New York, 1958.
5. F. D. Snell and C. T. Snell, *Colorimetric Methods of Analysis*, Vol. II, Van Nostrand, New York, 1949; Vol. IIA, Van Nostrand, Princeton, N.J. 1959.
6. R. P. Bauman, *Absorption Spectroscopy*, Wiley, New York, 1962.
7. N. L. Allport and J. E. Brocksopp, *Colorimetric Analysis*, Vol. II, 2nd ed., Chapman and Hall, London, 1963.
8. G. Charlot, *Colorimetric Determination of Elements*, Elsevier, Amsterdam, 1964.
9. E. J. Meehan, in *Treatise on Analytical Chemistry*, Part I, Vol. 5 (eds. I. M. Kolthoff and P. J. Elving), Interscience, New York, 1964.
10. J. R. Edisbury, *Practical Hints on Absorption Spectrometry*, Hilger and Watts, London, 1966; Plenum Press, New York, 1967.
11. W. West (ed.), *Chemical Applications of Spectroscopy*, 2nd ed., Part I, Wiley, New York, 1968.
12. E. I. Stearns, *The Practice of Absorption Spectrophotometry*, Wiley-Interscience, New York, 1969.
13. ASTM Committee, *Manual on Recommended Practices in Spectrophotometry*, 3rd ed., American Society for Testing and Materials, Philadelphia, 1969.
14. F. Grum, in *Techniques of Chemistry*, Volume I, Part IIIB (eds. A. Weissberger and B. W. Rossiter), Wiley, New York, 1972.
15. P. G. Jeffery, *Chemical Methods of Rock Analysis*, 2nd ed., Pergamon, Elmsford, N.Y., 1975.
16. R. E. Stanton, *Rapid Methods of Trace Analysis*, Arnold, London, 1966.
17. D. F. Boltz and M. G. Mellon, *Anal. Chem.*, **38**, 317R (1966); **40**, 255R (1968); **42**, 152R (1970); **44**, 300R (1972); **46**, 227R (1974); **48**, 216R (1976).
18. M. L. McGlashan, *Physicochemical Quantities and Units*, 2nd ed., Royal Institute of Chemistry, London, 1971.
19. "Spectrometry Nomenclature," *Anal. Chem.*, **39**, 1943 (1967).
20. H. G. Pfeiffer and H. A. Liebhafsky, *J. Chem. Educ.*, **28**, 133 (1951).
21. H. A. Liebhafsky and H. G. Pfeiffer, *J. Chem. Educ.*, **30**, 450 (1953).
22. D. D. Perrin, *Masking and Demasking of Chemical Reactions*, Wiley-Interscience, New York, 1970.
23. D. D. Perrin, *Organic Complexing Agents*, Wiley-Interscience, New York, 1964.
24. K. Ueno, in *M.T.P. International Review of Science, Physical Chemistry Series One*, Vol. 13: Analytical Chemistry, Part 2 (ed. T. S. West), Butterworths, London, 1973.

REFERENCES

25. G. F. Lothian, *Analyst*, **88**, 678 (1963).
26. N. T. Gridgeman, *Anal. Chem.*, **24**, 445 (1952).
27. C. M. Crawford, *Anal. Chem.*, **31**, 343 (1959).
28. G. H. Ayres, *Anal. Chem.*, **21**, 652 (1949).
29. A. Ringbom, *Z. Anal. Chem.*, **115**, 332 (1939).
30. L. S. Goldring, R. C. Hawes, G. H. Hare, A. O. Beckman, and M. E. Stickney, *Anal. Chem.*, **26**, 869 (1953).
31. R. Bastian, *Anal. Chem.*, **21**, 972 (1949).
32. C. F. Hiskey, *Anal. Chem.*, **21**, 1440 (1949).
33. C. N. Reilley and C. M. Crawford, *Anal. Chem.*, **27**, 716 (1955).
34. C. V. Banks, P. G. Grimes, and R. I. Bystroff, *Anal. Chim. Acta*, **15**, 367 (1956).
35. J. W. O'Laughlin and C. V. Banks, in *Encyclopedia of Spectroscopy* (ed. G. L. Clark), Reinhold, New York, 1960, pp. 19–33.
36. R. N. Hager, *Anal. Chem.*, **45**, 1131A (1973).
37. S. Shibata, M. Furukawa, and K. Goto, *Anal. Chim. Acta*, **46**, 271 (1969).
38. P. B. Sweetser and C. E. Bricker, *Anal. Chem.*, **25**, 253 (1953).
39. A. L. Underwood, *Anal. Chem.*, **26**, 1322 (1954).
40. L. Shapiro and W. W. Brannock, *Anal. Chem.*, **27**, 725 (1955).
41. J. B. Headridge, *Photometric Titrations*, Pergamon, London, 1958.
42. A. L. Underwood, in *Advances in Analytical Chemistry and Instrumentation*, Vol. 3 (ed. C. N. Reilley), Interscience, New York, 1964, pp. 31–104.
43. G. Schwarzenbach and H. Flaschka, *Complexometric Titrations*, 2nd English ed. (transl. H. M. N. H. Irving), Methuen, London, 1969.
44. M. Bose, *Anal. Chim. Acta*, **10**, 201, 209 (1954).
45. R. Fuge, *Chem. Geol.*, **2**, 289 (1967).
46. J. T. Hatcher and L. V. Wilcox, *Anal. Chem.*, **22**, 567 (1950).
47. M. Fleet, *Anal. Chem.*, **39**, 253 (1967).
48. R. E. Stanton, A. J. McDonald, and I. Carmichael, *Analyst*, **87**, 134 (1962).
49. M. Ziegler, O. Glemser, and E. Preisler, *Mikrochim. Acta*, 1526 (1956).
50. H. J. Cluley, *Analyst*, **76**, 523, 530 (1951).
51. W. A. Schneider and E. B. Sandell, *Mikrochim. Acta*, 263 (1954).
52. J. D. Burton, F. Culkin, and J. P. Riley, *Geochim. Cosmochim. Acta*, **16**, 151 (1959).
53. J. D. Burton and J. P. Riley, *Mikrochim. Acta*, 586 (1959).
54. S. Megregian, *Anal. Chem.*, **26**, 1161 (1954).
55. P. L. Sarma, *Anal. Chem.*, **36**, 1684 (1964).
56. W. H. Huang and W. D. Johns, *Anal. Chim. Acta*, **37**, 508 (1967).
57. H. Fischer, G. Leopoldi, and H. von Uslar, *Z. Anal. Chem.*, **101**, 1 (1935).
58. E. B. Sandell and J. J. Neumayer, *Anal. Chem.*, **23**, 1863 (1951).
59. G. C. B. Cave and D. N. Hume, *Anal. Chem.*, **24**, 1503 (1952).

60. J. P. Riley, *Anal. Chim. Acta*, **19**, 413 (1958).
61. R. H. Linnell and F. H. Raab, *Anal. Chem.*, **33**, 154 (1961).
62. R. A. Chalmers and M. A. Basit, *Analyst*, **93**, 629 (1968).
63. A. D. Wilson and G. A. Sergeant, *Analyst*, **88**, 109 (1963).
64. J. E. Portmann and J. P. Riley, *Anal. Chim. Acta*, **31**, 509 (1964).
65. V. Vasak and V. Sedivec, *Chem. List.*, **46**, 341 (1952).
66. D. Liederman, J. E. Bowen, and O. I. Milner, *Anal. Chem.*, **31**, 2052 (1959).
67. E. W. Fowler, *Analyst*, **88**, 380 (1963).
68. E. B. Sandell, *Anal. Chem.*, **20**, 253 (1948).
69. B. J. MacNulty and L. D. Woollard, *Anal. Chim. Acta*, **13**, 154 (1955).
70. W. J. Holland and J. Bozic, *Anal. Chem.*, **39**, 109 (1967).
71. M. R. Hayes and J. Metcalfe, *Analyst*, **87**, 956 (1962).
72. L. R. Uppstrom, *Anal. Chim. Acta*, **43**, 475 (1968).
73. L. Danielsson, *Talanta*, **3**, 138 (1959).
74. G. G. Karanovich, *Z. Anal. Khim.*, **11**, 417 (1956).
75. J. B. Pollock, *Analyst*, **81**, 45 (1956).
76. L. C. Covington and M. J. Miles, *Anal. Chem.*, **28**, 1728 (1956).
77. A. M. Cabrera and T. S. West, *Anal. Chem.*, **35**, 311 (1963).
78. H. A. Mottola and E. B. Sandell, *Anal. Chim. Acta*, **25**, 520 (1961).
79. M. B. Kalt and D. F. Boltz, *Anal. Chem.*, **40**, 1086 (1968).
80. J. G. Bergmann and J. Sanik, *Anal. Chem.*, **29**, 241 (1957).
81. R. E. Kitson, *Anal. Chem.*, **22**, 664 (1950).
82. R. S. Young and A. J. Hall, *Ind. Eng. Chem., Anal. Ed.*, **18**, 264 (1946).
83. R. Lundquist, G. E. Markle, and D. F. Boltz, *Anal. Chem.*, **27**, 1731 (1955).
84. K. J. McNaught, *Analyst*, **67**, 97 (1942).
85. L. F. Rader and F. S. Grimaldi, U. S. Geol. Surv. Prof. Paper 391-A (1961).
86. J. N. Pascual, W. H. Shipman, and W. Simon, *Anal. Chem.*, **25**, 1830 (1953).
87. L. J. Clark, *Anal. Chem.* **30**, 1153 (1958).
88. A. Claassen and A. Daamen, *Anal. Chim. Acta*, **12**, 547 (1955).
89. S. Christow. *Z. Anal. Chem.*, **125**, 278 (1943).
90. E. S. Pilkington and P. R. Smith, *Anal. Chim. Acta*, **39**, 321 (1967).
91. J. M. Howard and H. O. Spauschus, *Anal. Chem.*, **35**, 1016 (1963).
92. J. P. Riley and P. Sinhaseni, *Analyst*, **83**, 299 (1958).
93. A. R. Gahler, *Anal. Chem.*, **26**, 577 (1954).
94. G. F. Smith and W. H. McCurdy, *Anal. Chem.*, **24**, 371 (1952).
95. R. Belcher, M. A. Leonard, and T. S. West, *Talanta*, **2**, 92 (1959).
96. P. G. Jeffery, *Geochim. Cosmochim. Acta*, **26**, 1355 (1962).
97. W. B. Fortune and M. G. Mellon, *Ind. Eng. Chem. Anal. Ed.*, **10**, 60 (1938).
98. L. G. Saywell and B. B. Cunningham, *Ind. Eng. Chem. Anal. Ed.*, **9**, 67 (1937).

99. M. L. Moss and M. G. Mellon, *Ind. Eng. Chem. Anal Ed.*, **14**, 862 (1942).
100. J. T. Woods and M. G. Mellon, *Ind. Eng. Chem. Anal. Ed.*, **13**, 551 (1941).
101. F. Culkin and J. P. Riley, *Analyst*, **83**, 208 (1958).
102. F. Culkin and J. P. Riley, *Anal. Chim. Acta*, **24**, 413 (1961).
103. V. I. Kuznetzov and L. I. Bol'shakova, *Z. Anal. Khim.*, **15**, 523 (1960).
104. L. M. Skrebkova, *Z. Anal. Khim.*, **16**, 422 (1961).
105. W. H. Crouch, *Anal. Chem.*, **34**, 1689 (1962).
106. H. G. C. King, G. Pruden, and N. F. Janes, *Analyst*, **92**, 695 (1967).
107. W. H. Evans, *Analyst*, **93**, 306 (1968).
108. C. K. Mann and J. H. Yoe, *Anal. Chem.*, **28**, 202 (1956).
109. C. K. Mann and J. H. Yoe, *Anal. Chim. Acta*, **16**, 155 (1957).
110. R. F. Apple and J. C. White, *Talanta*, **8**, 419 (1961).
111. S. Abbey and J. A. Maxwell, *Anal. Chim. Acta*, **27**, 233 (1962).
112. F. Nydahl, *Anal. Chim. Acta*, **3**, 144 (1949).
113. C. L. Luke, *Anal. Chim. Acta*, **34**, 302 (1966).
114. E. B. Sandell, *Ind. Eng. Chem. Anal. Ed.*, **8**, 336 (1936).
115. F. S. Grimaldi and R. C. Wells, *Ind. Eng. Chem. Anal. Ed.*, **15**, 315 (1943).
116. P. G. Jeffery, *Analyst*, **82**, 558 (1957).
117. K. M. Chan and J. P. Riley, *Anal. Chim. Acta*, **36**, 220 (1966).
118. K. M. Chan and J. P. Riley, *Anal. Chim. Acta*, **39**, 103 (1967).
119. F. N. Ward and A. P. Marranzino, *Anal. Chem.*, **27**, 1325 (1955).
120. F. S. Grimaldi, *Anal. Chem.*, **32**, 119 (1960).
121. J. Esson, *Analyst*, **20**, 488 (1965).
122. R. Belcher, T. V. Ramakrishna, and T. S. West, *Talanta*, **9**, 943 (1962).
123. A. Lieberman, *Analyst*, **80**, 595 (1955).
124. A. J. Easton and J. F. Lovering, *Geochim. Cosmochim. Acta*, **27**, 753 (1963).
125. G. Norwitz and H. Gordon, *Anal. Chem.*, **37**, 417 (1965).
126. A. Claassen and L. Bastings, *Analyst*, **91**, 725 (1966).
127. D. E. Bodart, *Z. Anal. Chem.*, **247**, 32 (1969).
128. O. B. Michelson, *Anal. Chem.*, **29**, 60 (1957).
129. D. F. Boltz and M. G. Mellon, *Anal. Chem.*, **19**, 873 (1947).
130. C. H. Lueck and D. F. Boltz, *Anal. Chem.*, **28**, 1168 (1956).
131. J. C. Gage, *Analyst*, **80**, 789 (1955); **82**, 453 (1957).
132. L. J. Snyder, *Anal. Chem.*, **19**, 684 (1947).
133. R. Keil, *Z. Anal. Chem.*, **229**, 117 (1967).
134. R. M. Dagnall, T. S. West, and P. Young, *Talanta*, **12**, 583, 589 (1965).
135. D. E. Ryan, *Analyst*, **76**, 167 (1951).
136. F. S. Grimaldi and M. M. Schnepfe, U. S. Geol. Surv. Prof. Paper 575-C, 141 (1967).

137. J. H. Yoe and J. J. Kirkland, *Anal. Chem.*, **26**, 1335, 1340 (1954).
138. F. N. Ward and H. W. Lakin, *Anal. Chem.*, **26**, 1168 (1954).
139. R. W. Ramette and E. B. Sandell, *Anal. Chim. Acta*, **13**, 455 (1955).
140. R. E. Stanton and A. J. McDonald, *Analyst*, **87**, 299 (1962).
141. R. W. Burke and O. Menis, *Anal. Chem.*, **38**, 1719 (1966).
142. V. G. Brudz, V. I. Titov, E. P. Osiko, D. A. Drapkina, and K. A. Smirnova, *Z. Anal. Khim.*, **17**, 568 (1962).
143. K. L. Cheng, *Anal. Chem.*, **28**, 1738 (1956).
144. R. E. Stanton and A. J. McDonald, *Analyst*, **90**, 497 (1965).
145. Y. K. Chau and J. P. Riley, *Anal. Chim. Acta*, **33**, 36 (1965).
146. O. P. Case, *Ind. Eng. Chem. Anal. Ed.*, **16**, 309 (1944).
147. D. R. Shink, *Anal. Chem.*, **37**, 764 (1965).
148. P. Pakalns and W. W. Flynn, *Anal. Chim. Acta*, **38**, 403 (1967).
149. L. Shapiro and W. M. Brannock, *U. S. Geol. Surv. Circ.* **165** (1952).
150. L. Shapiro and W. M. Brannock, *U. S. Geol. Surv. Bull.* **1036-C** (1956); **1144-A** (1962).
151. P. G. Jeffery and A. D. Wilson, *Analyst*, **85**, 478 (1960).
152. R. E. D. Clark, *Analyst*, **61**, 242 (1936); **62**, 661 (1937).
153. H. Onishi and E. B. Sandell, *Anal. Chim. Acta*, **14**, 153 (1956).
154. R. M. Dagnall, T. S. West, and P. Young, *Analyst*, **92**, 27 (1967).
155. A. J. McDonald and R. E. Stanton, *Analyst*, **87**, 600 (1962).
156. N. N. Pavlova and I. A. Blyum, *Zavod. Lab.*, **28**, 1305 (1962); **32**, 1196 (1966).
157. R. A. Johnson and F. P. Kwan, *Anal. Chem.*, **23**, 651 (1951).
158. I. May and L. B. Jenkins, U. S. Geol. Surv. Prof. Paper 525-D, 192 (1965).
159. A. Weissler, *Ind. Eng. Chem. Anal. Ed.*, **17**, 695 (1945).
160. J. H. Yoe and A. R. Armstrong, *Anal. Chem.*, **19**, 100 (1947).
161. T. Rigg and H. A. Wagenbauer, *Anal. Chem.*, **33**, 1347 (1961).
162. P. G. Jeffery and G. R. E. C. Gregory, *Analyst*, **90**, 177 (1965).
163. R. S. Clarke and F. Cuttitta, *Anal. Chim. Acta*, **19**, 555 (1958).
164. N. T. Voskresenskaya, *Z. Anal. Khim.*, **11**, 623 (1956).
165. J. E. Currah and F. E. Beamish, *Anal. Chem.*, **19**, 609 (1947).
166. O. A. Nietzel and M. A. DeSesa, *Anal. Chem.*, **29**, 756 (1957).
167. T. R. Scott, *Analyst*, **75**, 100 (1950).
168. K. L. Cheng, *Anal. Chem.*, **30**, 1027 (1958).
169. J. M. Florence and Y. Farrar, *Anal. Chem.*, **35**, 1613 (1963).
170. E. Singer and M. Matucha, *Z. Anal. Chem.*, **191**, 248 (1962).
171. A. K. Majumdar and G. Das, *Anal. Chim. Acta*, **31**, 147 (1964).
172. P. G. Jeffery and G. O. Kerr, *Analyst*, **92**, 763 (1967).
173. K. L. Cheng, *Talanta*, **8**, 658 (1961).

REFERENCES

174. K. M. Chan and J. P. Riley, *Anal. Chim. Acta*, **34**, 337 (1966).
175. G. Telep and D. F. Boltz, *Anal. Chem.*, **23**, 901 (1951).
176. R. Jakubiec and D. F. Boltz, *Anal. Chim. Acta*, **43**, 137 (1968).
177. E. R. Wright and M. G. Mellon, *Ind. Eng. Chem. Anal. Ed.*, **9**, 251 (1937).
178. R. M. Sherwood and F. W. Chapman, *Anal. Chem.*, **27**, 88 (1955).
179. E. B. Sandell, *Ind. Eng. Chem. Anal. Ed.*, **18**, 163 (1946).
180. C. E. Crouthamel and C. E. Johnson, *Anal. Chem.*, **26**, 1288 (1954).
181. C. H. R. Gentry and L. G. Sherrington, *Analyst*, **73**, 57 (1948).
182. B. Reef and H. G. Doge, *Talanta*, **14**, 967 (1967).
183. P. G. Jeffery, *Analyst*, **81**, 104 (1956).
184. B. F. Quin and R. R. Brooks, *Anal. Chim. Acta*, **58**, 301 (1972).
185. E. B. Sandell, *Ind. Eng. Chem. Anal. Ed.*, **9**, 464 (1937).
186. B. L. Vallee, *Anal. Chem.*, **26**, 914, 1244 (1954).
187. I. Carmichael and A. J. McDonald, *Geochim. Cosmochim. Acta*, **22**, 87 (1961).
188. L. Greenland, *Geochim. Cosmochim. Acta*, **27**, 269 (1963).
189. J. H. Yoe and R. M. Rush, *Anal. Chem.*, **26**, 1345 (1954).
190. C. Huffman, H. H. Lipp, and L. F. Rader, *Geochim. Cosmochim. Acta*, **27**, 209 (1963).
191. K. L. Cheng, *Talanta*, **2**, 61 (1959).
192. A. K. Babko and V. T. Vasilenko, *Zavod. Lab.*, **27**, 640 (1961).
193. F. Culkin and J. P. Riley, *Anal. Chim. Acta*, **32**, 197 (1965).

CHAPTER

8

MOLECULAR FLUORIMETRY

The analysis of geological materials by fluorimetric methods is not nearly as widespread as the use of absorption spectrophotometry. There are several elements, however, including aluminum, beryllium, selenium, uranium, and some of the lanthanides, for which fluorimetry is one of the most sensitive and most satisfactory methods available.

Extensive discussion of the basic phenomenon and theory of fluorescence, of fluorimetric instruments, and of practical applications in inorganic analysis, can be found in specialized texts.[1-7] Useful information is contained in several articles and reviews,[8-10] and the progress of investigations in inorganic fluorimetry can be followed in the reviews compiled by Weissler[11] and his successors.[12] Only a brief outline is possible here.

8.1 INTRODUCTION AND THEORY

Most molecules possess a set of electronic states in which all electron spins are paired. These are described as singlet states, designated S, that of lowest energy being designated S_0. The absorption of ultraviolet or visible radiation may raise an electron to a molecular orbital of higher energy without altering the electron spin, that is, the molecule may undergo a transition from state S_0 to state S_1 (or a higher excited singlet state), as shown in Fig. 8.1.

During the time spent in the excited state, often 10^{-6} to 10^{-9} s, the molecule loses vibrational energy by colliding with neighboring molecules. Return to state S_0 may occur as a result of spontaneous emission of a quantum of light of rather lower energy (longer wavelength) than that which was absorbed. This process, molecular fluorescence, is one of several that may occur to the excited molecules. Other possible processes include (*a*) radiationless, or collisional, deactivation to reach S_0, (*b*) decomposition, (*c*) chemical reactions, and (*d*) transition to a triplet state T in which two electron spins are unpaired, the return to S_0 occurring by collisional deactivation or by emission of radiation (phosphorescence).

For reasons previously noted (p. 99) both the absorption and the fluorescence occur over a rather broad band of wavelengths, as illustrated in Fig.

INTRODUCTION AND THEORY

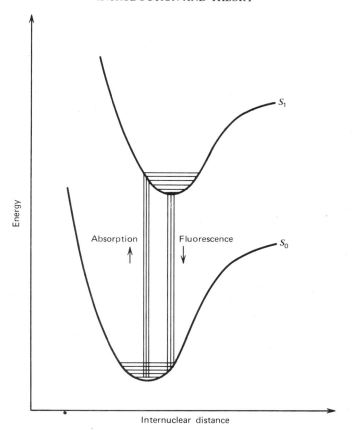

Fig. 8.1. Molecular electronic excitation and fluorescence.

8.2. The fluorescence excitation spectrum indicates the relative ability of radiation of different wavelengths to cause excitation to the state from which the observed fluorescence occurs. It is similar to the absorption spectrum, but may differ from the latter as a result of certain instrumental factors (p. 132), or if part of the absorption spectrum is attributable to excitation to another excited state. Most practical applications make use of a band of exciting radiation centered on the longest wavelength peak in the excitation spectrum. The use of higher energy (shorter wavelength) radiation increases the possibility of molecular decomposition.

The fluorescence emission spectrum is a plot of fluorescence power (or intensity) at various wavelengths, using a fixed wavelength or wavelength

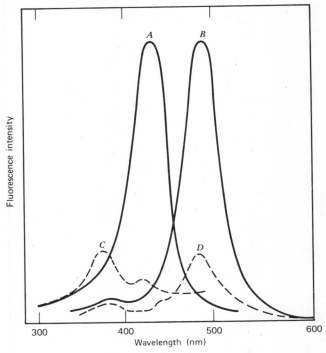

Fig. 8.2. Fluorescence excitation and emission spectra of 3,4′,7-trihydroxyflavone and its antimony complex in 1 M perchloric acid. A, Excitation, Sb + reagent; B, emission, Sb + reagent; C, excitation, reagent alone; D, emission, reagent alone. *Source*: Filer.[13] Reprinted by permission of *Analytical Chemistry*. Copyright by the American Chemical Society.

band for excitation. Changing the excitation wavelength changes the intensity, rather than the shape or position, of the fluorescence emission spectrum. The efficiency with which an absorbing species fluoresces is described by the fluorescence quantum efficiency Φ, defined as the ratio of the number of quanta emitted as fluorescence to the number of quanta absorbed. The quantum efficiency depends on the probability of occurrence of the other processes that compete with fluorescence—if collisional deactivation to state S_0 occurs readily, for example, then Φ will be low. The quantum efficiency is a property of a molecule under specified environmental conditions: solvent, ionic strength, pH, and temperature.

For several reasons the development of a relationship between the fluorescence signal and the concentration of the fluorescing species is more complex than in the case of absorption. (*a*) The amount of incident radiation *absorbed*

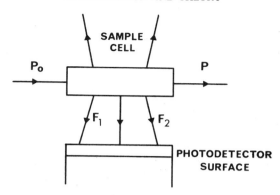

Fig. 8.3. Absorption and fluorescence by a sample. The fluorescent power F_1 from regions of the sample near the source is greater than the power F_2 from regions further from the source. The fluorescence from different parts of the cell is uniform when the absorbance is small.

varies with distance traversed through the sample, according to Beer's law. Consequently, fluorescent emission is greatest from the part of the sample nearest the source. (*b*) The sample is generally not perfectly transparent to the fluorescent radiation, that is, some reabsorption occurs. (*c*) The measured fluorescence signal depends on geometric factors involving the source, sample, and detector, such as the solid angle subtended by the photodetector surface at the sample cell.

Many situations have been analyzed in detail; a simple case is shown in Fig. 8.3, in which a transparent sample cell of length l cm contains a single fluorescent species at a concentration c mol l^{-1}, having a molar absorption coefficient ε cm^{-1}mol^{-1}l at the excitation wavelength. If absorption occurs in accordance with Beer's law, if the fluorescent power from each element of the sample is proportional to the power incident on that element, and if none of the fluorescence is reabsorbed by the sample, then the power transmitted by the sample is

$$P = P_0 10^{-\varepsilon c l} = P_0 10^{-A} \tag{8.1}$$

and the power absorbed is

$$P_0 - P = P_0(1 - 10^{-A}) \tag{8.2}$$

where A is the absorbance of the sample at the excitation wavelength.

The fluorescent power F reaching the detector is

$$F = K P_0 (1 - 10^{-A}) \tag{8.3}$$

where the constant K includes both the quantum efficiency of the fluorescence process and a geometric factor indicating the efficiency of detection of the fluorescence. (In general, each term in equation 8.3 is wavelength dependent).

For solutions of low absorbance (e.g., $A < 0.01$), equation 8.3 simplifies to

$$F = 2.303 K P_0 A = 2.303 K P_0 \varepsilon l c \qquad (8.4)$$

A linear relationship between F and c is therefore obtained, provided only a small fraction (e.g., $<2\%$) of the incident power is absorbed by the sample. In practice the dependence of F on concentration is often investigated over several decades of concentration and illustrated on a log-log plot. The proportionality between F and c, and hence the analytical utility of fluorimetry, is best at trace concentrations of the fluorescent species.

Where fluorimetry is applicable there are several potential advantages over absorptiometry. Selectivity can be obtained from the ability to vary two parameters, the excitation and emission wavelengths. Extremely high sensitivity can be obtained, detection of some substances being possible at levels below 1 ng ml^{-1}. Whereas detection of traces by absorptiometry involves the measurement of very small differences in the power of the incident and transmitted beams, the fluorescent radiation is measured directly, and the signal can be increased by increasing the source power (equation 8.3). The detectability of a fluorescent species is therefore largely dependent on the magnitude of any background signal from impurities or interfering substances.

Disadvantages of fluorimetry include the sensitivity of the fluorescence signal to changes in the environment of the fluorescent species and the possibility of the species undergoing photodecomposition. Fluorescence quenching, that is, the reduction of fluorescence by some competing deactivation process, caused by a specific interaction between the fluorescent species and some other substance present, is also an important problem, which must be investigated in each analytical situation. The presence of other strongly absorbing substances is also undesirable, since this reduces the radiant power available for absorption by the fluorescent species.

8.2 INSTRUMENTATION

The basic optical components of a fluorimeter are shown in Fig. 8.4. An excitation source of moderately high intensity is needed. A mercury-vapor lamp is useful if excitation can be achieved by one of the lines in the mercury spectrum (e.g., 365 nm). Alternatively, a mercury lamp can be modified by painting the inner surface with a crystalline phosphor that gives a broadband emission at longer wavelengths. A high-pressure xenon lamp,

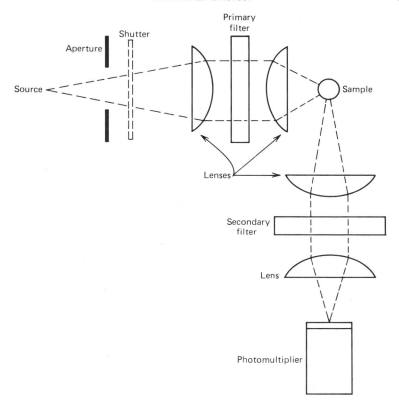

Fig. 8.4. Optical system of a filter fluorimeter.

emitting a continuous spectrum, is used if scanning of the excitation spectrum is necessary.

The excitation wavelength band is selected by the primary filter (in a filter fluorimeter) or monochromator (in a spectrofluorimeter), and the radiation is directed into the sample cell. Cells of pyrex glass are usually satisfactory, as wavelengths below 300 nm are seldom used for excitation. The fluorescence is generally (but not necessarily) studied at an angle of 90° to the direction of incident radiation. A secondary filter or monochromator is used to select the emission wavelength. The fluorescent emission passes to the photodetector, which is most commonly a photomultiplier tube, although some instruments employ a barrier-layer photocell.

Fluorescence emitted from the surface of opaque solid samples can be measured by placing the detector at a small angle to the direction of the exciting radiation. Reflected exciting radiation is prevented from reaching the detector by a suitable choice of wavelength for the secondary filter or monochromator. In geochemical work the determination of uranium by measurement of its fluorescence in a solid fluoride-carbonate matrix has found extensive application.

With simple single-beam optics such as those of Fig. 8.4 it is essential that the source and photomultiplier voltage supplies be adequately stabilized. More elaborate apparatus makes use of a double-beam arrangement, for example, one in which the photodetector receives light alternately from the sample and from a fluorescent standard. The latter may be a standard solution of the species being measured or of some other stable fluorescing species (e.g., quinine sulfate), or may be a standard fluorescing uranium glass.

Much useful analytical work can be carried out with a filter fluorimeter—the use of a broader band of exciting radiation increases P_0 (equation 8.4) and generally provides greater sensitivity than can be obtained with an instrument equipped with monochromators. However, the use of a spectrofluorimeter is necessary if excitation and emission spectra are being investigated and if exploratory work is being carried out on interferences and conditions of analysis.

It should be noted that most commercial spectrofluorimeters do not provide "true" excitation and emission spectra. The true spectra are modified instrumentally by the wavelength variation of the source power, the transmission characteristics of the monochromators, and the spectral characteristics of the photodetector. The observed excitation spectrum is the true absorption spectrum of the fluorescent substance multiplied by the spectral power distribution of the incident beam emerging from the excitation monochromator; the observed emission spectrum is the true emission spectrum multiplied by the transmission characteristics of the secondary monochromator and the spectral response of the detector. Only the more sophisticated instruments provide facilities for producing "corrected" spectra. The use of corrected spectra is not important in analytical work, but is valuable in studies of any new fluorescing system.

Detailed discussion of methods for the correction of excitation and emission spectra can be found elsewhere,[2] and fluorescence instrumentation in general is described more fully in several texts.[1–4,6,9]

8.3 APPLICATIONS

Many fluorimetric methods of analysis, particularly in biological fields, involve the determination of species that are themselves fluorescent. With the exception of certain methods involving the uranyl ion or lanthanide ions, however, most inorganic analyses are based on the reaction of a metal ion with an organic reagent to form a fluorescent complex. Organic fluorimetric reagents include flavonols, rhodamines, hydroxyanthraquinones, 8-hydroxyquinolines, azobenzenes and azonaphthalenes (particularly 2,2'-dihydroxy derivatives), salicylideneamines, and β-diketones.

Indirect fluorimetric methods include (a) measurement of quenching by the analyte of the fluorescence of another species, (b) displacement by the analyte of the organic ligand from a fluorescent metal-organic complex, and (c) use of the analyte to liberate a ligand that can form a fluorescent metal-ligand complex. Examples of these processes can be seen in (a) the determination of osmium, using the reduction of the fluorescence of 5-amino-4,6-bis(methylthio)pyrimidine by osmium tetroxide,[14] (b) the determination of fluoride, using the removal of aluminum from fluorescent aluminum complexes,[15,16] and (c) the determination of cyanide and sulfide by their release of 8-hydroxyquinoline-5-sulfonate from its nonfluorescent palladium complex, the ligand then combining with magnesium to form a fluorescent species.[17]

Solid-state fluorimetric determination of uranium in fluoride or fluoride-carbonate matrices is discussed by Grimaldi et al.[18] and by Sandell.[19] Short summaries have been given of studies of the fluorescence of lanthanides in various solid matrices and in solution.[2,6,19]

Fluorimetric procedures for a number of elements are described in works such as those of White and Argauer[2] and Sandell.[19] These include such widely used methods as those for aluminum with Pontachrome Blue Black R, beryllium with morin, and selenium with 2,3-diaminonaphthalene. Selected fluorimetric methods, with references, are shown in Table 8.1. When combined with suitable extractions to ensure the absence of interfering metals, many of these procedures allow the determination of trace inorganics at concentrations in the range 1 ng ml^{-1} to 1 μg ml^{-1}. A more complete tabulation, with sensitivities for each method, has been given by Guilbault.[6]

TABLE 8.1 Fluorimetric Methods for Trace Elements

Element	Reagent	Excitation wavelengths/nm	Emission wavelengths/nm	Ref.
Ag	1,10-Phenanthroline/eosin (quenching by Ag)	540	580	20
Al	8-Hydroxyquinoline-5-sulfonic acid	375	485	21
	Pontachrome Blue Black R[a]	330	635	22, 23
	Acid Alizarin Garnet R[b]	470	575	2
	Salicylidene-o-aminophenol	410	520	24
	8-Hydroxyquinoline	365	520	25
	Lumogallion[c]	485	576	26

(continued)

TABLE 8.1 (continued)

Element	Reagent	Excitation wavelengths/nm	Emission wavelengths/nm	Ref.
Au	Rhodamine B	550	575	27
B	Benzoin	405	480	28–30
	4'-Chloro-2-hydroxy-4-methoxybenzophenone	365	490	31
Be	Morin[d]	470	570	19, 32–36
	Quinizarin[e]	545	640	37
Cu	1,1,3-Tricyano-2-amino-1-propene	304	510	38
Eu	1,1,1,5,5,5,-Hexafluoropentane-2,4-dione	312	614	39
	2-Thenoyltrifluoroacetone	390	615	40
F	Zirconium/calcein blue	350	410	41
Ga	Rhodamine B	550	575	42, 43
	8-Hydroxyquinoline	365	520	44, 45
Ge	Rezarson[f]	—	610	46
In	8-Hydroxyquinoline	—	—	47
	2-(2'-Pyridyl)benzimidazole	335	411	48
Li	8-Hydroxyquinoline	365	540	49
Mg	N,N'-bis(salicylidene)ethylenediamine	355	440	50
	N,N'-bis(salicylidene)-2,3-diaminobenzofuran	485	545	51
	2,2'-Dihydroxyazobenzene	470	580	52
Re	Ethylrhodamine B	560	590	53
Sb	3,4',7-Trihydroxyflavone	422	475	13
	Morin[d]	420	500	54
Se	2,3-Diaminonaphthalene	365	525	55–57
Sn	Morin[d]	420	500	54
Ta	Butylrhodamine B	560	595	58
Tb	1,1,1,5,5,5-Hexafluoropentane-2,4-dione	312	544	39
Th	3,4',7-Trihydroxyflavone	—	520	59
Tl	Rhodamine B	546	600	60, 61
U	Fluoride	365	550	18, 19, 62–64
	Phosphate	280	493	65, 66
	Rhodamine B	555	575	67
W	Carminic acid	515	585	68
Zn	Dibenzothiazolylmethane	365	415	69
Zr	Flavonol	400	465	70

[a] 2,2'-Dihydroxyazonaphthalene-4-sulfonic acid (Na or Zn salt).
[b] 2,2',4'-Trihydroxyazobenzene-5-sulfonic acid (Na salt).
[c] 5-Chloro-3-[(2,4-dihydroxyphenyl)azo]-2-hydroxybenzenesulfonic acid.
[d] 2',4',3,5,7-Pentahydroxyflavone.
[e] 1,4-Dihydroxyanthraquinone.
[f] 5-Chloro-3-[(2,4-dihydroxyphenyl)azo]-2-hydroxybenzenearsonic acid.

References

1. G. G. Guilbault (ed.), *Fluorescence. Theory, Instrumentation and Practice*, Dekker, New York, 1967.
2. C. E. White and R. J. Argauer, *Fluorescence Analysis*, Dekker, New York, 1970.
3. D. M. Hercules (ed), *Fluorescence and Phosphorescence Analysis*, Wiley-Interscience, New York, 1966.
4. C. A. Parker, *Photoluminescence of Solutions*, Elsevier, New York, 1968.
5. J. D. Winefordner, S. G. Shulman, and T. C. O'Haver, *Luminescence Spectrometry in Analytical Chemistry*, Wiley-Interscience, New York, 1972.
6. G. G. Guilbault, *Practical Fluorescence*, Dekker, New York, 1973.
7. R. S. Becker, *Theory and Interpretation of Fluorescence and Phosphorescence*, Wiley-Interscience, New York, 1969.
8. A. Weissler and C. E. White, in *Handbook of Analytical Chemistry* (ed. L. Meites), McGraw-Hill, New York, 1963.
9. W. Wotherspoon, G. K. Oster, and G. Oster, in *Techniques of Chemistry*, Vol. I, Part IIIB (eds. A. Weissberger and B. W. Rossiter), Wiley-Interscience, New York, 1972.
10. G. G. Guilbault, in *M.T.P. International Review of Science*, Physical Chemistry, Series One, Vol. 12, Analytical Chemistry, Part 1 (ed. T. S. West), Butterworths, London, 1973.
11. C. E. White and A. Weissler, *Anal..Chem.*, **36**, 116R (1964); **38**, 115R (1966); **40**, 114R (1968); **42**, 57R (1970); **44**, 182R (1972); A. Weissler, *Anal. Chem.*, **46**, 500R (1974).
12. C. M. O'Donnell and T. N. Solie, *Anal. Chem.*, **48**, 175R (1976).
13. T. D. Filer, *Anal. Chem.*, **43**, 725 (1971).
14. S. Burchett and C. E. Meloan, *Anal. Lett.*, **4**, 471 (1971).
15. H. W. Willard and C. A. Horton, *Anal. Chem.*, **24**, 862 (1952).
16. W. A. Powell and J. H. Saylor, *Anal. Chem.*, **25**, 960 (1953).
17. J. S. Hanker, A. Gelberg, and B. Witten, *Anal. Chem.*, **30**, 93 (1958).
18. F. S. Grimaldi, I. May, M. H. Fletcher, and J. Titcomb, *U. S. Geol. Surv. Bull.* **1006** (1954).
19. E. B. Sandell, *Colorimetric Determination of Traces of Metals*, 3rd ed., Interscience, New York, 1959.
20. D. N. Lisistsyna and D. P. Shcherbov, *Z. Anal. Khim.*, **25**, 2310 (1970).
21. D. E. Ryan and B. K. Pal, *Anal. Chim. Acta*, **44**, 385 (1969).
22. A. Weissler and C. E. White, *Ind. Eng. Chem. Anal. Ed.*, **18**, 530 (1946).
23. L. H. Simons, P. H. Monaghan, and M. S. Taggart, *Anal. Chem.*, **25**, 989 (1953).
24. R. M. Dagnall, R. Smith, and T. S. West, *Talanta*, **13**, 609 (1966).
25. F. S. Grimaldi and H. Levine, *U. S. Geol. Surv. Bull.* **992**, 39 (1953).

26. T. Shigematsu, Y. Nishikawa, K. Hiraki, and N. Nagano, *Jap. Anal.*, **19**, 551 (1970).
27. J. Marinenko and I. May, *Anal. Chem.*, **40**, 1137 (1968).
28. C. E. White, A. Weissler, and D. Busker, *Anal. Chem.*, **19**, 802 (1947).
29. C. E. White and D. E. Hoffman, *Anal. Chem.*, **29**, 1105 (1957).
30. G. Elliot and J. A. Radley, *Analyst*, **86**, 62 (1961).
31. B. Liebich, D. Monnier, and M. Marcantonatos, *Anal. Chim. Acta*, **52**, 305 (1970).
32. C. W. Sill and C. P. Willis, *Anal. Chem.*, **31**, 598 (1959).
33. E. B. Sandell, *Anal. Chim. Acta*, **3**, 89 (1949).
34. I. May and F. S. Grimaldi, *Anal. Chem.*, **33**, 1251 (1961).
35. C. W. Sill and C. P. Willis, *Geochim. Cosmochim. Acta*, **26**, 1209 (1962).
36. C. W. Sill, C. P. Willis, and J. K. Flygare, *Anal. Chem.*, **33**, 1671 (1961).
37. M. H. Fletcher, C. E. White, and M. S. Sheftel, *Ind. Eng. Chem. Anal. Ed.*, **18**, 179 (1946).
38. K. Ritchie and J. Harris, *Anal. Chem.*, **41**, 163 (1969).
39. D. E. Williams and J. C. Guyon, *Anal. Chem.*, **43**, 139 (1971).
40. R. Belcher, R. Perry, and W. I. Stephen, *Analyst*, **94**, 26 (1969).
41. T. L. Har and T. S. West, *Anal. Chem.*, **43**, 136 (1971).
42. H. Onishi, *Anal. Chem.*, **27**, 832 (1955).
43. A. I. Chuvileva and I. A. Blyum, *Zavod. Lab.*, **35**, 1153 (1969).
44. J. W. Collat and L. B. Rogers, *Anal. Chem.*, **27**, 961 (1955).
45. E. B. Sandell, *Anal. Chem.*, **19**, 63 (1947).
46. D. P. Shcherbov, R. N. Plotnikova, and I. N. Astaf'eva, *Zavod. Lab.*, **36**, 528 (1970).
47. R. Bock and K.-G. Hackstein, *Z. Anal. Chem.*, **138**, 337 (1953).
48. L. Bark and L. Rixon, *Anal. Chim. Acta*, **45**, 425 (1969).
49. C. E. White, M. H. Fletcher, and J. Parks, *Anal. Chem.*, **23**, 478 (1951).
50. C. E. White and F. Cuttitta, *Anal. Chem.*, **31**, 2083 (1959).
51. R. M. Dagnall, R. Smith, and T. S. West, *Analyst*, **92**, 20 (1967).
52. H. Diehl, R. Olsen, G. Spielholtz, and R. Jensen, *Anal. Chem.*, **35**, 1144 (1963).
53. I. A. Blyum and N. A. Brushtein, *Zavod. Lab.*, **36**, 1032 (1970).
54. D. P. Shcherbov, I. N. Astaf'eva, and R. N. Plotnikova, *Zavod. Lab.*, **39**, 546 (1973).
55. C. A. Parker and L. G. Harvey, *Analyst*, **87**, 558 (1962).
56. J. H. Watkinson, *Anal. Chem.*, **38**, 92 (1966).
57. I. I. Nazarenko and I. V. Kislova, *Zavod. Lab.*, **37**, 414 (1971).
58. I. A. Blyum and T. I. Shumova, *Z. Anal. Khim.*, **25**, 511 (1970).
59. T. D. Filer, *Anal. Chem.*, **42**, 1265 (1970).
60. H. Onishi, *Bull. Chem. Soc. Jap.*, **30**, 827 (1957).

REFERENCES

61. A. D. Matthews and J. P. Riley, *Anal. Chim. Acta*, **48**, 25 (1969).
62. J. T. Byrne, *Anal. Chem.*, **29**, 1408 (1957).
63. F. A. Centanni, A. M. Ross and M. A. DeSesa, *Anal. Chem.*, **28**, 1651 (1956).
64. G. R. Price, R. J. Ferretti, and S. Schwartz, *Anal. Chem.*, **25**, 322 (1953).
65. C. W. Sill and H. E. Peterson, *Anal. Chem.*, **19**, 646 (1947).
66. A. Danielsson, B. Roennholm, L.-E. Kjellstroem, and F. Ingman, *Talanta*, **20**, 185 (1973).
67. G. Leung, Y. S. Kim, and H. Zeitlin, *Anal. Chim. Acta*, **60**, 229 (1972).
68. G. F. Kirkbright, T. S. West, and C. Woodward, *Talanta*, **13**, 1637 (1966).
69. R. R. Trenholm and D. E. Ryan, *Anal. Chim. Acta*, **32**, 317 (1965).
70. W. C. Alford, L. Shapiro, and C. E. White, *Anal. Chem.*, **23**, 1149 (1951).

CHAPTER

9

EMISSION SPECTROCHEMICAL ANALYSIS

9.1 INTRODUCTION

The radiation emitted by excited atoms has been used in qualitative analysis since 1861, when Bunsen and Kirchhoff discovered cesium spectroscopically. A further 16 elements were discovered by this means during the following 50 years. However, the large-scale development of emission spectrography for quantitative analysis occurred only after the pioneering work of Goldschmidt and his co-workers at Göttingen in the 1930s. From then until the 1960s, the emission spectrograph was the analyst's mainstay for routine instrumental analysis of trace elements in geological materials. More recently it has suffered some decline in popularity because of competition from atomic absorption (Chapter 10) and X-ray fluorescence spectroscopy (Chapter 12). Nevertheless, the emission spectrograph is still an important tool, and new developments are ensuring its continued use in trace analysis.

The spectrochemical method depends on the fact that atoms of elements may be excited in an electric arc or laser beam, and will emit radiation of characteristic wavelengths. This radiation is passed through a monochromator (prism or grating optics) and is dispersed into a spectrum. The intensity of each spectral line is related to the population of excited atoms in the radiation source, and may be used for quantitative analysis. Line intensities can be measured either photographically, as in the so-called "optical" instruments, or photoelectrically, as in direct-reading spectrometers. Direct readers usually have grating optics, whereas optical spectrographs may have either gratings or prisms.

9.2 THE ORIGIN OF ATOMIC SPECTRA

Absorption and emission of electromagnetic radiation by monatomic vapors arise from transitions between discrete electronic energy levels of the atoms. The quantity of energy, ΔE, absorbed or emitted in such a transition, is related to the frequency v (or wavelength λ) by the equation

$$\Delta E = hv = \frac{hc}{\lambda}$$

where h is Planck's constant and c is the speed of light.

An atom with its electronic configuration in the state of lowest energy is said to be in the ground state. Absorption of energy from a suitable source produces an excited atom, which may subsequently emit radiation to return to the ground state or to reach another excited state of lower energy. Quantum mechanics indicates that not all of the possible transitions between the different energy levels are allowed. The selection rules governing these transitions are discussed in texts that deal with atomic spectra and atomic structure in some detail.[1-3]

For maximum sensitivity in emission spectroscopy, the ratio of excited atoms to ground-state atoms should be as high as possible. If E_n is the excitation potential of state n (i.e., the excitation energy above the ground state), the equilibrium number of atoms in state n at temperature T is N_n, given by

$$\frac{N_n}{N_0} = \frac{g_n}{g_0} \exp -\frac{E_n}{kT} \qquad (9.1)$$

where N_0 is the number of atoms in the ground state, k is the Boltzmann constant, and g_n, g_0 are the statistical weights of state n and the ground state, respectively. Equation 9.1 indicates that low temperatures are sufficient to cause significant excitation to states with low excitation potentials. For example, for the $^2S_{1/2} - {}^2P_{3/2}$ transition of cesium at 852.1 nm ($E_n = 1.454$ eV), N_n/N_0 is about 1.2×10^{-3} at 2000°C. To observe lines from transitions of higher energy, higher temperatures are needed. For the line from the $^1S_0 - {}^1P_1$ transition of zinc at 213.9 nm ($E_n = 5.795$ eV), N_n/N_0 is about 4×10^{-13} at 2000°C, 3×10^{-9} at 3000°C, and 10^{-5} at 5000°C. It is clear that only a small proportion of zinc atoms will be excited even at 5000°C, and that this proportion changes radically with relatively small temperature changes. This highlights one of the greatest problems in emission spectroscopy: unless the arc temperature is stabilized, there are serious fluctuations in the population of excited atoms and hence in the spectral line intensities.

It should also be noted that, at the temperatures generally used for spectrochemical excitation, many elements are appreciably ionized. Where the ionization potential is low and the temperature is high, the emission spectrum may consist of a mixture of lines of the neutral atom M and the ion M^+. In some cases (scandium, yttrium, and many of the lanthanides), ion lines are intense enough to be the most useful for quantitative analysis.

For a detailed discussion of the theory of spectrochemical excitation, the work of Boumans[4] should be consulted.

9.3 SOURCES AND ELECTRODES

A wide variety of electrical methods has been used to excite atomic emission. These include the d.c. arc, a.c. arc, high-voltage spark, plasma jet, plasma torch, and radiofrequency and microwave plasmas. The d.c. arc, high-voltage spark, and radiofrequency plasma are most widely used. Each of these types of discharge uses different instrumentation and excites the sample by a different mechanism.

9.3.1 The Direct Current Arc

When d.c. excitation is used, the sample is packed into a small graphite or carbon electrode, and a current of 5–20 A is passed between the sample electrode and a pointed counterelectrode. A temperature of 3000°C to 8000°C is attained, sufficient to volatilize all the constituents of silicate rocks and other materials. Anode excitation (in which the sample is made the anode) is preferred for most silicate rock analysis, but cathode excitation is often used for soils.[5] When cathode excitation is used, the spectral lines are uneven and show enhancement near the cathode. Use of the cathode layer, selected by masking off the rest of the spectral line, often results in greater sensitivity. The respective merits of cathode and anode excitation have been discussed by Ahrens and Taylor[6] and by Brooks and Boswell.[7]

Figure 9.1 shows some of the many different shapes of electrode used in spectrochemical analysis. Necked electrodes help to maintain heat and assist in volatilization of the sample. There is now a trend toward the use of smaller electrodes because of the tendency of the d.c. arc to wander around the rim

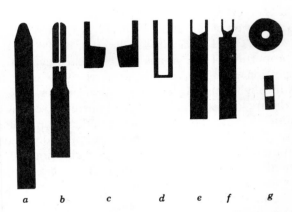

Fig. 9.1. Typical electrodes used in spectrochemical analysis. *a*, rounded upper electrode; *b*, vacuum cup; *c*, Teflon cup for vacuum cup; *d*, porous cup; *e*, crater electrode; *f*, necked crater electrode; *g*, rotating disk electrode.

of the electrode and produce instability. In our laboratory we use 3-mm solid graphite electrodes with a 1.5-mm hole drilled to a depth of about 6 mm. Such an electrode holds about 20 mg of sample. The use of larger electrodes, holding up to 100 mg of sample, results in a higher intensity of the spectral lines, but also increases the background signal and the arc instability.

The two main disadvantages of the d.c. arc are its instability and the formation of cyanogen from the combination of the carbon of the electrodes with atmospheric nitrogen. Molecular bands of CN introduce an intense interference in the spectral range 350–420 nm and can mask many important spectral lines. In recent years both problems have been largely overcome by the development of the Stallwood jet[8] and its modifications. The original Stallwood jet consisted of a quartz envelope enclosing the arc. An appropriate gas mixture was forced upward through a hole in the envelope, excluding nitrogen and thereby eliminating the CN band emission. The gas flow stabilized the arc and partially eliminated the outer fringe in which self-absorption can occur. One problem with the original Stallwood jet was the necessity of removing and cleaning the quartz envelope at frequent intervals. This difficulty was obviated by the modification of Margoshes and Scribner[9] in which the quartz envelope was eliminated and an annular flow of gases protected and stabilized the arc.

The use of smaller electrodes and the modified Stallwood jet has had a marked effect on the precision of emission spectrography. Quantitative analysis of many trace elements in geological materials is possible with a precision of 5% to 10%.

9.3.2 The High-Voltage Spark

The high-voltage spark is the most commonly used alternative to the d.c. arc. The sample is in the form of a solid pellet and the counterelectrode is made of carbon or graphite. The high-voltage spark is an intermittent discharge of alternating polarity, in which currents as large as 1000 A pass for periods of about 100 μs. Between discharges there are intervals of some milliseconds, preventing overheating of the whole sample. The localized heating, however, is very intense; temperatures of the order of 10,000°C are achieved. A greater variety of spectral lines is therefore excited than is the case in the d.c. arc.

Because each spark touches and vaporizes only a small part of the surface of the sample, it is desirable that the sample be as homogeneous as possible. The spark source has been extensively used for metallurgical analysis, but has only recently begun to be used for geological samples, usually with a direct-reading spectrometer. In favorable cases precision of 1% to 5% can be achieved.

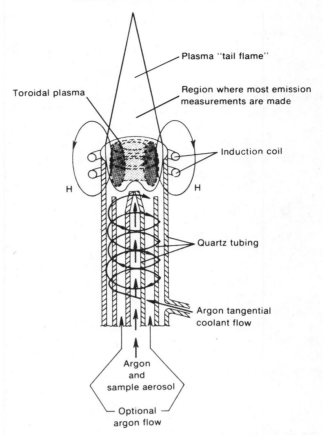

Fig. 9.2. Inductively coupled radiofrequency plasma. *Source*: Fassel and Kniseley.[11] Reprinted with permission from *Analytical Chemistry*. Copyright by the American Chemical Society.

9.3.3 Inductively Coupled Radiofrequency Plasma

This is probably the most notable recent development in excitation sources for emission spectrometry. This source depends on the interaction between the magnetic field of an oscillating radiofrequency current and the charged species present in the plasma, as shown in Fig. 9.2. Ionization of an argon gas stream in the radiofrequency field, produces a toroidal plasma in which temperatures of over 5000°C can be attained. The sample is introduced as an aerosol in the inert gas stream and is efficiently vaporized and atomized in the plasma. The high stability possible with the radiofrequency plasma has made it possible to achieve analytical results with a precision approaching 1%. This source is considered to have great potential for multielement emission spectrometry.[10-12]

9.4 OPTICAL SPECTROGRAPHS

The term "optical spectrograph" is generally used for prism or grating instruments using a photographic plate to record the spectra. Figure 9.3a shows a schematic drawing of an optical spectrograph.[13]

Radiation from the arc is focused on to a slit by a lens system and is reflected through the optical path of the instrument. A collimating lens directs the radiation on to a glass or quartz prism that has a reflecting surface at the back. The radiation is refracted twice before being collected at the camera. The camera is angled to minimize chromatic aberration caused by the different focal lengths of radiation of different wavelengths.

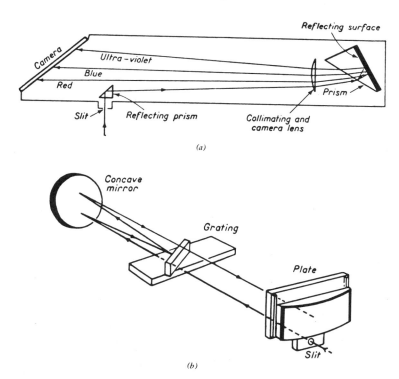

Fig. 9.3. Optical arrangements in optical spectrographs. (a) Prism instrument; (b) grating instrument. *Source*: Taylor and Ahrens.[13]

The choice of prism depends on the elements that are to be determined. One of the most important considerations is the provision of adequate dispersion for the work contemplated. Because most geological samples have very complex spectra, it is essential that the lines be separated as far as possible. The dispersing ability of an optical system is indicated by the reciprocal dispersion

$$\mu = \frac{\Delta\lambda}{\Delta s}$$

where $\Delta\lambda$ is the wavelength difference between two lines separated by a distance Δs on the photographic plate in the focal plane of the spectrograph. With prism optics μ is a function of wavelength.

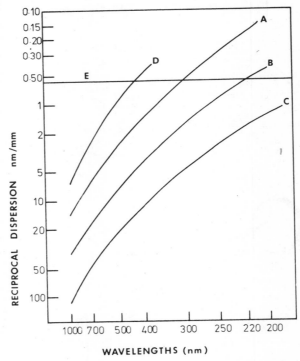

Fig. 9.4. Reciprocal dispersion (nm/mm). *A*, For quartz optics instrument (170 cm focal length); *B*, quartz optics instrument (60 cm focal length); *C*, quartz optics instrument (18 cm focal length); *D*, glass optics instrument (170 cm focal length); *E*, grating instrument (3 m focal length, 576 grooves/mm). *Source*: Hilger and Watts Ltd.[14] Reproduced by permission of Rank Hilger, Westwood, Margate, Kent, England. Copyright by Rank Hilger, 1962.

Figure 9.4 gives reciprocal dispersions for five types of spectrograph.[14] Although none of these instruments is now manufactured, they illustrate very well the basic principles involved. Spectrograph A is a 170-cm-focal-length instrument with quartz optics. Reciprocal dispersion is satisfactorily low in the ultraviolet (0.18 nm/mm at 220 nm) but becomes unacceptable in the red (6 nm/mm at 700 nm). Such an instrument would be unsuitable for determining traces of the alkali metals, for which the most sensitive lines lie near the red end of the visible spectrum. An instrument with glass optics (Spectrograph D in Fig. 9.4) has a much lower reciprocal dispersion in the red (2.5 nm/mm at 700 nm) but would be unusable below 400 nm because of absorption by glass in the ultraviolet. Spectrographs with interchangeable quartz-glass optics can be used if necessary. Figure 9.4 also shows the reciprocal dispersion for a 3-m grating instrument (E). In this case μ is independent of wavelength. The instrument is more satisfactory than glass prism instruments above 450 nm and is superior to quartz instruments above 300 nm. The advantage of better dispersion is slightly offset by the need for two 250 mm by 10 mm plates to cover the same range as one plate in the prism instruments.

A typical grating spectrograph is shown schematically in Fig. 9.3b. Radiation enters a slit at the base of the plate holder, passes under the grating and is reflected by a concave mirror back to the grating. The light returns to the mirror and is reflected over the grating on to the photographic plate. Diffraction of light of wavelength λ by a grating is governed by the equation

$$n\lambda = d(\sin \alpha \pm \sin \beta)$$

where n is the spectral order ($n = 1, 2, 3, \ldots$); d is the spacing between the grooves ruled on the grating; α and β are the respective angles made to the grating normal by the incident and reflected radiation. The positive sign applies when α and β are on the same side of the normal, and the negative sign when they are on opposite sides. It is clear that light of submultiples of the primary (first-order) wavelength may also appear on the photographic plate. For example, second-order diffraction of a line of wavelength 200 nm overlaps a 400 nm line in the first order. Such interference problems can be eliminated by using suitable filters or order-sorting prisms.

A problem encountered with air-path spectrographs is the absorption of wavelengths below about 200 nm by air. Vacuum spectrographs have been designed for work with those elements (mostly nonmetals) that have persistent lines in the range 150 to 200 nm. The vacuum spectrographs are usually designed for use as direct readers because of the difficulty of obtaining photographic emulsions sensitive to wavelengths below 200 nm. In conjunction with an a.c. spark source, the vacuum spectrograph can be used to determine elements such as nitrogen (174 nm), phosphorus (178 nm), sulfur (181 nm), tin (190 nm), selenium (196 nm), and arsenic (197 nm).

The choice of a suitable spectrograph for analysis of geological materials depends on several factors. The complexity of the emission spectra makes it almost mandatory to use an instrument with low reciprocal dispersion. In general, grating instruments are more satisfactory but are considerably more expensive than prism spectrographs. If cost is not a consideration, it is useful to have the flexibility provided by a system capable of being used with both d.c. arc and high-voltage spark sources.

9.5 PHOTOGRAPHIC MEASUREMENT OF SPECTRAL LINE INTENSITIES

9.5.1 Characteristic Curves

The exposure E of the plate at wavelength λ is equal to the integral of the intensity I of the radiation of wavelength λ emitted over the time of exposure t

$$E = \int I(\lambda, t) dt$$

In the idealized case where I is independent of time,

$$E = It$$

Development of the photographic plate causes a blackening that is related to the exposure and can be measured with a microdensitometer. In this instrument a narrow beam of light passes through the plate on to a photocell, producing a reading d on a galvanometer scale. If power P_0 is transmitted through an unexposed part of the plate, giving a galvanometer reading d_0, and if power P is transmitted through an exposed part (e.g., at the center of a spectral line), then the optical density D is given by

$$D = \log_{10} \frac{P_0}{P} = \log_{10} \frac{d_0}{d}$$

If D is plotted as a function of the logarithm of the exposure, a *characteristic curve* (or Hurter and Driffield curve) is obtained, as shown in Fig. 9.5. Little increase of density occurs (region A–B) until a certain critical level of exposure is reached. Density increases linearly with log E over the region B–C, but tends toward a limiting value at D. The slope of the linear portion (tan θ) is known as the *gamma* of the emulsion and is a measure of its contrast.

The main problem encountered by the spectrographer in making photographic intensity measurements is the need to operate in the linear portion of the characteristic curve. Unfortunately, this is not always possible in trace analysis. However, it is possible to prefog the plates so that the point B in Fig. 9.5 is reached. A more widespread procedure is to modify the density

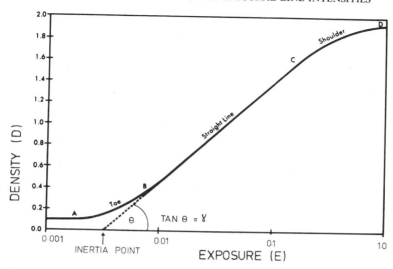

Fig. 9.5. Characteristic curve for a photographic emulsion. Note the toe (AB) and shoulder (CD). Extrapolation of the linear portion (BC) to the x axis gives the inertia point, which corresponds to the energy the plate must receive before density increases appreciably. *Source*: after Mitteldorf.[15]

function $\log(d_0/d)$ in such a way as to extend the linear region of the characteristic curve. Such modifications include the Seidel function,[16] $\log(d_0/d - 1)$, and the partial Seidel functions[17,18] $[\log(d_0/d) + \log(d_0/d - 1)]/2$ and $50[\log(d_0/d) + \log(d_0/d - 1) + 1.2]$. It is convenient to use a galvanometer scale calibrated in one of these modified-density functions.[18]

9.5.2 Plate Calibration and Internal Standards

To obtain spectral-line intensities it is necessary to calibrate the emulsion (i.e., to determine the gamma at each wavelength used) and to ensure that the exposure of the line falls within the useful range. Both of these requirements are met by using a *step sector*. This is a device that rotates in front of the spectrograph slit and causes each spectral line to appear on the plate as a series of steps (usually seven), corresponding to exposures $E, E/2, E/4, \ldots, E/64$. The spectrographer can then choose two or three steps suitable for density measurement. It is then possible to find, for each line, the exposure corresponding to a certain standard density. The exposure found in this way is proportional to the intensity of the spectral line. Photographic methods of photometry are discussed more extensively by Strock[19] and by Ahrens and Taylor.[6]

The total amount of radiant energy emitted at the spectral line of an analyte, depends on the concentration of the analyte in the sample, on the mass of sample consumed in the arc, and on various physical conditions prevailing in the arc itself. Because of the great difficulty in reproducing these conditions precisely from one arcing to the next, photographic analysis is most often carried out by a technique using an *internal standard*. A known proportion of another element (the internal standard) is added to the sample, and the analyte spectral-line intensity is measured relative to that of an internal standard line. Ideally, any disturbance in the arc, any variation from one arcing to the next, or any variation in the mass of sample vaporized and excited, should affect the internal standard in the same way as the analyte, and the line intensity ratio should be unaffected. Ahrens and Taylor[6] have noted the following criteria for choosing a good internal standard for a given analyte: the concentration of the internal standard in the original sample should be negligible, the standard and the analyte should volatilize at the same rate, both spectral lines should have similar wavelengths and excitation potentials, the standard line should be free of self-absorption, and the standard should be available in a high state of purity.

In analysis of the highest precision, it is wise to choose the best internal standard for each element. For example, rubidium (420.2 nm) may be used as an internal standard for potassium (404.7 nm), beryllium (249.4 nm) for silicon (252.8 nm), and bismuth (289.8 nm or 306.7 nm) for lead (283.3 nm). For a wide range of elements of moderate volatility, palladium is a good general internal standard and is frequently used. The internal standard can be added to the carbon powder, which is then mixed with the samples in a constant proportion before arcing.

The intensities of the analyte and standard lines are each corrected for background adjacent to the line. The ratio of corrected intensities I/I_s is then referred to a working curve of the form shown in Fig. 9.6. The linear part of the working curve, given by the relationship

$$\log c = k_1 \log \frac{I}{I_s} + k_2$$

often extends over at least two decades of concentration. At high concentrations there is a characteristic shoulder due to self-absorption of the analysis line, and a "toe" at low concentrations can result from traces of the analyte in the matrix used to prepare the standards.

Reading spectrographic plates and drawing self-calibration curves are extremely tedious processes. Several short cuts, which can be used without causing too much loss of precision, are described elsewhere.[6,18]

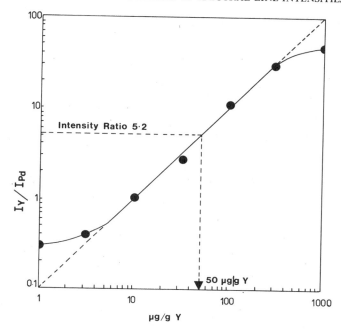

Fig. 9.6. Typical working curve for yttrium in a silicate rock. The ratio of the intensity of the yttrium line (377.7 nm) to that of the internal standard palladium (301.0 nm) is plotted as a function of yttrium concentration. The curve shows a toe due to impurities in the matrix used for the standards, and a shoulder due to self-absorption of the analysis line.

One of the greatest problems in emission spectroscopy is the *matrix effect*. The intensity of a spectral line depends not only on the concentration of the element concerned, but also on the bulk chemical composition of the sample. Intensity differences of up to an order of magnitude can be obtained for the same element in different matrices. One solution to this problem is to prepare an artificial material similar in bulk composition to the material being analyzed, but lacking the analyte element(s). The analyte is added in increments to this artificial material to form the standards. It is usually impossible to find matrix materials of sufficient purity to eliminate the "toe" when determining common elements, but it can be removed by subtracting from each point on the graph, the minimum intensity ratio that will straighten the curve. Because of the logarithmic scale, this procedure has a large effect on points near the toe, and a negligible effect at points corresponding to higher concentrations.

9.6 QUALITATIVE AND QUANTITATIVE ANALYSIS WITH THE D.C. ARC

9.6.1 Qualitative Analysis

For qualitative analysis of rocks, the finely ground sample is mixed with carbon powder in the ratio one of sample to three or four of powder. The two are mixed thoroughly and loaded into a graphite electrode. The sample is then consumed in the d.c. arc at 5–15 A and the spectrum is recorded on the photographic plate. An iron reference spectrum of smaller width than the analysis spectrum should be superimposed. The various elements are identified by the position of the spectral lines relative to those of the iron spectrum. This can be achieved with a comparator in which two plates (the analysis plate and a standard plate) can be viewed simultaneously in an eyepiece. The two plates are adjusted so that their iron reference spectra coincide. The standard plate contains spectra of individual elements each with a superimposed iron spectrum. The eyepiece is then moved in a horizontal direction and scans the spectra to detect coincident lines. The identity of the spectral lines and possible line interferences are established by reference to standard wavelength tables [20–22] and tables of interferences.[6,23]

If the analysis spectra are stepped, it is possible to obtain a *relative* semiquantitative value for elemental concentrations by recording the highest step number that is just visible. Hence a scale of values from 1 to 7 can be assigned.

A simple alternative to the comparator is a spectrum projector in which an image of the spectrum is projected on one of a series of cards belonging to an atlas of spectral lines. The cards contain a photograph of an iron reference spectrum and show the positions of major spectral lines of the elements. When using the projector, the instrument is adjusted so that the iron spectra on the plate and card exactly coincide and the spectrum is then scanned by lateral movement of the plate in the optical field.

9.6.2 Quantitative Analysis

For quantitative analysis the internal standard is added, either to the carbon powder or to the spectroscopic buffer if this is being used to reduce the matrix effect. A suitable buffer material, such as calcium carbonate or a salt of an alkali metal, is added in such a proportion that it becomes a major constituent of the matrix. This procedure improves accuracy, but the extra dilution of the sample reduces the sensitivity.

The reproducibility of spectrographic data depends largely on the arcing step. Errors in weighing, photographic plate development, and measurement

TABLE 9.1 Persistent Lines and Detection Limits of the Elements

Element	Line (nm)	Limit of detection (μg/g) Open arc	Limit of detection (μg/g) Stallwood jet	Element	Line (nm)	Limit of detection (μg/g) Open arc	Limit of detection (μg/g) Stallwood jet
Ag	328.1	1	0.5	Mn	257.6	5	1
Al	309.3	10	5	Na	330.2	500	500
As	235.0	100	50	Nb	309.4	10	5
Au	242.8	<5	<5	*Nd	430.4	20	5
B	249.8	2	2	Ni	341.5	20	5
Ba	455.4	5	2	P	253.6	100	50
Be	234.9	0.1	0.5	Pb	283.3	20	5
Be	313.0	0.5	0.02	Pd	324.3	10	5
Bi	306.8	5	2	*Pr	422.5	20	5
Ca	422.7	1	0.2	Pt	306.5	<5	<5
Cd	326.1	20	10	Rb	420.2	<1000	1000
Ce	418.7	1000	500	Re	346.1	50	10
*Ce	418.7	10	5	Rh	343.5	10	<5
Co	345.4	20	5	Ru	343.7	100	10
Cr	283.6	10	5	Sb	259.8	50	50
Cs	455.5	500	500	Sc	402.4	1	0.5
Cu	324.8	1	0.5	Se	206.3	1000	500
*Dy	404.6	5	2	Si	288.2	1	0.5
*Er	369.3	5	2	*Sm	442.4	20	5
Eu	459.4	5	2	Sn	317.5	20	5
Fe	302.1	5	2	Sr	407.8	50	5
Ga	417.2	20	5	Ta	331.1	1000	1000
Ga	294.4	5	5	*Tb	350.9	50	10
*Gd	342.3	10	2	Te	238.6	500	100
Ge	265.1	5	5	Th	401.9	500	500
Hf	319.4	50	10	Ti	334.9	10	1
Hg	253.6	500	50	Tl	377.6	50	20
*Ho	389.1	5	2	*Tm	346.2	5	2
In	410.2	20	5	U	424.2	500	500
In	325.6	10	<5	V	309.3	20	5
Ir	322.1	50	10	W	430.2	100	100
K	404.4	<1000	1000	Y	324.2	5	2
*La	394.9	5	2	*Yb	328.9	1	0.5
Li	323.3	500	500	Zn	330.3	50	10
*Lu	261.5	5	2	Zr	349.6	50	10
Mg	279.6	0.2	0.2				

* Lithium carbonate matrix. All other elements with graphite matrix. *Source:* Mitteldorf.[15]

of line densities are usually much less significant. In favorable cases, a precision of 5% can be attained by skilled operators. The accuracy of the results, however, depends largely on the operator's success in reducing errors due to the matrix effect.

Table 9.1 lists limits of detection for many elements in the d.c. arc.[15] The data were obtained by using a 3.4-m grating spectrograph with a grating ruled to 600 lines/mm. The spectral lines are the so-called R.U. (*raies ultimes*), persistent lines that are the most sensitive for each element.

9.7 DIRECT-READING SPECTROMETERS

The direct-reading spectrometer is a modification of the grating spectrograph in which photographic recording of spectral information is replaced by a photoelectric system with an automatic readout. Direct readers were developed more than 30 years ago and were initially used mainly for metallurgical analysis. In recent years, however, these instruments have found increasing use in the analysis of geological materials.

The progress of direct-reading emission spectrometry has been stimulated by developments in the computer field. Not only have analog computers been built into the instruments themselves, but external digital computers have served to handle the vast amount of data that these instruments can produce in a short time. The emergence of fully computerized direct-reading spectrometers has helped to maintain the importance of emission spectroscopy in the face of competition from techniques such as atomic absorption spectroscopy. Although the use of optical spectrographs has declined, a great deal of work is now carried out with direct readers that can measure up to 30 trace elements simultaneously, with precision and accuracy adequate for many purposes and with a speed unsurpassed by any other technique.

9.7.1 Instrumentation

Typical instrumentation for geochemical analysis has been described by Scott et al.[24] The basic design is similar to that of an optical grating instrument, except that the plate holder is replaced by a series of exit slits with a photomultiplier situated behind each one. The spectral lines impinge on the photomultipliers and produce photoelectric currents that charge a bank of capacitors. The voltage accumulated across each capacitor provides a manual readout. Alternatively, the voltages may be read out automatically and the data fed into a computer, which obtains a value for the intensity of each spectral line, removes background, obtains the analyte/internal standard ratio, refers this to a previously stored working curve, and gives the concentration in several alternative forms of readout. The usual forms are punch tape for further processing in an external computer, or typed information from an electronic typewriter.

A number of problems beyond the capability of a built-in analog computer have been solved by using an external digital computer. For example, in the technique developed by Scott et al.[24] background was measured by 4 channels of a 49-channel system and extrapolated values were subtracted from each of 45 analysis lines. The external computer was also used to correct for self-absorption and interelement interferences. In the analysis of geological materials, the matrix effect is largely a function of the concentration of a few major elements such as calcium and potassium. The digital computer was therefore used to apply matrix corrections to all analysis lines. Good examples of the use of the direct reader for rapid determination of many trace elements in geological materials can be found in the work of Scott et al.,[24] Tennant and Sewell,[25] and Timperley.[26]

Although the d.c. arc is the source most favored for analysis of geological samples, the a.c. spark can also be used successfully. A common technique is to fuse the sample with lithium tetraborate (containing the internal standard) in a small graphite crucible of known mass. The fused material and crucible are then ground together in a small mill and pressed into a pellet for sparking. Excellent sample homogeneity is achieved, and there is a substantial saving of time, compared with manual operation.

9.7.2 The Tape Machine

A useful accessory for use with the direct reader and a.c. spark source is the tape machine designed by Danielsson et al.[27,28] The sample is fed in powder form on to an adhesive tape that passes through a spark gap at constant speed. High precision and good sensitivity are obtained, and the machine is in use in several geochemical and mining laboratories.[29]

9.8 SPECTROCHEMICAL ANALYSIS OF LIQUIDS

Emission spectrography has been less widely used for liquid samples than for solids. However, several types of electrode and several nebulizing devices suitable for dealing directly with solutions have been developed.

9.8.1 Porous Cup and Vacuum Cup Electrodes

Porous cup and vacuum cup electrodes are illustrated in Fig. 9.1. The porous cup[30] is a long hollow electrode with a thin porous base, less than 1 mm in diameter. The porous cup becomes the upper electrode and is sparked with a lower pointed electrode. During sparking the liquid seeps through the base and is excited by the a.c. spark. The vacuum cup[31] is also designed for use with the a.c. spark. It consists of a narrow electrode with a vertical capillary, which meets at right angles a shorter capillary leading to the side near the base. A small reservoir of liquid in a teflon cup fits over the electrode so that the horizontal capillary dips into the solution. The liquid is drawn into the spark, partly by capillary action and partly by the partial vacuum induced by the spark. The use of both these devices has declined with the development of atomic absorption and of other emission techniques for solutions.

9.8.2 The Rotating Disk Electrode

The rotating disk electrode[32,33] has been extensively used with direct readers. It is a simple device (Fig. 9.1) in which a graphite wheel rotates against the surface of a solution and carries with it a thin even film of liquid. An a.c. spark is struck against the wheel, and the spectrum is recorded in the normal manner. An advantage of the rotating disk is that it can be used with slurries, concentrated solutions, and viscous liquids such as oils, more readily than other types of electrode.

9.8.3 The Plasma Jet

In the late 1950s, Margoshes and Scribner[34] and Korolev and Weinstein[35] developed a method for direct nebulization of liquids into a gas-stabilized d.c. arc. The arc is struck between a lower graphite ring and an upper tungsten electrode. The solution is nebulized vertically through the graphite ring. The arc is enclosed in a teflon-lined chamber capped with a second ring maintained at the same voltage as the tungsten electrode. A fast stream of inert gas, such as argon, surrounds the arc, cooling the outer part, which then becomes less conducting. This causes the arc to narrow and become hotter. As a result of the electrical and thermal "pinching" of the arc, temperatures up to 10,000°C can be obtained. The high temperature ensures that good sensitivity is achieved for refractory elements such as beryllium. Disadvantages of the plasma jet include its acoustic noise, its consumption of large volumes of inert gas, and the need for currents of 20 to 25 A, which are beyond the range of many conventional source units. There is also a danger of the nebulizing jet becoming clogged because of its proximity to the arc. This last problem has been overcome in systems containing an aerosol desolvation chamber. The application of such a system to the analysis of geological materials has been described by Golightly and Harris.[36]

TABLE 9.2 Emission Spectroscopic Detection Limits with the Inductively Coupled Radiofrequency Plasma[a]

Concentration detection limit (μg/ml)	Element
≤0.0001	Ba, Ca, Sr
0.0001–0.0005	Be, Na, Y
0.0005–0.001	Cr, Cu, Er, Eu, Mg, Mn, Yb
0.001–0.005	Ag, Al, B, Cd, Co, Dy, Fe, La, Mo, Rh, Sc, Th, Ti, W, Zn, Zr
0.005–0.01	Ce, Gd, Hf, Ho, Lu, Nb, Ni, Pb, Pd, Si, Tm, V
0.01–0.1	As, Au, Bi, Ga, In, Nd, P, Pr, Pt, Se, Sm, Ta, Te, U
0.1–0.3	Ge, Hg, Sb, Sn, Tb, Tl

[a] After Fassel and Kniseley.[11]

9.8.4 Inductively Coupled Radiofrequency Plasma

The nature of this source has been described in Section 9.3.3 and Fig. 9.2. The amount of solvent that can be introduced into the plasma is limited by the poisoning action of the molecular gases released. For this reason, the solution is converted into an aerosol with an ultrasonic nebulizer, and the aerosol passes through a heated desolvation chamber before being carried into the plasma.[37]

Table 9.2 summarizes some detection limits obtained on solutions aspirated into the inductively coupled plasma. Although only a limited amount of work on geological materials has yet been carried out with this type of apparatus,[38] it is clear that excellent sensitivity can be obtained. Its future in geochemical analysis may depend on whether the need for extra sensitivity justifies the time spent in taking samples into solution.

9.9 THE LASER MICROPROBE

In the laser microprobe, a high-intensity pulsed laser beam is focused on the sample, vaporizing it rapidly. Above the sample are two electrodes that form an arc across the cloud of vaporized material, the arc being triggered by the increased conductivity of the air gap between the electrodes. The principle of the laser microprobe is illustrated schematically in Fig. 9.7.

Because the laser beam can be concentrated to a small spot, about 50 μm in diameter, minimal destruction of the sample occurs. It is suitable for the analysis of small or valuable samples, such as archaeological specimens. A modification of this approach is Vogel's microspark apparatus,[39] which

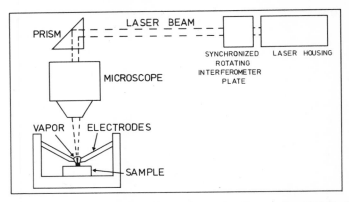

Fig. 9.7. Schematic diagram of the laser microprobe. *Source*: after Mitteldorf.[15]

uses a spark source and is capable of scanning a field as small as 5 μm in diameter. Geological applications of the laser microprobe are described in the book edited by Andersen.[40]

9.10 COMPARATIVE EVALUATION AND RECENT DEVELOPMENTS

Emission spectroscopy, atomic absorption spectroscopy, and X-ray fluorescence analysis are the most frequently used methods for routine trace analysis of geological materials. In some respects these methods are complementary and in other respects they are competitive. Table 9.3 summarizes the main characteristics of these methods, as they are applied to geological materials using conventional instruments.

It is beyond the scope of this volume to review all the recent developments in emission spectroscopy. Several successors to the earlier books of Nachtrieb[41] and Ahrens and Taylor[6] have recently appeared,[42–45] and detailed emission spectrochemical methods are published by the American Society for Testing and Materials.[46,47] The nineteenth and final volume of a long-running series of spectrochemical abstracts has also been published.[48] The annual reports on analytical atomic spectroscopy published by the Society for Analytical Chemistry[49–51] contain a large amount of information on applications of atomic absorption, emission, and fluorescence spectroscopy, and very comprehensive biennial reviews on emission spectroscopy have been prepared by Barnes.[52] It is evident from the recent literature that new and important developments are continuing to appear, and that emission spectroscopy will remain a valuable tool in geochemical and geological trace analysis for many years.

TABLE 9.3 Advantages and Disadvantages of Some Instrumental Methods

	Photographic spectrography	Direct-reading spectrometry	Atomic absorption spectrophotometry	X-ray fluorescence
Capital cost	Low	Extremely high	Low	High
Sensitivity	Good (1–10 μg/g range for many elements)	Good (1–10 μg/g range for many elements)	Very good (0.01–1 μg/ml for most elements in solution, 0.1–10 μg/g based on mass of solid sample)	Fair (1–10 μg/g for many elements of atomic number 15–70)
Speed of analysis	Limited by speed of photographic intensity measurement	Unsurpassed for multielement work	Fast for single elements in solution	Relatively slow for multielement trace analysis
Precision	Average (5–10%)	Fairly precise (2–5%)	Precise (1–2%)	Average (5–10%)
Matrix effect	Considerable	Considerable	Slight or easily controlled	Considerable
Accuracy	Fair	Fair	Good	Fair
Main advantage	Permanent record	Very high speed for multielement analysis	Good precision and ease of operation	Nondestructive
Required operator skill	High	Low if spark source used	Low	Fairly high

References

1. H. E. White, *Introduction to Atomic Spectra*, McGraw-Hill, New York, 1934.
2. G. Herzberg, *Atomic Spectra and Atomic Structure*, 2nd ed., Dover Publications, New York, 1944.
3. H. G. Kuhn, *Atomic Spectra*, Longmans Green, London, 1962.
4. P. W. J. M. Boumans, *Theory of Spectrochemical Excitation*, Plenum, New York, 1966.
5. R. L. Mitchell, *The Spectrochemical Analysis of Soils, Plants and Related Materials*, Commonwealth Agricultural Bureau, Farnham Royal, England, 1964.
6. L. H. Ahrens and S. R. Taylor, *Spectrochemical Analysis*, 2nd ed., Addison-Wesley, Reading, Mass., 1961.
7. R. R. Brooks and C. R. Boswell, *Anal. Chim. Acta*, **32**, 339 (1965).
8. B. J. Stallwood, *J. Opt. Soc. Am.*, **44**, 171 (1954).
9. M. Margoshes and B. F. Scribner, *Appl. Spectrosc.*, **18**, 154 (1964).
10. D. G. Peters, J. M. Hayes, and G. M. Hieftje, *A Brief Introduction to Modern Chemical Analysis*, Saunders, Philadelphia, 1976, p. 469.
11. V. A. Fassel and R. N. Kniseley, *Anal. Chem.*, **46**, 1110A (1974).
12. J. D. Winefordner, J. J. Fitzgerald, and N. Omenetto, *Appl. Spectrosc.*, **29**, 369 (1975).
13. S. R. Taylor and L. H. Ahrens, in *Methods in Geochemistry* (eds. A. A. Smales and L. R. Wager), Interscience, New York, 1960, p. 81.
14. Hilger and Watts Ltd., *Catalogue CH 403/4*, London, 1962.
15. A. J. Mitteldorf, in *Trace Analysis—Physical Methods* (ed. G. H. Morrison), Interscience, New York, 1965.
16. H. Kaiser, *Spectrochim. Acta*, **2**, 1 (1941).
17. M. Honerjäger-Sohm and H. Kaiser, *Spectrochim. Acta*, **2**, 396 (1944).
18. C. R. Boswell and R. R. Brooks, *Appl. Spectrosc.*, **19**, 147 (1965).
19. L. W. Strock, in *Trace Analysis* (eds. J. H. Yoe and H. J. Koch), Wiley, New York, 1957.
20. G. R. Harrison, *MIT Wavelength Tables*, Wiley, New York, 1939.
21. W. F. Meggers, C. H. Corliss, and B. F. Scribner, *Tables of Spectral Line Intensities. I. Arranged by Elements*, Natl. Bur. Stand. Monogr. 145, Part 1 (1975).
22. W. F. Meggers, C. H. Corliss, and B. F. Scribner, *Tables of Spectral Line Intensities. II. Arranged by Wavelengths*, Natl. Bur. Stand. Monogr. 145, Part 2 (1975).
23. J. Kroonen and D. Vader, *Line Interference in Emission Spectrographic Analysis*, Elsevier, Amsterdam, 1963.
24. R. O. Scott, J. C. Burridge, and R. L. Mitchell, *Bol. Geol. Min.*, **80**, 446 (1969).
25. W. C. Tennant and J. R. Sewell, *Geochim. Cosmochim. Acta*, **33**, 640 (1969).
26. M. H. Timperley, *Spectrochim. Acta*, **29B**, 95 (1974).
27. A. Danielsson, F. Lundgren, and G. Sundkvist, *Spectrochim. Acta*, **15**, 122 (1959).

28. A. Danielsson and G. Sundkvist, *Spectrochim. Acta*, **15**, 126, 134 (1959).
29. R. L. Mitchell, *Bol. Geol. Min.*, **80**, 395 (1969).
30. C. Feldman, *Anal. Chem.*, **21**, 1041 (1949).
31. T. H. Zink, *Appl. Spectrosc.*, **13**, 94 (1959).
32. C. E. Harvey, *Spectrochemical Procedures*, Applied Research Laboratories, Glenvale, Calif., 1950.
33. L. G. Young, *Analyst*, **87**, 6 (1962).
34. M. Margoshes and B. F. Scribner, *Spectrochim. Acta*, **15**, 138 (1959).
35. V. V. Korolev and E. E. Weinstein, *J. Anal. Chem. (USSR)*, **15**, 686 (1960).
36. D. W. Golightly and J. L. Harris, *Appl. Spectrosc.*, **29**, 233 (1975).
37. G. W. Dickinson and V. A. Fassel, *Anal. Chem.*, **41**, 1021 (1969).
38. S. Greenfield, I. L. Jones, H. M. McGeachin, and P. B. Smith, *Anal. Chim. Acta*, **74**, 225 (1975).
39. R. S. Vogel, *Process Dev. Quart. Rep.*, MCW-1479, U.S. Atomic Energy Commission, Washington, D.C., 1962.
40. C. A. Andersen (ed.), *Microprobe Analysis*, Wiley-Interscience, New York, 1973.
41. N. H. Nachtrieb, *Principles and Practice of Spectrochemical Analysis*, McGraw-Hill, New York, 1950.
42. M. Slavin, *Emission Spectrochemical Analysis*, Wiley-Interscience, New York, 1971.
43. N. W. H. Addink, *DC Arc Analysis*, Macmillan, London, 1971.
44. J. Mika and T. Török, *Analytical Emission Spectroscopy*, Vol. 1, Butterworths, London, 1973.
45. E. L. Grove (ed.), *Analytical Emission Spectroscopy*, Vol. 1, Dekker, New York; Part I, 1971; Part II, 1972; Part III (in prep.).
46. American Society for Testing and Materials, *Methods for Spectrochemical Analysis*, 5th ed., Philadelphia, 1968.
47. American Society for Testing and Materials, *Annual Book of ASTM Standards*, Part 32, Philadelphia, 1972; 1973.
48. E. H. S. Van Someren and F. T. Birks, *Spectrochemical Abstracts*, Vol. 19, 1972–1973, Adam Hilger, London, 1974.
49. D. P. Hubbard (ed.), *Annual Reports on Analytical Atomic Spectroscopy 1972*, Society for Analytical Chemistry, London, 1973.
50. C. Woodward (ed.), *Annual Reports on Analytical Atomic Spectroscopy 1973*, Society for Analytical Chemistry, London, 1974.
51. C. Woodward (ed.), *Annual Reports on Analytical Atomic Spectroscopy 1974*, Society for Analytical Chemistry, London, 1975.
52. R. M. Barnes, *Anal. Chem.*, **44**, 122R (1972); **46**, 150R (1974); **48**, 106R (1976).

CHAPTER

10

ATOMIC ABSORPTION SPECTROPHOTOMETRY

The impact of the development of atomic absorption spectrophotometry on the field of trace metal analysis since the late 1950s can hardly be overemphasized. With its particular combination of sensitivity, specificity, precision, and economy, atomic absorption has made possible the determination of a wide variety of elements in geological materials on an unprecedented scale.

Observations of the basic phenomenon, the absorption of radiation by atoms, can be traced back as far as the reports of Wollaston in 1802 and Fraunhofer in 1817, concerning the dark lines in the solar spectrum. About 1860, Bunsen and Kirchhoff associated certain spectral lines with particular elements and explained the so-called Fraunhofer lines as being due to absorption of specific wavelengths from the sun's continuum by a surrounding cooler region containing various atomic vapors. For most of the next century, however, the majority of investigations into elemental spectroscopy concentrated on the emission of radiation from atoms and small molecules excited by various means.

The absorption and reradiation (fluorescence) of light of characteristic wavelengths by atomic vapors was studied extensively by Wood and others between 1902 and 1930. Quantitative analytical use of the absorption phenomenon was for a long time confined to mercury, the estimation of which was investigated by several workers.[1-3]

Development of the atomic absorption technique for other elements did not gather momentum until attention was drawn by Walsh[4] and by Alkemade and Milatz[5,6] to some of its advantages. Appropriate components of apparatus had already been developed in other fields of study, and it was not long before the potential of atomic absorption in the analysis of real samples was demonstrated. The successful application of atomic absorption to such practical problems as the determination of magnesium in agricultural materials and of zinc and other elements in plants, by Allan[7] and David,[8] respectively, was an important factor in encouraging the widespread development and use of the method in the years following 1958. A personal account by Walsh of the early developments and current status of atomic absorption has recently been given.[9]

A book devoted to geological applications of atomic absorption has been published;[10] several other books deal with the subject in a more general way.[11-21]

10.1 THEORY

The absorption and emission of radiation by monatomic vapors has been discussed briefly in Chapter 9 (p. 138). Detailed accounts of atomic structure and atomic spectra can be found in several specialized texts.[22-24]

When atoms are produced, for example, in the dissociation of compounds in a flame, the distribution of atoms among the various electronic states at equilibrium at temperature T is governed by the Boltzmann equation

$$\frac{N_i}{N} = \frac{g_i \exp(-E_i/kT)}{\sum g_i \exp(-E_i/kT)} \tag{10.1}$$

where N_i, N denote the number of atoms in state i and the total number of atoms, respectively; g_i is the statistical weight of state i; E_i is the energy of state i above the ground state; k is the Boltzmann constant. An alternative form of this expression allows the relative numbers of atoms in two states, designated by the subscripts 1 and 2, to be calculated:

$$\frac{N_2}{N_1} = \frac{g_2}{g_1} \exp\left[\frac{-(E_2 - E_1)}{kT}\right] \tag{10.2}$$

In many cases the excited states are sufficiently far above the ground state that, at the temperatures used in flame spectroscopy (rarely much above 3000 K), almost all atoms are in the ground state. Table 10.1 shows for several elements the ratio of the number of atoms in the first excited state to that in the ground state at 2000 K and 3000 K.

It is apparent that under these conditions only the ground electronic state is significantly populated, and the most readily measurable absorptions of

TABLE 10.1 Values of N_2/N_1 for Several Resonance Lines

Element	Transition	Wavelength (nm)	g_2/g_1	N_2/N_1 2000 K	N_2/N_1 3000 K
K	$^2S_{1/2} - {}^2P_{3/2}$	766.5	2	1.7×10^{-4}	3.8×10^{-3}
Na	$^2S_{1/2} - {}^2P_{3/2}$	589.0	2	9.9×10^{-6}	5.9×10^{-4}
Ca	$^1S_0 - {}^1P_1$	422.7	3	1.2×10^{-7}	3.7×10^{-5}
Zn	$^1S_0 - {}^1P_1$	213.9	3	7.3×10^{-15}	5.6×10^{-10}

radiation are those originating from ground-state atoms. Absorption lines involving ground-state atoms are often described as resonance lines.

The equilibrium proportion of ground-state atoms is relatively insensitive to changes of temperature. In the case of sodium, for example, this proportion changes from approximately $(1 - 9.9 \times 10^{-6})$ to $(1 - 5.9 \times 10^{-4})$ in going from 2000 K to 3000 K; that is, it is virtually unchanged at a value of almost unity. The proportion in the first excited state, however, increases by a factor of about 60 over this temperature range. Atomic absorption should therefore be less influenced than atomic emission by short-term temperature fluctuations in the medium where the atoms are produced. However, if such fluctuations alter substantially the efficiency with which atoms are produced from compounds of the element (see Section 10.3), then absorption and emission are likely to be similarly affected.

A further consequence of the Boltzmann expression (equation 10.2) is that the proportion of atoms in the excited state is higher when the energy involved in the transition is lower (i.e., when the resonance line is of longer wavelength). This is also apparent from a comparison of the values of N_2/N_1 for potassium and zinc in Table 10.1. Elements such as potassium, cesium, and lithium, with resonance lines at long wavelengths, therefore, tend to be determined with greater sensitivity in emission than those with short-wavelength resonance lines, such as cadmium and zinc.

When comparing the relative sensitivities and limits of detectability for any given element in flame atomic absorption and emission, several different factors, mainly instrumental, must be considered. These have been discussed by Alkemade[25] and by Winefordner and Vickers.[26,27]

Absorption and emission of radiation occur not at an exact wavelength, but rather over a narrow range of wavelengths. Broadening of atomic spectral lines is caused by any influence capable of perturbing the atom and its electronic energy levels. These influences may be summarized briefly as follows.

1. Natural broadening, which is due to the finite lifetime of the atom in the excited state, and is only about 10^{-5} nm for most resonance lines.

2. Doppler broadening, caused by the absorbing or emitting atoms having a range of velocity components along the line of observation, is dependent on wavelength, temperature and atomic weight. The half-width from this effect (i.e., the width at half the peak height) is in the range 10^{-2} to 10^{-3} nm for many resonance lines at flame temperatures.

3. Collisional broadening (pressure, or Lorentz broadening) is due to the perturbation of the absorbing or emitting atoms by molecules of foreign gases. The extent of the broadening is dependent on temperature, wavelength, atomic weight of the absorbing or emitting species, pressure and molecular

THEORY

weight of the foreign species, and on a collisional cross-section parameter, which is itself a function of the conditions and species involved. For most elements, under flame conditions, collisional half-widths are estimated to be no more than 10^{-2} nm.[28]

4. Other effects, such as resonance broadening (perturbation by other atoms of the absorbing or emitting species itself) and broadening from electric fields and magnetic fields, are generally negligible by comparison with Doppler and collisional broadening.

In addition, hyperfine splitting of atomic spectral lines, occurring in those isotopes with nonzero nuclear spin, causes some lines to consist of a number of separate but closely spaced components, each of which is subject to the broadening influences described above. This effect is discussed in more detail elsewhere (Ref. 17, Chapter VIII).

In most cases, therefore, absorption of radiation by an atomic vapor at 2000 K to 3000 K occurs over a wavelength range of 10^{-2} to 10^{-3} nm. This range is so narrow that it is generally a disadvantage to attempt to use a continuous source of radiation, as this makes an unreasonable imposition on the spectral slit width and monochromator. The usual practice is to employ a source giving a sharp line of appropriate wavelength: the monochromator then needs only to be able to separate this line from any nonabsorbable neighboring lines emitted by the source. The line source most commonly used is a hollow-cathode lamp containing the element being determined (see p. 166), giving lines that are rather narrower than the absorption lines. This is illustrated schematically in Fig. 10.1.

The integrated energy I_A absorbed per unit time per unit cross-sectional area of the atomic vapor is given by

$$I_A = \int I_\lambda [1 - \exp(-k_\lambda l)] d\lambda \qquad (10.3)$$

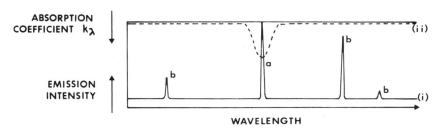

Fig. 10.1 Wavelength dependence of (i) the emission from a hollow cathode lamp, and (ii) the absorbing ability of an atomic vapor of the same element, as produced in a flame. a, Resonance line; b, nonresonance lines.

where I_λ is the spectral intensity of the source at wavelength λ, k_λ is the absorption coefficient of the atomic vapor at this wavelength, and l is the absorption pathlength. The integration is carried out over the wavelength interval over which the radiation is absorbed.

The fraction of radiation absorbed by the atomic vapor, α, is given by

$$\alpha = \frac{\int I_\lambda [1 - \exp(-k_\lambda l)] d\lambda}{\int I_\lambda d\lambda} = \frac{I_A}{I_o} \tag{10.4}$$

The denominator is the energy transmitted per unit time per unit cross-sectional area of the flame gases when no atomic vapor is produced in the light path (e.g., when a blank solution is aspirated into a flame). The evaluation of the integrals in equation 10.4 under different conditions has been discussed by Zeegers et al.[29] Expressions have been given that relate the absorption coefficient at the peak of the line (k_{max}), or the average absorption coefficient over the source linewidth ($\bar{k}$), and N, the number of atoms per cubic centimeter capable of absorbing light of wavelength λ. The exact form of these expressions depends on the assumptions made concerning the source-line and absorption profiles. For example, if the source linewidth is negligible compared to that of the absorption line, and if only Doppler broadening of the latter needs to be considered, then k_{max} is given by

$$k_{max} = \frac{2\lambda_0^2 (\pi \ln 2)^{1/2}}{\Delta \lambda_D} \frac{e^2}{mc^2} Nf \tag{10.5}$$

where $\Delta \lambda_D$ is the Doppler half-width, m and e are the mass and charge, respectively, of the electron, λ_0 is the wavelength at the line center, and f is the oscillator strength, which can be regarded as the average number of electrons per atom that the incident radiation can cause to undergo this particular transition.

Of importance in analytical spectroscopy is the fact that the absorbance of the atomic vapor, defined by

$$A = -\log(1 - \alpha) = \log \frac{I_o}{I_o - I_A} \tag{10.6}$$

is proportional to N over a usefully wide range of atom concentrations. At very high concentrations, other factors, such as resonance broadening, lead to a loss of this proportionality. (The analytical curves obtained in practice may suffer from curvature for a variety of other reasons, as discussed in Section 10.4.2).

Equation 10.6 can be rewritten

$$A = \log \frac{I_o}{I_T} \tag{10.7}$$

where I_T is the energy transmitted per unit time per unit cross-sectional area when the light path contains the atomic vapor. In a conventional spectrophotometer the quantity actually measured is the response of a photodetector, which is related to the power (energy per unit time) incident on its sensitive surface. Instrumentation is available that gives the output in the form of either "percent transmission," $(P_T/P_o) \times 100\%$ or absorbance, $\log(P_o/P_T)$, where P_o and P_T are the powers received by the detector in the absence and presence, respectively, of the absorbing vapor.

10.2 INSTRUMENTATION

The essential features of a conventional atomic absorption spectrophotometer are shown schematically in Fig. 10.2. These features include: (a) a radiation source emitting a sharp line at the wavelength of the element being determined, (b) a device (commonly a flame) for converting the sample into an atomic vapor, (c) a monochromator or other device to isolate the required wavelength, (d) a detector (generally a photomultiplier), (e) an amplifier, and a readout device (meter, recorder, digital readout, or printout facility).

From the earliest discussion of atomic absorption equipment,[4] it was recognized that a potential problem arises from excited atoms in the flame emitting light at the analytical wavelength. By modulating (chopping) the output of the source, either electronically or mechanically, and tuning the amplifier to the same frequency, measurement of the continuous signal emanating from the flame can be avoided.

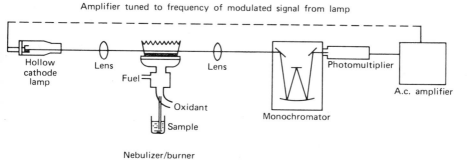

Fig. 10.2. Basic components of an atomic absorption spectrophotometer. (Modulation of the signal from the hollow cathode lamp is usually carried out electronically, rather than mechanically.)

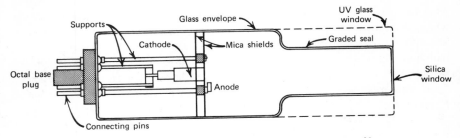

Fig. 10.3. Diagram of hollow cathode lamp. *Source*: Price[20].

10.2.1 Radiation Sources

Hollow Cathode Lamps

By far the most commonly used radiation source at the present time is the hollow cathode lamp (Fig. 10.3). The hollow cylindrical cathode is made of, or contains, the metal whose spectrum is required. The body of the lamp is made of an ultraviolet-transmitting glass, with the end-window of the same material if the resonance line lies above about 250 nm. For lower wavelengths a quartz window is used, attached to the body by a graded seal. The lamps are filled with neon or argon at a pressure of about 10 torr. A discharge at about 300 to 400 V and 3 to 30 mA occurs chiefly inside the cathode, where inert-gas ions sputter metal atoms off the cathode. Collisional excitation of the metal atoms leads to the emission of the characteristic line spectrum of the metal. Some lines of the fill gas are also emitted. For most elements, the lamp noise level is of the order of 0.1% to 0.2%; however, the stability of lamps of the more volatile elements (e.g., As, Se) is less satisfactory.

Several variants of the simple hollow cathode lamp may be mentioned. Multielement lamps have been developed in an attempt to provide greater convenience and economy. The range of elements that can be brought together in this way is limited by the need for them to be of similar volatility (e.g., Ca/Mg or Cu/Ag). Other limitations, such as diminished resonance line intensity and the greater inconvenience of lamp failure, make the value of multielement lamps disputable.

Demountable lamps have also been employed, allowing different cathodes to be used in a single lamp. This economy is offset by the need for facilities for lamp evacuation and inert gas purging.

High-brightness lamps, devised by Sullivan and Walsh,[30] use the hollow cathode as a source of the metal vapor, which is then excited by an auxiliary discharge of about 500 mA in front of the cathode. The object of this arrangement is to provide increased intensity without a high temperature and its accompanying problems of line-broadening and self-absorption. The advantage of these lamps has been reduced by recent improvements in the design of regular hollow cathode lamps.

Hollow cathode lamps have been operated in a pulsed mode both for atomic absorption[31] and atomic fluorescence.[32] Pulsing at several hundred hertz with a pulse duration of the order of several microseconds allows peak currents of several hundred milliamperes to be used.

Vapor-Discharge Tubes

Spectra of several elements (alkali metals, mercury) are conveniently produced in vapor discharge lamps. The metal whose spectrum is required is contained in a glass or silica tube with an inert gas at low pressure. A gas discharge between the sealed-in electrodes vaporizes the metal and leads to the emission of the metal spectrum. The spectral lines produced in this way tend to suffer from broadening and self-reversal, and the tubes should be operated at the minimum current that gives a stable output.

Electrodeless Discharge Lamps

Electrodeless discharge lamps (or tubes) have not found wide use in atomic absorption, but have been extensively studied in the development of atomic fluorescence spectroscopy, where high-intensity sources are required.[33–35] A lamp consists of a sealed quartz tube containing milligram quantities of a metal or metal iodide, together with an inert gas at low pressure. The lamp is placed in the intense field of a microwave source, using an antenna of the "A" type or of the "cavity" type. A Tesla coil initiates a discharge by causing some ionization of the inert gas. Electrons energized by the high-frequency component of the microwave field excite atoms of the metal. The discharge is mainly produced from near the tube wall, minimizing self-absorption, and consists largely of the emission spectrum of the metal. Although the effective electron temperature in the tube is very high, the gas is at low pressure and temperature ($< 400°C$), and the broadening of the atomic lines is limited.

Notable improvements in the stability and reproducibility of operation of electrodeless discharge lamps have been demonstrated by Browner et al.[36–38] using careful temperature control of the lamp environment. Multi-element lamps containing up to five elements have also performed well under controlled temperature conditions, provided the elements chosen have similar peak-performance temperatures.[37] A review of work on electrodeless discharge lamps has recently been published.[39]

10.2.2 Sample Atomization

Flame Atomization

Most routine atomic absorption measurements use a flame as the means of sample atomization. A detailed discussion of flames and burner design has been given by Mavrodineanu and Boiteux.[40] Both turbulent-flow and laminar-flow burners have been used.

In turbulent-flow burners, used in some of the early work in atomic absorption, the fuel and oxidant are not mixed until they reach the flame. Such burners are also described as "total-consumption" burners, as all the sample is sprayed directly into the flame. The apparent advantage of using the total sample does not seem to be confirmed in practice: there is evidence that sample vaporization is incomplete when total consumption burners are used. Turbulent-flow burners are inherently safe, and are almost mandatory if flames with high burning velocity (e.g., oxygen/acetylene) are being used.

Most flame atomizers for atomic absorption, however, are based on a laminar-flow premix burner, in which the flow of oxidant (often air or nitrous oxide) aspirates the sample into a spray chamber. Impact of the sample with a glass bead produces a spray of fine droplets, the largest of which settle out and are drained to waste. The remainder of the aerosol, generally only approximately 10% to 20% of the sample, is mixed with the fuel gas (e.g., acetylene, propane, or hydrogen) and passes into the burner. The construction of a spray chamber and burner head is illustrated in Fig. 10.4.

A typical burner head is designed to give a flame approximately 4 mm wide and 10 cm long. These dimensions result from the need to provide a long absorption path, the need to use the incident radiation as effectively as possible by passing it through the region of the flame with the greatest atom concentration, and the need for a finite beam width to fall on the entrance slit of the monochromator.

Laminar-flow burners usually give a very stable flame, almost devoid of the acoustic noise that accompanies the operation of some turbulent-flow burners, and provide greater freedom than the latter from chemical interferences in analysis (see Section 10.3.3). The risk of a laminar-flow burner "burning back" is negligible if the manufacturer's directions are followed strictly.

When the sample aerosol passes into the flame, the droplets (initially up to about 10 μm in diameter) are first desolvated, leading to the formation of minute solid particles. In the hotter regions of the flame, these may be decomposed into constituent molecules and atoms. The extent of the decomposition into the desired vapor is largely determined by the thermal stability of the compounds formed in the solid particle and by the flame

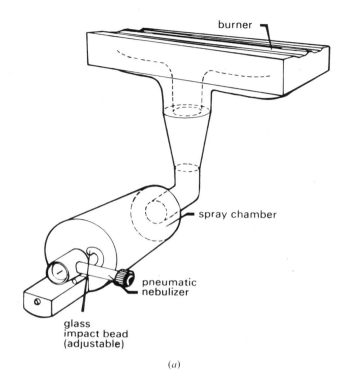

(a)

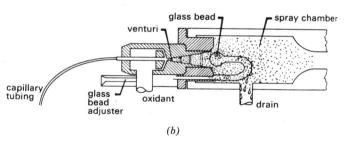

(b)

Fig. 10.4. (a) Construction of spray chamber and burner head. (b) Construction of pneumatic nebulizer and spray chamber. *Source*: Amos et al.[41] Reproduced by permission of Varian Techtron Pty. Ltd., Australia.

TABLE 10.2 Characteristics of Premixed Flames Used in Atomic Absorption

Oxidant	Fuel	Mixture	Approx. max. temperature (°C)	Ref.
Air	Propane	Stoichiometric	1925	40
Air	Acetylene	Lean	2145	42
Air	Acetylene	Stoichiometric	2150	42
Air	Acetylene	Rich	2060	42
Nitrous oxide	Acetylene	Lean	2670–2740	42, 43
Nitrous oxide	Acetylene	Stoichiometric	2795	42
Nitrous oxide	Acetylene	Rich	2580–2750	43, 44

temperature and flame chemistry. The atomization process is discussed further in Section 10.3.

Many different flames have been investigated for use in atomic absorption. Approximate maximum temperatures of those most commonly used are given in Table 10.2.

In addition to the flames listed above, use has also been made of several diffusion flames, in which a fuel such as hydrogen is mixed with an inert diluent (argon or nitrogen). The oxidant consists simply of oxygen in the air entrained in the flowing gases above the burner. The argon/hydrogen diffusion flame is much more transparent at low wavelengths than hydrocarbon flames and has found particular application in the determination of arsenic and selenium.[45,46]

The comparatively low temperature of diffusion flames increases the danger of chemical interferences (see Section 10.3.3). However, this problem is overcome in a variant of the normal sample aspiration technique for those elements (e.g., As, Sb, Se, Te, Ge) which form volatile hydrides. The sample is treated with a reducing agent to convert the analyte into a vapor of the corresponding hydride; this is then flushed into the diffusion flame where atomization readily occurs. Early work by this technique[47] used reagents such as zinc in an acidified reducing medium for hydride formation, but the advantages of sodium borohydride solutions have recently been demonstrated.[48–51]

Investigations have also been made of "separated" flames, in which the different reaction zones are physically separated from one another, along a tube.[52,53] In some arrangements, the flowing flame gases are protected by an inert gas sheath.

A useful feature of the separated air-acetylene flame is that in the interconal zone, where many rather stable oxides are reduced to atoms, the band emission from species such as OH, CO, and C_2 is much less than in the unseparated flame. Absorption by this flame at wavelengths below 200 nm is also much less in the separated flame, giving an advantage in the determination of arsenic and selenium.

Flameless Atomization

The early use of atomic absorption in the determination of mercury, the monatomic vapor of which can readily be produced at low temperatures, has already been noted. For other elements the nebulizer/flame arrangement has become the standard method for atomizing samples, although the fundamental inefficiency of the process is well recognized.[54] A high proportion of the sample is not nebulized finely enough to be carried into the flame. The part that does enter the flame has been diluted with fuel gas and oxidant by four or five orders of magnitude. A further order-of-magnitude dilution occurs when the gases react and expand to flame temperature.

Flameless atomization has been studied in an attempt to make more efficient use of the sample, and several devices have been produced that allow the determination of many elements in quantities in the nanogram to picogram range, using sample volumes of only 1–20 microliters. Most types of apparatus for flameless atomization generate the atomic vapor by very rapid electrical heating of a suitable material (e.g., graphite or tantalum) on which a small quantity of the sample has been deposited from a microliter syringe.

Electrothermal atomizers have mainly been of the furnace type or the heated rod (or filament) type. The former was initially developed by L'vov[55,56] who used an electrically heated graphite tube. An electrode containing the sample was introduced into a hole in the furnace and heated to atomize the sample. Temperatures of about 2500°C could be achieved in the furnace within a few seconds. By enclosing the system in a chamber filled with inert gas, oxidation of both the graphite furnace and the atomic vapor was largely prevented.

Among the many other electrothermal atomizers are wire loops of platinum or tungsten,[57] tantalum boats,[58] and graphite rods.[59,60] Flameless atomization following electrolytic deposition of metals on mercury[61] or tungsten[62] has also been demonstrated.

For many elements the concentration detection limit is about one order of magnitude lower than that obtained by flame atomization, and, as sample volumes of only a few microliters are needed, the absolute (mass) detection limits are very much lower. In general, the precision attainable by electrothermal atomization is rather poorer than with flame atomizers, and a great deal more study is being made of the basic processes occurring during operation of the electrothermal devices. The variety of apparatus with which different workers have experimented has given rise to conflicting reports on the extent and mechanism of sample matrix interference effects; work in this area is continuing. Flameless atomization techniques have been reviewed by Kirkbright[63] and more recent work has been summarized.[64,65]

In the case of mercury, for which poor sensitivity is obtained by flame atomic absorption, flameless atomization techniques have become widely used. Monatomic mercury vapor can be produced by chemical reduction of mercury compounds (e.g., by stannous chloride) in the cold. Mercury can also be electrodeposited on copper[66] or amalgamated on to a metal such as silver[67] or gold[68] for subsequent release on heating. The mercury vapor is aspirated into a long-path absorption tube maintained at, or slightly above, room temperature. A comprehensive review[69] of more than 400 papers on mercury determinations by flameless atomic absorption and fluorescence, covering the period through 1974, is a good entry point into the literature on this subject. This review lists many applications to various categories of materials of geological interest.

Apart from mercury determinations, the amount of work using flameless atomization for analysis of geological materials has been limited. Although some work has been reported on solid materials,[70] the greatest advantages would appear to be in water analysis, where the greater sensitivity of the flameless atomizers may obviate the need for a solvent extraction step, and in the analysis of crude petroleum and lubricating oils, where it is possible to avoid making dilutions with organic solvents.

10.2.3 Monochromators

It is important that the absorption line should be separated from any adjacent lines of the source. With very simple source spectra, such as that from a sodium vapor lamp, a filter may be enough. However, in general, it is necessary to select the required wavelength with a good monochromator designed for use between approximately 190 and 850 nm.

The spectra emitted by hollow cathode lamps of some elements (e.g., transition elements such as Cr, Fe, Ni) contain several closely spaced lines, not equally well absorbed. To the extent that nonabsorbing or more weakly absorbing lines of the analyte (or lines from the inert filler gas) are admitted to the detector, bending of analytical curves and reduced sensitivity will result. This makes it preferable to use a diffraction grating capable of separating two lines about 0.1 nm apart when using the minimum effective monochromator slit width.

Atomic absorption without the use of a conventional monochromator can be performed by using the "resonance" system of Sullivan and Walsh.[30] This type of system has not, however, been adopted in standard commercial atomic absorption instruments.

10.2.4 Detectors

Although a barrier-layer photocell and galvanometer can be combined with filter systems in detecting the resonance radiation of the alkali metals, photomultipliers are almost invariably preferred in instrumentation for more general use. Standard cesium-antimony photomultipliers have adequate sensitivity in the 190 to 600-nm range, the sensitivity often being greater than 10 mA/watt between approximately 210 and 570 nm, and falling off sharply beyond these wavelengths. For work with potassium (766 nm), rubidium (780 nm), and cesium (852 nm) other types of photomultiplier, such as those with gallium arsenide cathodes, are much more sensitive.

10.2.5 Amplifier and Readout Systems

The output of the detector passes to an a.c. amplifier, the signal from which is rectified before being fed to a meter. Continuous signals arising from flame emission are rejected when the amplifier is tuned to the source modulation frequency. There are limitations, however, on the ability of such a system to eliminate the unwanted continuous signal. Fluctuations in the continuous signal may contain an appreciable a.c. component close to the modulation frequency. Furthermore, photomultiplier shot noise is a function of the total incident power, and the photomultiplier may even become saturated if the flame emission is very intense.

Photomultiplier output is a linear function of the transmission, but many instruments make provision for readout to be given both as percent transmission and as absorbance. In the latter case, direct conversion to concentration units is readily achieved if the absorbance-concentration relationship is linear. Units are also available that provide direct concentration readout even with nonlinear analytical curves.

Signal-averaging may be used to increase the reliability of the output, and the speed of analysis is greatly facilitated in instruments equipped with digital readout and printout systems. Scale-expansion facilities are also often provided and are particularly useful at levels approaching the detection limit (in practice, this is often at an absorbance of the order of 0.001). Scale-expansion combined with "backing-off" of part of the signal may also be useful in improving precision at higher absorbances.

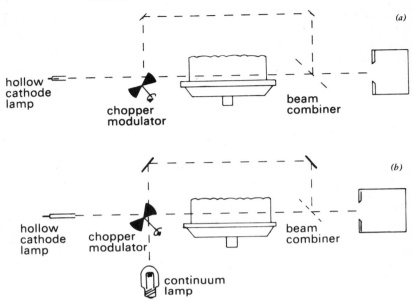

Fig. 10.5. Double-beam optical systems (schematic). (a) Reducing the effect of variation in source intensity. (b) Correcting for the effect of nonatomic background absorption. *Source*: Amos et al.[41] Reproduced by permission of Varian Techtron Pty. Ltd., Australia.

10.2.6 Optical Systems

Although much atomic absorption work has been done with single-beam instrumentation, the effect of any drift or fluctuations in the light source can be greatly reduced by using a double-beam system (Fig. 10.5a) in which modulation and switching of the beam is achieved by a rotating mirror. Such a system is not entirely analogous to the double-beam optics of solution absorptiometry (p. 108) in that the reference beam does not pass through a "cell" containing everything except the analyte: instead, it bypasses the flame completely. (No useful purpose is served in passing the reference beam through a second flame, as the noise-time relationships of the two flames are certain to differ). As a result of improvements in the stability of hollow cathode lamps, a large part of the fluctuation may originate in the flame, and the double-beam system provides only a small advantage. A further consideration in double-beam systems is the loss of at least 50% of the energy of the radiation that would otherwise be permitted to pass through the flame.

However, in many situations it is advisable to check and correct for the contribution of molecular absorption to the total absorption signal (p. 178).

In these cases it is a major convenience to have a double-beam system in which light from the element hollow cathode lamp and from a continuous source can be passed alternately through the sample vapor (Fig. 10.5b).

10.3 ASPECTS OF ATOM PRODUCTION

The concentration of analyte atoms n (in atoms cm^{-3}) produced in a flame depends on the concentration of analyte c (in mol l^{-1}) in the aspirated solution, according to the relation[26,71]

$$n = 1.0 \times 10^{19} \frac{F\varepsilon\beta}{Qe_f} c \qquad (10.8)$$

where F is the rate at which solution flows into the nebulizer (in cm^3 min^{-1}); ε is the aspiration efficiency, i.e., the fraction of the analyte that passes into the flame as gaseous species of any kind; β is the atomization efficiency, i.e., the efficiency of producing atoms from the gaseous species; Q is the flow rate of the unburnt gases into the flame (in cm^3 s^{-1}); e_f is the flame gas expansion factor, reflecting the change in the temperature of the gases from room temperature and the change in the number of gaseous moles accompanying the flame reaction.

The factor ε accounts for the transfer efficiency of the nebulizing device and for the efficiency with which solid particles entering the flame are decomposed to give gaseous species containing the analyte. The atomization efficiency β is the ratio of the concentration of free analyte atoms in the flame to the total concentration of analyte in all gaseous forms (molecular species, atoms, ions). This takes into account such factors as the incomplete dissociation of analyte compounds passing into the flame, ionization of analyte atoms, and formation of compounds between the analyte atoms and other species present in the flame. A comprehensive discussion of the factors that can cause β and ε to vary with analyte concentration has been given by Zeegers et al.[29]

Atomic absorption analyses are made by comparing the absorbances of a set of standard solutions with those of the samples; therefore, it is clear that the factor $F\varepsilon\beta/Qe_f$ of equation 10.8 must be identical for standards and samples. Any of the several causes of deviation from this condition will lead to errors in the analysis. These and other sources of so-called "interference effects" are examined in the following sections.

10.3.1 Aspiration and Nebulization

The flow rate of the solution into the nebulizer is largely dependent on the solution viscosity; the size distribution of the droplets formed in the nebulizer varies with the surface tension. Species containing the analyte enter the flame at a rate that therefore depends on the physical properties of the solution. There are significant differences in the aspiration and nebulization behavior (i.e., the factors F and ε of equation 10.8) between solutions with different solvents, different sample macroconstituents, or different temperatures.

Attempts have been made to improve the aspiration efficiency by heating the flame gases before they enter the spray chamber or by heating the spray chamber itself. However, the increased sensitivity is usually offset by a decrease in the stability of nebulizer operation, and such devices have not found general favor with instrument manufacturers.

Solvents

Early in the development of atomic absorption techniques it was reported that aspiration of solutions in organic solvents gave better sensitivity (often two- to threefold greater) than aspiration of the corresponding aqueous solutions.[72-75] Because of the lower viscosity and surface tension of the solvents used, a greater proportion of the sample reaches the flame. Improved sensitivity is obtained, provided the atomization efficiency β is not adversely affected. The flame characteristics are modified by organic solvents, in some cases so seriously that they should not be used for analytical purposes. The most suitable organic solvents appear to be aliphatic esters and ketones such as isobutyl acetate and 4-methyl-2-pentanone (methyl isobutyl ketone). The latter is particularly useful, being the basis of schemes for preliminary separation and concentration of trace amounts of many elements by solvent extraction (pp. 57, 187, 193–195).

Some water-miscible organic solvents (e.g., methanol, ethanol, and acetone) may be used in aqueous solutions containing up to approximately 25% of the organic solvent. With higher proportions the flame tends to become too unstable. Organic solvents added to a flame act as an additional fuel and alter the absorption by the flame itself. A reduced level of the normal fuel gas is usually necessary, and the zero absorbance (100% transmission) level should be set while aspirating a solvent blank, as usual.

Sample Macroconstituents

Large differences in macroconstituent composition (e.g., salts, acids) between samples and standards can also lead to differences in solution flow rate and atomizer efficiency, resulting from differences in physical properties. Macroconstituent differences may also lead to differences in behavior after the droplets enter the flame. Both the size-distribution of droplets and the atom distribution in the flame (vertically and laterally) may therefore be affected.

It is often recommended that samples and standard solutions be matched as closely as possible with respect to acid concentrations, for example, and with respect to major matrix salts, the purity of which must be checked before they are incorporated in the standard solutions. Typical observations of the effects of different acid media in the determination of several transition elements include those of Perrault[76] and Maruta et al.[77]

Solution temperature

Otherwise identical solutions at different temperatures give small, but significant, differences in absorbance, again as a result of different physical properties. Experiments in the authors' laboratory showed, for example, that an aqueous solution containing 2.5 μg Cu/ml gave an absorbance 10% greater at 60°C than at 25°C. This is worth noting in cases where residues from fusion or acid-dissolution have been redissolved with warming: the resulting solutions must be allowed to come to room temperature before the analysis is carried out.

10.3.2 Influence of Flame Type

A number of elements (including the alkali metals, Ag, Au, Cd, Cu, Mn, Ni, Pb, Sb, Tl, and Zn) are well atomized in relatively cool flames, such as air/coal-gas and air/propane. Where the compounds are completely dissociated in low-temperature flames, the sensitivity is usually a little better than in the air/acetylene flame. This appears to be the case for the alkali metals, in particular. However, as some of the above elements are incompletely atomized and are subject to chemical interferences (p. 179) in low-temperature flames, the air/acetylene flame is generally preferred.

The air/acetylene flame allows the determination of many elements in addition to those noted above, including Ba, Ca, Co, Cr, Fe, Mg, Mo, Sn, and Sr. In some cases the extent of atomization is very sensitive to flame composition: Cr and Mo, for example, are atomized better in a fuel-rich air/acetylene flame, even though the temperature is slightly lower than that of the stoichiometric flame.

Elements that form particularly stable oxides (e.g., the lanthanides, Al, B, Be, Si, Ti, V, W, Y, Zr) require the higher temperature and better reducing conditions of the nitrous oxide/acetylene flame to give satisfactory atomization.[78-80] In addition, where the degree of atomization in air/acetylene is strongly dependent on flame conditions, height of observation, and the presence of other substances in the sample, the nitrous oxide/acetylene flame is a useful alternative (e.g., for Ba, Ca, Mg, Mo, Sn, Sr).

Although chemical interferences are often reduced or eliminated in a higher temperature flame, the neutral atom population may also decrease as a result of ionization. The prevention of interferences from this source is discussed in the next section.

The flame chemistry of several elements (e.g., As, Se, Sn) is such as to give satisfactory atomization when hydrogen is used as a fuel, as in the air/hydrogen and argon/hydrogen diffusion flames. Greater sensitivity may be achieved than with air/acetylene, but often at the expense of greater susceptibility to interferences. However, interferences are of smaller concern if a gas-liberation method of sampling (p. 170) is used.

10.3.3 Interferences

Differences in behavior between samples and standard solutions, arising from different physical properties of the solutions, have already been described. It is also necessary to consider the possibility of spectral interference effects, and of interferences related to various phenomena that may occur in the flame and that affect the atomization efficiency.

Spectral Interferences

1. *Atomic line interference.* The simplicity of atomic absorption spectra and the narrow widths of the source emission lines both contribute to making atomic absorption relatively free of the problem of atomic line interference. A few cases are known, however, where a resonance line from a hollow cathode of one element may also be absorbed by a second element. There is mutual interference,[81] for example, between lines of gallium (403.298 nm) and manganese (403.307 nm). The cobalt line at 253.649 nm is capable of causing interference[82] in the determination of mercury at 253.652 nm. Other examples have been noted by Fassel et al.[83] The danger of atomic line interference increases when multielement hollow cathode lamps are used. For example, a silver/copper lamp emits a strongly absorbed copper line (327.45 nm) near the most strongly absorbed silver line (328.07 nm). In the analysis of silver care must be taken to exclude the copper line by using a narrow spectral bandpass, otherwise the detector will record a contribution from the copper line, and aspiration of a solution containing both silver and copper will give an absorption consisting of components from both elements. Other examples of this type of problem have been discussed by Jaworowski and Weberling.[84]

2. *Molecular absorption.* Under some conditions, particularly when solutions of high salt content are aspirated into a flame or vaporized in a flameless atomizer, a significant concentration of molecular species may be produced in the light path. Molecular absorption may occur, leading to an analytical interference if the absorption band extends over the absorption line of the analyte. Light losses observed under these conditions were initially attributed to light scattering by salt particles, but strong evidence has been

produced to indicate that scattering is often insignificant by comparison with molecular absorption, both in flame and nonflame situations.[56,85-88]

Molecular absorption is most pronounced in low-temperature flames and nonflame atomizers, or when long-path burners are used. However, in the air/acetylene flame, Willis[89] observed significant broadband absorption in the 210 to 320 nm range from 5% solutions of salts such as NaCl, $CaCl_2$, and K_2SO_4. Kahn[90] noted that a 10% NaCl solution in an air/acetylene flame gave the same absorption at 213.9 nm as a zinc solution at a concentration of 0.04 μg/ml.

Correction for nonatomic absorption can be made by measuring the light loss at a nearby line where atomic absorption does not occur.[89] A suitable line is not always available, and the preferred procedure involves measuring the light loss from a continuum source at the analyte wavelength.[85,90,91] Atomic absorption makes a negligible contribution to this loss. The absorbance from the continuum source is subtracted from that observed from the hollow cathode source to obtain the value arising from purely atomic absorption. Recent commercial instrumentation makes provision for automatic subtraction of nonatomic "background" absorption by means of a double-beam system in which light beams from a continuous source (e.g., a hydrogen or deuterium lamp) and from the atomic hollow cathode lamp are alternately passed through the absorbing medium.

In trace-element analysis of solutions resulting from the treatment of geological materials, either by fusion or acid-dissolution methods, a check for molecular absorption is essential, as the solution may contain high concentrations of salts of K, Na, Mg, Ca, Al, and Fe. The importance of this check was illustrated by Fletcher[92] in the determination of Co, Ni, Pb, Cu, and Zn in standard rocks.

Chemical Interferences

During the period of several milliseconds following the introduction of aerosol droplets into a flame, many processes may occur: solvent evaporation, melting and vaporization of minute solid particles, chemical decomposition of particles and gaseous molecules, excitation and ionization of the gaseous species, recombination of atomic vapor to form oxides, hydroxides and other molecular species, and condensation of gaseous species to reform solid particles. The distribution of analyte atoms in the flame is influenced by the rate and extent of the above processes (which in turn depend on flame parameters such as velocity, composition and temperature-distribution), and on the rate of diffusion of the various species through the flame to the surroundings. Maximum production of atoms in the light path can only be achieved if all atom-forming processes go to completion. Otherwise, changes in the flame parameters or in the chemical composition of the solution droplets entering the flame may alter the factors ε and β (equation 10.8).

Changes in the absorption signal caused by concomitants in the flame, can be loosely described as chemical interferences. They are further classified as depressions or enhancements by comparison with an absorption signal obtained under some specified reference conditions. Enhancements, however, can be considered simply as the removal of a depressive effect existing under the reference conditions.

It is convenient to consider these interferences under the following headings: (a) stable compound formation, (b) analyte occlusion, (c) atom redistribution, (d) ionization effects. The first two of these types of interference have been discussed in detail by Alkemade.[93] It should be noted that in many cases the observed behavior may result from a combination of these effects, their relative importance varying in different parts of the flame. It is also important to note that, because these effects alter the efficiency of atom production, flame atomic emission and atomic absorption are affected in the same way.

Detailed summaries of the literature on interference studies can be found in a number of specialist atomic absorption texts[10,20,21] and in recent review articles.[64,65,94]

1. *Stable compound formation.* In some cases the addition of the interferent to the solution results in the formation of a compound that is particularly stable or slow to volatilize, thereby reducing the yield of ground-state atoms in the light path. The effects of various oxyanions in depressing the atomization of alkaline earth elements in air/acetylene flames have been extensively studied by users of flame emission and atomic absorption. Workers in the latter field have paid particular attention to the determination of calcium, magnesium, and strontium in the presence of anions such as phosphate, silicate, and sulfate.[95-101] Evidence from double-nebulizer experiments indicates that stable compounds are formed in the solid phase upon evaporation of the solvent, rather than as a result of gas-phase combination.

The depressive effect of aluminum on magnesium absorption is also well known[7,102] and is probably due to the formation of spinel ($MgAl_2O_4$), particles of which have been isolated from an air/acetylene flame when solutions containing both elements have been aspirated.[103] Because the magnitude of this kind of interference depends on flame composition and temperature and on the position of observation in the flame, quantitative disagreement among different observers is common.

The methods used for overcoming these depressive interferences include the use of a higher-temperature flame and the addition of a "releasing agent" to sample and standard solutions. Strontium and lanthanum salts are effective in removing the effect of phosphate on calcium, for example, by combining preferentially with the phosphate. Similarly, calcium, strontium, barium, and lanthanum salts remove the interference of aluminum on magnesium by forming more stable aluminates. In some situations chelating agents such as EDTA and 8-hydroxyquinoline also act as releasing agents, either by hindering stable oxide formation or by their reducing action on the dried solid particles in the flame.

There are several instances of an enhancement of atom production resulting from the formation of compounds that are more volatile than those formed from the reference solution. In the nitrous oxide/acetylene flame, for example, addition of small amounts of fluoride to an aqueous solution enhances the absorption from Zr, Hf, Ta, and Ti.[79,80,104,105] Metals with oxides of high dissociation energy are more readily volatilized if the metals in solution are bonded to elements other than oxygen, as in fluorocomplexes and metallocenes.[105,106] The importance of considering solution equilibria in explaining some chemical interferences has been emphasized.[107]

Depressive interferences related to stable compound formation become significant at interferent:analyte mole ratios smaller than unity and often reach a plateau near a simple stoichiometric mole ratio. The effects are generally symmetrical with respect to the analyte and interferent;[93] that is, if B depresses atomization of A, then A depresses atomization of B (assuming that B is being determined under the same conditions).

2. *Analyte occlusions.* Other interference effects are not related to the formation of specific compounds. When the solid particle remaining after evaporation of the solvent consists of analyte atoms dispersed in a matrix of the concomitant, the volatility of this matrix has an important bearing on the yield of free analyte atoms. Depression or enhancement may be observed, depending on whether the matrix is less or more volatile than that resulting from desolvation of the reference solution.

An explanation of this kind has been used to account for the depression of chromium and molybdenum absorption in the presence of high iron concentrations in the air/acetylene flame.[108] Several observations indicate that specific compound formation is not involved. The depression levels off only at very high iron concentrations, and furthermore iron absorption signals are little affected by a high concentration of chromium. It is suggested that, at high iron/chromium ratios, the chromium atoms are occluded in an iron matrix (b.p. 3000°C) which is incompletely volatilized. In the reverse situation where iron is determined in the presence of an excess of chromium, the formation of the more volatile chromium matrix (b.p. 2480°C) does not affect the eventual release of the iron atoms.

Interferences of this type can often be overcome by using the nitrous oxide/acetylene flame, or by adding salts such as ammonium chloride or sodium sulfate to the solution. This helps to ensure that the analyte atoms become trapped in matrix particles that volatilize completely before reaching the light beam.

It has been pointed out[109] that a description of matrix volatility in terms of boiling point and heat of vaporization may be an oversimplification. The rate of evaporation of small particles is a complex function of factors such as drop size, surface tension, heat-transfer parameters, and diffusion coefficients of the evaporating species.

3. *Atom redistribution.* Detailed studies[110,111] on atomization in nitrous oxide/acetylene flames have shown that some concomitants cause interference by changing the spatial distribution of free analyte atoms. Matrices containing rather involatile acids (e.g., $HClO_4$, H_3PO_4) or salts, give rise to particles that undergo slower lateral diffusion, the atom population being enhanced in the center of the flame and depleted at the edge. This effect is seen, for example, when calcium is measured in the presence of phosphoric acid, observations being made 5 to 10 mm above the burner.[111] The overall calcium atom population at this height appears to be unchanged, but the lateral redistribution may cause a change in the population intercepted by the light beam. Although this type of interference has been indicated in only a few cases, it provides a further reason for careful matching of the major constituents in sample and standard solutions.

4. *Ionization effects.* The equilibrium between neutral and singly ionized atoms of the same species can be written

$$M \rightleftharpoons M^+ + e^- \tag{10.9}$$

and can be characterized by an ionization equilibrium constant, K_i,

$$K_i = \frac{P_{M^+} P_{e^-}}{P_M} \tag{10.10}$$

where P_M, P_{M^+}, P_{e^-} are partial pressures of M, M^+ and free electrons, respectively. (In flames, free electrons arise both from the ionization of introduced metal atoms and from the flame combustion process itself). The value of K_i can be found from the Saha equation

$$\log K_i = -\frac{5040 E_i}{T} + \frac{5}{2} \log T + \log \frac{g_+}{g} - 6.182 \tag{10.11}$$

where E_i is the ionization energy (in electron volts) and g_+, g are the electronic partition functions of the ion and atom, respectively.

TABLE 10.3 Extent of Ionization of Selected Metals in Different Flames

Flame	Ionization			
	Predominant (>90%)	Extensive (>50%)	Significant (>10%)	Slight (>1%)
Air/propane	—	—	Cs, Rb	K
Air/acetylene	—	Cs, Rb	K, Sr	Na, Li, Ca
N_2O/acetylene	Cs, Rb, K	Na, Li, Ba, Sr, Eu	Ca, Al, Yb	Mg

If the temperature of the flame is high enough, a significant proportion of the metal atoms may ionize, depending on E_i. The effect of ionization in diminishing the ground-state population of neutral atoms is particularly important for alkali metals and alkaline earths in high-temperature flames. Although the position of the ionization equilibrium varies with the metal-atom concentration, ionization being greater at low concentrations, an approximate indication can be given of the extent of ionization of selected metals under analytical conditions in different flames (Table 10.3).

This equilibrium has been extensively studied in both flame emission and atomic absorption. It is clearly a potential source of interference effects, and may also be a cause of curvature of analytical curves (p. 186). In the presence of a second metal that ionizes more readily than the analyte, the increased free electron concentration suppresses ionization of the analyte and thereby increases the concentration of neutral analyte atoms. In the air/acetylene flame, for example, the atomic absorption from low concentrations of sodium is increased by the presence of an excess (e.g., 100–1000 μg/ml) of potassium. Similarly, the alkali metals suppress ionization of the alkaline earths in the nitrous oxide/acetylene flame.[79,80,112,113] Because many elements are appreciably ionized in the nitrous oxide/acetylene flame, the use of an "ionization buffer" (e.g., at least 1000 μg/ml of potassium) is often recommended.

10.4 ANALYTICAL TECHNIQUES

This section deals with a number of matters relevant to the practice of atomic absorption. It is not, however, intended to provide details of the operation of atomic absorption instruments. Information on specific instruments and components can be found in the manuals of the manufacturers. Notes on such matters as optimum performance of hollow cathode lamps, alignment of burners in the optical path, adjustment of nebulizers and gas flows, cleaning of spray chambers and burner heads, and the use of different readout modes can be found in specialized works on atomic absorption (e.g., Refs. 19–21).

10.4.1 Standard Solutions

Atomic absorption standard solutions are generally prepared from relatively concentrated stock solutions with a metal concentration of 1000 μg/ml, for example. Analytical grade reagents are suitable. Where metals are used, they should be as free as possible from surface oxide coatings. Details of the preparation of stock solutions suitable for most purposes are given in Table 10.4. Organometallic compounds suitable for the preparation of standards in nonaqueous solvents have been listed by Price.[20]

Most of these solutions may be stored in polyethylene bottles for long periods. Dilute standards are prepared from the stock solutions by addition of the same solvent medium as is contained in the stock solution, together with any necessary additives (releasing agents, ionization suppressors, matrix-matching salts, etc.). Analyte-impurity levels in the additives must be carefully noted to ensure that they can be tolerated: in some cases a special atomic-absorption grade reagent is advisable (e.g., La_2O_3 or $LaCl_3$ with very low calcium and magnesium levels). For many applications it is convenient to use multielement dilute standard solutions, provided due consideration is given to undesirable chemical reactions during storage. It is recommended that solutions of approximately 20 μg/ml or less be renewed every few weeks and solutions of approximately 2 μg/ml or less be renewed every few days (see also Chapter 2).

10.4.2 Analytical Curves

Analytical curves should be developed by aspirating at least three (preferably four) standard solutions giving absorbances of approximately 0.05 to 0.80. Most curves of absorbance versus concentration are linear or slightly curved toward the concentration axis. The most recent instrumentation provides a direct concentration readout for sample solutions, with facilities to take account of a certain amount of curvature in the absorbance-concentration relation.

Some of the most important causes of bending of analytical curves are summarized below.

1. *Failure to isolate a single, narrow absorbing line.* In this case, the detector also responds to unabsorbable radiation. This may consist of lines from other metal components of the hollow cathode lamp or from the fill gas, or neighboring unabsorbed (or weakly absorbed) lines of the analyte. The

TABLE 10.4 Stock Solutions for Atomic Absorption Standards[a]

Element	Reagent	Mass/g	Initial dissolution
Aluminum	Metal foil	1.000	25 ml conc. HCl + few drops conc. HNO_3
Antimony	Metal shot	1.000	10 ml conc. HCl + few drops conc. HNO_3 + 10 g tartaric acid
Arsenic	As_2O_3	1.320	50 ml conc. HCl or 25 ml 1 M NaOH
Barium	$BaCO_3$	1.438	20 ml 1 M HCl
Beryllium	$BeSO_4 \cdot 4H_2O$	19.64	Water
Bismuth	Metal	1.000	50 ml conc. HNO_3
Boron	H_3BO_3	5.714	Water
Cadmium	CdO	1.142	20 ml 5 M HCl
Calcium	$CaCO_3$	2.497	100 ml 1 M HCl
Chromium	Metal	1.000	50 ml conc. HCl
Cobalt	Metal	1.000	50 ml 6 M HNO_3
Copper	Metal	1.000	50 ml 5 M HNO_3
Germanium	Metal chips	1.000	20 ml 1:1 $HCl:HNO_3$
Gold	Metal	1.000	15 ml conc. HCl + 5 ml conc. HNO_3
Indium	Metal	1.000	10 ml conc. HCl + 5 ml conc. HNO_3
Iron	Metal	1.000	20 ml 5 M HCl + 5 ml conc. HNO_3
Lead	Metal	1.000	50 ml 2 M HNO_3
Lithium	$Li_2SO_4 \cdot H_2O$	9.215	Water
Magnesium	Metal ribbon	1.000	50 ml 5 M HCl
Manganese	Metal	1.000	50 ml conc. HCl
Mercury	Metal	1.000	20 ml 5 M HNO_3
Molybdenum	$(NH_4)_6Mo_7O_{24} \cdot 4H_2O$	1.829	Water
Nickel	Metal	1.000	50 ml 5 M HNO_3
Niobium	Metal powder	1.000	10 ml HF + 5 ml conc. HNO_3
Palladium	$(NH_4)_2PdCl_4$	2.668	Water
Platinum	$(NH_4)_2PtCl_6$	2.275 *	Water
Potassium	KCl (dry)	1.905	Water
Rhodium	$(NH_4)_3RhCl_6 \cdot 1.5H_2O$	3.858	Water
Ruthenium	$[Ru(OH)(NO)(NH_3)_4]Cl_2$	2.839	Water
Selenium	Element	1.000	15 ml conc. HCl + 5 ml conc. HNO_3
Silicon	$Na_2SiO_3 \cdot 5H_2O$	7.6[b]	Water
Silver	$AgNO_3$	1.575	Water
Sodium	NaCl (dry)	2.542	Water
Strontium	$SrCO_3$	1.684	20 ml 1 M HCl
Tantalum	$TaCl_5$	1.980	Water
Tellurium	Metal	1.000	15 ml conc. HCl + 5 ml conc. HNO_3
Thallium	Tl_2SO_4	1.235	Water
Tin	Metal	1.000	200 ml conc. HCl + 5 ml conc. HNO_3
Titanium	Potassium titanyl oxalate	7.394	Water
Tungsten	WO_3	1.260	0.5 M KOH
Vanadium	Ammonium metavanadate	2.296	20 ml "100-volume" (30% w/v) H_2O_2
Zinc	Metal	1.000	50 ml 5 M HCl

[a] All solutions when made up to 1 liter contain 1000 μg/ml as the metal. Much of the information in this table is extracted from Price (Ref. 20, pp. 82–83).
[b] Requires gravimetric standardization.

problem can sometimes be avoided by careful choice of slit-width and monochromator spectral bandpass, but in some cases the lines cannot be resolved. Where some of the light admitted is not capable of being absorbed, the analytical curve approaches asymptotically a limiting absorbance that depends on the intensity of the unabsorbable radiation. Where several unresolved analyte lines are included the analytical curves tend toward an inclined asymptote. Bending of analytical curves may also be caused by broadening and self-absorption of the line source, which can occur when hollow cathode lamps are operated at very high currents.

2. *Defects in the optical system.* Because of the finite width of the light beam (usually rather variable over the pathlength through the flame), not all the radiation passes through regions with uniform atom concentration. This leads eventually to bending of the analytical curve, although the problem is smaller with optical systems that keep the beam narrow through the flame. At high absorbances (e.g., > 1.0) stray light may become a significant part of the total light detected. As this component is not absorbable, this also causes bending of the analytical curve.

3. *Ionization.* Where the ionization of analyte atoms makes a significant contribution to the total electron concentration in the flame, the degree of ionization changes with the total concentration of analyte introduced into the flame. A simple analysis of equations 10.9 and 10.10 shows that, as the total analyte concentration (atoms and ions) in the flame increases, the proportion of atoms lost by ionization becomes smaller. This causes analytical curves to bend toward the absorbance axis. Such behavior has been observed with sodium and potassium in the air/acetylene flame and with barium, strontium, and europium in the nitrous oxide/acetylene flame. Both the curvature and the loss of sensitivity resulting from ionization can be avoided by the use of ionization-suppressing salts (p. 183).

4. *Failure to match physical properties of solutions.* If a standard stock solution, prepared in a particular solvent medium, is not diluted with the same solvent, then the series of diluted standards may show a regular change of physical properties that will result in variable aspiration efficiency. This is of particular importance in dealing with mixed organic/aqueous solvents. Depending on the way the dilution is carried out, the analytical curve may bend in either direction. The problem is simply avoided by proper matrix matching of all standard solutions.

10.4.3 Sensitivities and Detection Limits

The term *sensitivity*, used in a quantitative sense, refers to the concentration (in solution) of an element that produces an absorption of 1% (absorbance of 0.0044) of the radiation transmitted by a solvent blank alone. The *detection limit* may be taken as the concentration corresponding to twice the standard deviation of a series of at least 10 readings taken close to the blank level.[20] The detection limit is therefore dependent on both the sensitivity and the level of noise from all instrumental sources. Because the noise levels in the readout are usually well below 1% of the radiation transmitted by the solvent blank, detection limits are often lower than the sensitivities by a factor between about 3 and 10.

Some variation is found in sensitivities quoted by workers with different instruments. The values given in Table 10.5, however, give an approximate guide for many of the most frequently used absorption lines.

In cases where a high concentration of analyte makes a less-sensitive determination desirable, several alternatives are available. (a) Dilution of the sample is the most obvious method, but is tedious if many samples are involved. (b) The effective pathlength through the atomic vapor can be reduced by rotation of the burner; with the burner slot perpendicular to the optical path, the absorbance of a given solution is lowered by approximately one order of magnitude. (c) In favorable cases it may be possible to select a less strongly absorbing line of the analyte. Some of the alternative lines are given in Table 10.5.

In many situations, however, better sensitivity than that shown in Table 10.5 is needed. Some workers analyzing freshwater samples have concentrated the solution by evaporation. A small improvement resulting from the addition of certain organic solvents has already been noted (p. 176). Optical methods of improving sensitivity, such as multiple passage of radiation through the flame, or the use of long-path flame absorption cells, have found only limited application. Sensitivities for most elements are improved by at least one order of magnitude when electrothermal atomizers are used, and elements forming volatile hydrides are determined with better sensitivity by the vapor-generation technique (p. 170) than by direct aspiration of sample solutions.

Alternatively, a chemical preconcentration technique can be used. Solvent extraction systems employing chelating agents such as APDC have found extensive application.[114-120] With careful choice of chelating agent, organic solvent, and pH of the aqueous phase, concentration factors of at least 100 can be achieved. The fact that alkali and alkaline earth elements generally remain in the aqueous solution is an added benefit, as any molecular absorption from salts of these elements may be eliminated.

Ion-exchange resins and chelating resins have also been used for analyte preconcentration by some workers.[121,122]

10.4.4 Method of Standard Additions

In situations where interferences cannot be completely controlled by close matching of standards and samples (e.g., if the full composition of the

TABLE 10.5 Sensitivity of Some Atomic Absorption Lines[a]

Element	Wavelength (nm)	Sensitivity[b]	Flame[c]	Element	Wavelength (nm)	Sensitivity[b]	Flame[c]
Ag	328.07	0.03	AA(l)[d]	Mn	279.48	0.02	AA(s)
	338.28	0.06			403.08	0.26	NA(r)
Al	309.27, 28[e]	0.75	NA(r)	Mo	313.26	0.2	AA(r)
	396.15	1.1			313.26	0.5	AP(l)
As	193.70	0.6	AH	Na	589.00	0.003	AA(l)
	193.70	1.3	AA(j)		589.00	0.01	AP(l)
	197.20	1.0	AH		589.59	0.008	AP(l)
Au	242.80	0.1	AA(l)		330.23, 30[e]	1.7	AP(l)
	267.60	0.2		Nb	334.91	19	NA(r)
B	249.68, 77[e]	8.4	NA(r)		358.03	21	
Ba	553.55	0.2	NA(r)	Nd	492.45	6.3	NA(r)
	455.40[f]	2.0		Ni	232.00	0.05	
Be	234.86	0.02	NA(r)		341.48	0.25	AA(l)
Bi	223.06	0.2	AA(l)[d]		352.45	0.25	
	306.77	0.3		Os	290.91	1.2	NA(r)
Ca	422.67	0.013	NA(l)	Pb	217.00	0.11	
	422.67	0.03	AA(s)		283.30	0.23	AA(l)
Cd	228.80	0.01	AA(l)[d]	Pd	244.79	0.09	
Ce	520.01, 04[e]	30	NA		247.64	0.13	AA(l)
	569.70	39		Pr	495.14	18	NA(r)
Co	240.73	0.05	AA(l)	Pt	265.95	1.2	AA(s)
	304.40	0.85		Rb	780.02	0.03	AP
Cr	357.87	0.06	AA(r)		780.02	0.5	AA(l)
	357.87	0.1	NA(r)		794.76	0.08	AP
	425.44	0.17	AA(r)	Re	346.05	9.5	NA(r)
Cs	852.11	0.04	AP	Rh	343.49	0.1	AA(l)
	852.11	0.5	AA(l)	Ru	349.89	0.7	AA(r)
	894.35	0.05	AP	Sb	217.58	0.3	
	455.54	0.21	AP		206.83	0.35	AA(s)
Cu	324.75	0.04			231.18	0.6	
	327.40	0.14	AA(l)	Sc	391.18	0.27	NA(s)

Element	Wavelength	Sensitivity	Flame	Element	Wavelength	Sensitivity	Flame
Dy	421.17	0.67	NA(r)	Se	196.03	0.47	AH
Er	400.80	0.46	NA(r)		196.03	0.8	AA(j)
Eu	459.40	0.34	NA(r)	Si	251.61	1.5	NA(r)
Fe	248.33	0.05	AA(s)	Sm	429.67	6.6	NA(r)
	371.99	0.41		Sn	224.61	0.8	NA(r)
Ga	294.36, 42^c	0.72	NA		235.48	1.1	NA(r)
	294.36, 42^c	2.4	AA		286.33	1.4	NA(r)
Gd	368.41	19	NA		286.33	8	AA
Ge	265.12, 16^c	1.3	NA	Sr	460.73	0.04	NA(s)
	265.12, 16^c	5	AA(j)		460.73	0.2	AA(s)
Hf	307.29	10	NA(r)	Ta	271.46	11	NA(r)
Hg	253.65	2.2	AA(l)	Tb	432.65	8	NA(r)
Ho	410.38	0.76	NA(r)	Te	214.28	0.26	AA(l)
In	303.94	0.17	AA(l)	Ti	364.27	1.4	NA(r)
Ir	208.88	2.3	AA(s)		365.35	1.5	
K	766.49	0.01		Tl	276.79	0.2	AA(s)
	769.90	0.03	AP(l)	Tm	371.79	0.27	NA
	404.41	3.3		U	358.49	110	NA
La	550.13	40	NA(r)	V	318.54	0.7	NA(s)
Li	670.78	0.01	AP		318.34, 39^c	1.1	
	670.78	0.03	AA	W	255.14	5.8	NA(r)
Lu	335.96	7.9	NA	Y	410.24	2.3	NA
Mg	285.21	0.003	AA(s)	Yb	398.80	0.07	NA(s)
	285.21	0.003	NA(s)	Zn	213.86	0.01	AA(s)
	202.50	0.09	AA(s)	Zr	360.12	9.1	NA(r)
	279.55^f	0.2	NA(s)				

[a] Data from various sources, including Varian Techtron Hollow Cathode Lamp Data, Varian Techtron (1973); Price;[20] authors' laboratory.
[b] Defined as concentration of element, in μg/ml, producing 1% absorption.
[c] Flames: AA, air/acetylene; NA, nitrous oxide/acetylene; AH, argon/hydrogen; AP, air/propane; r, rich; j, just luminous; s, stoichiometric; l, lean.
[d] Similar sensitivity in air/propane flame.
[e] Doublet, not normally separated.
[f] Ion line; no ionization buffer should be used.

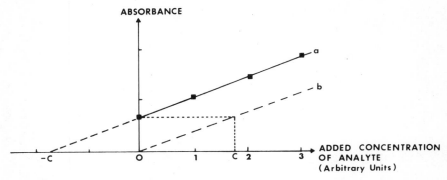

Fig. 10.6. Method of standard additions. Absorbance is plotted as a function of the added analyte concentration (line a) and extrapolated to zero absorbance at concentration $-C$. Alternatively, use is made of a parallel line (b), and the concentration C is obtained from the absorbance of the original solution.

samples is unknown), it is possible to use the method of standard additions. The sample is divided into at least three equal aliquots, and known different amounts of analyte are added to all but one. The solutions are then made up to the same volume. The measured absorbance is plotted as a function of the added concentration of analyte in the solutions (Fig. 10.6, line a). It is assumed that the additions alone make a contribution given by the parallel line b. The concentration C of analyte in the unspiked solution is found from the point on line b having the same absorbance as the unspiked solution, or alternatively, by extrapolating line a back to zero absorbance, where the abscissa is $-C$.

Molecular absorption, if it occurs, should make a similar contribution to the absorbance of each solution. It is preferable to correct all absorbances for this effect before plotting.

The method should be used with great caution if the cause of the interference is unknown, as it is based on the assumption that the interferent acts on the addition in exactly the same way as it does on the analyte in the original sample. This assumption does not hold when the absorbance of a given concentration of analyte is very sensitive to the interferent:analyte ratio.

10.4.5 Indirect Methods of Analysis by Atomic Absorption

Several indirect methods have been devised for cases where the direct determination of the species sought is inconvenient or impossible. Use has been made of chemical processes such as precipitation, heteropolyacid formation, and complex formation, and also of flame interference effects.

Chloride[123] and sulfate,[124] for example, can be determined by precipitation with excess silver ions and barium ions, respectively. The excess of precipitant can be found by atomic absorption, or the precipitate can be separated and determined after dissolution in a suitable reagent (e.g., ammonia for silver chloride, ammonium EDTA solution for barium sulfate). The applicability of this technique depends on the specificity of the precipitation/dissolution procedure and on the solubility product of the precipitate.

Several oxyanions (e.g., phosphate, silicate, arsenate, titanate, vanadate) form complex heteropolyacids with molybdate or phosphate-molybdate mixtures.[125-127] The acids, such as silicomolybdic acid and molybdotitanophosphoric acid, can be extracted into organic solvents and the molybdenum determined by atomic absorption. Sensitivity is gained both by the extraction process and by the stoichiometry of the reactions forming the heteropolyacids which contain, for example, 12 molybdenum atoms per silicon atom, or 11 molybdenum atoms per 2 titanium atoms. Such amplification procedures for indirect atomic absorption analysis have been reviewed by Johnson et al.[128]

Indirect determinations of chelating agents such as EDTA and nitrilotriacetic acid (NTA) may be made by forming an appropriate metal complex and precipitating the excess metal[129] or extracting the complex into an organic solvent. Alternatively, some compounds form extractable complexes with metal chelates. Iodide can be determined by measuring cadmium atomic absorption after extraction into benzene of the ion-pair from *tris*-(1,10-phenanthroline)cadmium(II) and iodide.[130] Cyanide can be determined by forming dicyano-*bis*(1,10-phenanthroline)iron(II), which can be extracted into chloroform.[131]

Atomic absorption interference effects have been used as the basis for indirect determination of certain species. These effects include the depression of calcium, magnesium, or strontium absorption in air/acetylene flames by phosphate, silicate, sulfate, and other oxyanions[132-135] and the enhancement of zirconium or titanium absorption by fluoride.[136] The use of such interferences to determine species such as phosphate, silicate, and fluoride is of limited application, as the effects are seldom specific, and the extent of the interference is very dependent on flame conditions and the position of the optical path.

Summaries of recent work on indirect flame atomic absorption determinations have been given by Pinta,[137] Winefordner and Vickers,[64] and Hieftje et al.[65]

10.5 APPLICATIONS TO THE ANALYSIS OF GEOLOGICAL MATERIALS

The fact that atomic absorption can be used for more than 60 elements, many having detection limits well below 1 μg/ml in solution, has led to extensive use being made of this technique in the analysis of geological and related materials. The sensitivity, specificity, and precision that can be achieved often help to outweigh the factors that may be disadvantages: the need for the sample to be in solution and the capability of determining only one element at a time. This section outlines the work that has been carried out in applying atomic absorption to various types of geological samples: waters, both fresh and saline; ores, minerals, rocks, soils, and coal ash; crude petroleum.

10.5.1 Water Analysis

Atomic absorption is particularly well suited to freshwater analysis, as samples can often be aspirated directly into a flame. Sample preparation may need to include filtration (e.g., through a 0.45 μm millipore filter) to remove suspended matter. Alternatively, if the analysis is to include suspended matter, digestion with nitric or hydrochloric acid may be carried out. In the determination of elements such as calcium and magnesium, the usual steps must be taken to eliminate interferences from anions such as phosphate and silicate, if these are present. The other main factor influencing sample preparation is the required sensitivity: this must often be adjusted by one of the methods outlined on p. 187.

Some reservations have been expressed concerning the reliability of results given by electrothermal atomizers in the analysis of freshwater.[138,139] Because of uncertainties surrounding the presence or absence of matrix interference effects, the standard-addition technique has been recommended,[138] although this greatly reduces the convenience of the method.

A selection of papers dealing with the analysis of nonsaline water (i.e., lake, river, bore and well waters, purified water, and industrial waste waters) is summarized in Table 10.6. Additional discussion and references can be found in various articles and reviews.[64,65,94,140–142]

Of the metallic elements present in seawater, only Na, Mg, Ca, K, Sr, Li, and Rb are normally found in sufficient concentration for direct analysis by flame atomic absorption. Other elements can be determined directly if their concentrations are artificially elevated, for example, as a result of pollution.

An early paper[168] reported the direct determination of Cu, Fe, Mn, Ni, and Zn in seawater. However, the concentrations reported were close to, or below, the generally accepted detection limits for these elements, and later workers have found it necessary to use solvent extraction or chelating resins to obtain higher concentrations and more reliable results. Electrothermal atomizers are also being increasingly used, both for direct analysis and in conjunction with extraction methods. Some of the work on seawater is summarized in Table 10.7.

TABLE 10.6 Atomic Absorption Analysis of Freshwater

Elements determined	Special treatment, extraction, or analytical procedure	Ref.
Ag, Al, Be, Cd, Cu, Fe, Ni, Pb, Zn	Extract with mixture of chelating agents in ethyl propionate	143
Ag	Ion exchange; elute and extract with APDC/MIBK	144
Ag	Ash suspended matter; dissolve, extract with tri-iso-octyl thiophosphate/MIBK; Ta boat atomizer	145
Al, Cr, Cu, Fe, Hg, Mn, Pb	Graphite atomizer	146
Al	Extract with mixture of chelating agents in benzene	147
Al	Extract with oxine/MIBK	148
As, Sb, Se	Hydrides decomposed in silica tube furnace	149
Au	Chelate with polyschiff base; elute, extract into MIBK	150
Ca, Cu, Fe, K, Mg, Na	—	151
Ca, Cu, Fe, Mg, Mn	Acidify with HCl, autoclave to dissolve suspended matter	152
Ca, Cu, Fe, Mg, Na, Ni, Zn	Extract with APDC/MIBK where necessary	153
Ca, Cu, K, Li, Mg, Mn, Na, Sr, Zn	—	154
Cd, Cu, Fe, Ni, Pb, Zn	Ion exchange on Dowex A-1	121
Cd, Cu, Fe, Ni, Zn	Extract with DEDTC/MIBK	155
Cd	Rf-heated carbon bed atomizer	156
Co, Cr, Cu, Fe, Li, Mn, Rb, Sr, Zn	Evaporate to concentrate where necessary	157
Co, Cr, Cu, Fe, Mn, Ni, Pb, Zn	Extract with DEDTC/MIBK	158
Co, Ni, Pb	Extract with APDC/MIBK	159
Co, Ni	Coprecipitate with $Fe(OH)_3$; extract Fe; extract Co, Ni with APDC/MIBK	160
Cr	Convert to Cr(VI); extract with APDC/MIBK	161
Hg	Solution reduction	162
Hg	Amalgamate on silver; electrothermal atomization	67
Hg	Extract with sulfide-treated polyurethane; solution reduction	163
Mo	Extract with APDC/MIBK	164
Mo	Extract with oxine/MIBK	165
Se	Graphite atomizer	166
V	Extract with dichloro-oxine/MIBK	167

TABLE 10.7 Atomic Absorption Analysis of Seawater

Elements determined	Special treatment, extraction, or analytical procedure	Ref.
Ca, K, Mg, Na, Rb, Sr	Direct aspiration	169
Cd, Co, Cu, Ni, Zn	Extract with chelating resin	122
Cd, Cu, Fe, Mn, Pb, Zn	Extract with APDC/MIBK	170
Cd	Coprecipitate with $SrCO_3$; extract with DEDTC/MIBK	171
Co, Cu, Fe, Ni, Pb, Zn	Extract with APDC/MIBK	172
Cr	Oxidize with $KMnO_4$; extract with APDC/MIBK	173
Cr	Vitreous carbon rod atomizer	174
Cu, Fe	Extract with APDC/MIBK; graphite furnace atomizer	175
Cu	Extract with APDC/ethyl acetate	118
Cu	Extract on chitosan column; elute with 1,10-phenanthroline; graphite furnace atomizer	176
Cu	Concentrate by deposition on Hg drop; graphite atomizer	61
Fe, Mn	Extract with DEDTC/MIBK	177
Fe	Graphite atomizer	178
Hg	Solution reduction; amalgamate on gold	179
Hg	Collect ionic Hg on adsorption colloid; solution reduction	180
K, Li, Mg, Rb, Sr	Direct aspiration	181
Li, Rb, Sr	Direct aspiration	182
Li	Direct aspiration	183
Mo	Extract on chitosan or p-aminobenzylcellulose columns; graphite furnace atomizer	184
Ni	Extract with dimethylglyoxime/chloroform	185
V	Extract on chitosan column; homogenized column material atomized in graphite furnace	186

If seawater is aspirated directly, it is desirable to employ a burner specifically designed for solutions with a high dissolved-solid content, otherwise salts deposit in the burner slot and change the flame characteristics. Standards should be prepared with a matching salt content, for example, by making use of an artificial seawater preparation.[187] Nonatomic absorption should be checked, particularly at wavelengths below 300 nm, and corrections made if necessary.

Except in a few cases where sample dilution may be possible, the trace analysis of samples containing very high dissolved-salt concentrations, such as oil-field brines and solutions from sylvite and rock-salt specimens, involves a preliminary extraction. The analysis of oil-field brines is of particular interest because of the possibility of obtaining information on the origin of the brine and thereby helping to predict the location of oil. Procedures suitable for determining various elements in brines are indicated in Table 10.8.

APPLICATION TO THE ANALYSIS OF GEOLOGICAL MATERIALS 195

TABLE 10.8 Atomic Absorption Analysis of Brines

Elements determined	Procedure	Ref.
Cd, Co, Cu, Fe, Mn, Ni, Pb	Extract with APDC/MIBK	117
Co, Cr, Mo, Ni, V	Extract with APDC/MIBK	119
Co, Cu, Fe, Mn, Ni, Pb, Zn	Extract on chelating resin	188
Co, Cu, Mo, Ni	Extract with oxine/chloroform	119
Cr	Extract with diphenylthiocarbazone/MIBK	189
Cu, Fe, Mn, Ni	Extract with cupferron/MIBK	189
Mo	Extract dithiol complex into MIBK	189
Sr	Dilute	190
V	Extract with cupferron/MIBK	191

10.5.2 Analysis of Solid Geological Samples

Although some semiquantitative work has been done by flame atomization of finely powdered solids from slurries or combustible pellets, or by direct introduction of solids into electrothermal atomizers, almost all quantitative analyses of solid geological materials by atomic absorption involve an initial chemical attack on the sample. The procedures vary according to the nature of the sample. Several of the most commonly used procedures for dealing with some of the more important types of sample are noted below. Many of the general methods indicated here are equally applicable to the determination of major and trace constituents; in the case of major elements the sample solutions are usually diluted so that the atomic absorption measurements are made on solutions with "trace" concentrations of these elements.

Early applications of atomic absorption to mining and geochemical samples have been reviewed by Slavin,[192] and some methods appropriate for geochemical exploration have been given in detail by Ward et al.[193] A number of procedures for silicate rocks and minerals, sulfide minerals, quartz, limestones, and recent sediments, have been set out by Angino and Billings, who reviewed the literature through 1966, and have added further material in a revised edition of their book.[10] More recent work is discussed in later general books on atomic absorption.[17,20,21]

A selection of papers dealing with the determination of many elements in solid geological materials is given in Table 10.9. More extensive bibliographies of recent atomic absorption work are given by Slavin,[194] Winefordner and Vickers,[64] and Hieftje et al.[65]

Siliceous Samples

For some geochemical exploration work, attack with boiling nitric acid is suitable.[193] The use of mixtures of nitric acid with hydrofluoric acid,[247]

TABLE 10.9 Atomic Absorption Analysis of Solid Geological Materials

Elements	Material	Ref.
Ag	Lead concentrates	195
	Ores	196–199
	Platiniferous materials	200
	Quartz	201
	Rocks, soils, sediments	202–206
	Sulfides	120, 207–209
Al	Bauxite	112, 210
	Limestones	211
	Rocks	212, 213
	Silica, silicates	214–219
	Soils and extracts	220, 221
	Standard rocks	222, 223
	Sulfides	224
As	Coal	225
Au	Ores	196, 226–228
	Platiniferous materials	200
	Quartz	201
	Rocks	204, 229
	Sulfides	120
Ba	Silicates	112
Bi	Ores, rocks, soils	230, 231
Ca	Bauxite	210
	Coal ash	232
	Limestones	211, 233
	Silica, silicates	215–219, 233–235
	Soils and extracts	96, 220
	Standard rocks	222, 223, 236
	Sulfides	224
Cd	Ores	197, 237, 238
	Rocks	239
	Silicates	240, 241
	Sulfides	208
Co	Silicates	234, 241
	Soil extracts	242
	Standard rocks	92
Cr	Bauxite	210
	Ilmenite	243
	Rocks, soils	206
	Silicates	215, 244
	Standard rocks	223, 245
Cs	Silicates	246
Cu	Lavas	76
	Ores	192, 247, 248
	Platiniferous materials	200
	Quartz	201
	Rocks, soils	206
	Silicates	233, 241, 244, 249

TABLE 10.9 (*continued*)

Elements	Material	Ref.
Fe	Soils and extracts	114
	Standard rocks	92
	Sulfides	208, 224
	Bauxite	210
	Coal ash	232
	Feldspar	250
	Limestones	211, 233
	Obsidian	251
	Platiniferous materials	200
	Quartz	201
	Rocks	213
	Silica, silicates	215–219, 233, 234, 241
	Soils and extracts	220, 252
	Standard rocks	222, 223, 236
	Sulfides	224
Ga	Limonite	253
	Ores	254
Ge	Limonite	253
Hg	Coal	68, 255–259
	Rocks, soils, sediments, and minerals	68, 260–270
K	Bauxite	210
	Coal ash	232
	Obsidian	251
	Rocks	271
	Silicates	215, 217–219, 234
	Soils and extracts	96, 220
	Standard rocks	223, 236
	Sulfides	224
Li	Rocks	272
	Silicates	244, 246, 273
Mg	Bauxite	210
	Coal ash	232
	Gypsum, limestone	233, 274
	Silica and silicates	215–219, 233, 235, 238, 275
	Soils and extracts	96, 220, 276
	Standard rocks	222, 223, 236
	Sulfides	224
Mn	Bauxite	210
	Limestones	233
	Obsidian	251
	Rocks	213
	Silicates	215, 218, 233
	Soils and extracts	252, 277
	Standard rocks	222, 223, 236
	Sulfides	224

TABLE 10.9 (*continued*)

Element	Materials	Ref.
Mo	Rocks, sediments, soils	164, 278, 279
	Siliceous ores	280, 281
Na	Bauxite	210
	Coal ash	232
	Limestone	282
	Obsidian	251
	Rocks	271
	Silicates	215, 217–219, 234
	Soils and extracts	96, 220
	Standard rocks	223, 236
	Sulfides	224
Nb	Ores, rocks	283
Nd	Phosphate rock	284
Ni	Lavas	76
	Ores	192
	Platiniferous materials	200
	Silicates	241, 244
	Standard rocks	245
Pb	Ores	192, 285
	Rocks, soils	206
	Silicates	241, 244, 286, 287
	Sulfides	208, 224
Pt	Basic rocks	288, 289
	Platiniferous materials	200
Rare earths	Minerals	290
Rb	Feldspar	250
	Rocks	234, 291, 292
	Silicates	246
	Soils	293
Re	Ores, concentrates	294
Rh	Chromite concentrates	295
	Platinum concentrates	296
Sb	Galena	297
	Ores, rocks, soils	298–300
	Sulfides	301
Se	Galena	302
	Rocks, soils	303, 304
Si	Bauxite	112, 210
	Limestones	211
	Silicates	215, 217–219
	Sulfides	224
Sn	Cassiterite	305, 306
	Ores, rocks	307
	Silicates	308
	Sulfides	308

TABLE 10.9 (continued)

Element	Materials	Ref.
Sr	Coal ash	309
	Phosphate rock	310
	Rocks	233, 311
	Silicates	233, 234, 244
	Soils	99
	Standard rocks	223, 245
Te	Copper ores	312
	Rocks, soils	303
Ti	Bauxite	112, 210
	Rocks	213
	Silica, silicates	215, 216, 218, 219, 313
	Standard rocks	222, 223
	Sulfides	224
Tl	Silicates	314
	Sulfides	301
V	Copper ores	315
	Silicates	313, 316
W	Ores, concentrates	317
	Silicates	318, 319
Y	Phosphate rock	284
Zn	Lavas, obsidian	76, 251
	Ores	192, 197, 248
	Quartz	201
	Rocks, soils	206, 320
	Silicates	244, 249
	Sulfides	208, 224

perchloric acid,[192] or both[248] in ore and concentrate analysis has also been described. Acid mixtures based on hydrofluoric acid are used where complete extraction of trace elements from siliceous material is required. Silicate decomposition prior to atomic absorption analysis has been studied extensively by Langmyhr et al.[215,321] If silicon itself is not being determined, it is possible to use an unsealed vessel for digestion by hydrofluoric acid[215] or a hydrofluoric/nitric acid mixture,[322] the complete removal of silicon and fluoride being facilitated by subsequent treatment with perchloric or sulfuric acid.[214,236,323] Sulfuric acid should not be used where sulfate interference may occur, or in the determination of elements with insoluble sulfates, such as barium.[112] The residue from acid digestion is usually taken up in hydrochloric acid.

Fusions have been carried out with carbonate fluxes[112,236] or with lithium fluoride/boric acid,[324] but lithium metaborate is more commonly used.[112,218,219,325,326] Standard solutions are prepared with similar concentrations of lithium borate and nitric acid to those of the samples. A potential disadvantage of the fusion methods is the relatively high dissolved-solids content of the resulting solutions: the possibilities of aspiration problems and molecular absorption need to be borne in mind. A lower concentration of added metal ions results from the use of a 10:1 mixture of boric oxide and lithium carbonate.[327]

Schemes of silicate analysis based on a lithium metaborate fusion followed by digestion with hydrofluoric acid, have been described by Abbey et al.[328,329] Excess fluoride in the solution is complexed by the addition of boric acid.

Sulfide Minerals and Ores

The decomposition of sulfide minerals and ores can be carried out with nitric and/or sulfuric acids[207] or with a mixture of hydrofluoric, nitric, and hydrochloric acids in a teflon bomb.[224] The latter procedure ensures the complete digestion of both sulfide species and gangue. Boric acid is added to the solutions before the analysis is performed.

Limestone

Analysis of high-purity limestones requires only initial dissolution in dilute hydrochloric acid. However, if a total analysis of calcareous material is needed, the samples should be ignited at 1100°C and treated with HF/H_2SO_4.[233] Alternatively, initial dissolution in hydrochloric acid can be followed by ignition of the residue and treatment with $HF/HClO_4$, the final solution being made up with HCl.[282] The sequence, ignition, HCl digestion, and HF treatment has also been followed for siliceous limestones.[211]

Miscellaneous Ores

Although many ores can be attacked satisfactorily by one of the techniques described above, there are some cases where special methods of treatment must be devised. The determination of tin in cassiterite and other ores, for example, has been carried out by grinding the sample with ammonium iodide and heating the mixture to 500°C, vaporizing SnI_4 which is collected in HCl.[305] A similar method has been used for antimony in geological materials.[298,300]

Gallium and germanium have been determined in limonite[253] by treatment of the samples with titanous chloride and HCl, followed by extraction of the gallium and germanium into MIBK. Osmiridium has been analyzed for a number of precious metals by a method involving dry chlorination of the samples at 700°C in the presence of sodium chloride.[200]

APPLICATIONS TO THE ANALYSIS OF GEOLOGICAL MATERIALS 201

Coal Ash

Many trace elements can be determined in coal ash by using one of the standard treatments for silicate rocks, either HF/HClO$_4$ digestion[232,309] or lithium metaborate fusion.[219] A procedure for trace analysis of 10 elements in coal, involving wet digestion with HNO$_3$ and then HF in a teflon bomb, has also been described.[330]

Recent Sediments

Detailed procedures for determining a wide variety of elements (Na, K, Mg, Ca, Co, Fe, Mn, Mo, Ni, Zn) in various fractions of recent sediments are described by Angino and Billings.[10] The methods mainly employ HF/H$_2$SO$_4$ digestion, with hydrogen peroxide being used to aid in the oxidation of organic matter. Although HNO$_3$/H$_2$SO$_4$ mixtures have been used in sediment digestion for the determination of heavy metals, the use of HNO$_3$/HCl appears to be more satisfactory, particularly for lead.[331]

Soils

In the analysis of soils, either the total amount of a trace element or the amount extractable in one of a number of standard extracting solutions may be required. Determination of the total content of an element involves preliminary drying and ignition to destroy organic matter, followed by digestion with an acid mixture similar to one of those used for silicate rocks, such as HF/HNO$_3$/HClO$_4$.

More frequently, in soil analysis the amount of an element extractable in a buffer or chelating agent is sought in the hope that the fraction of the element so extracted will have some agricultural or biogeochemical significance. The various buffers used to release exchangeable ions from soils are described elsewhere.[332] Exchangeable calcium and magnesium have been determined following extraction with 1 M ammonium chloride[96] or 0.5 M acetic acid.[333] Sodium and potassium were also extracted with ammonium chloride.[96] An oxalate buffer has been used to determine free silicon,[334] iron, and aluminum.[335] Cobalt has been extracted with 0.5 M acetic acid[242] and 1 M ammonium acetate has been used for nickel.[336] The extraction of manganese from soil has been studied with four different extractants.[337] Copper, zinc, and manganese have been extracted with an EDTA solution.[338]

10.5.3 Analysis of Crude Oils

The determination of trace metals in crude oils and refinery feed stocks is important because some metals that are often present, such as nickel, vanadium, copper, and iron, are capable of poisoning the catalysts used in refining processes. Sources of crude oil have also been characterized by the relative concentrations of several trace elements, particularly nickel and vanadium, and the information has been useful in identifying sources of oil pollution.

To give a solution suitable for aspiration into a flame, it is necessary to dilute the oil with an organic solvent such as p-xylene, n-heptane, dioxan, or MIBK. Standards are prepared by dissolving suitable organometallic compounds in a metal-free oil base, and diluting them in the same way as the samples. A mixed solvent system, however, has enabled inorganic compounds to be used as standards[339] in the determination of lubricating-oil additives. The labor of sample dilution can be avoided and greater sensitivity can be obtained by the use of electrothermal atomizers.

Elements that have been determined in crude oils by atomic absorption include Ag,[340] Cu,[340–342] Fe,[341] Ni,[340,343–346] and V.[112,345,347,348]

10.5.4 Current and Future Developments

Atomic absorption is now firmly established as one of the most important techniques for precise determination of many elements in solution at concentrations of 0.1 to 100 μg/ml. Developments in electrothermal atomization and gas-liberation methods have extended the accessible concentration ranges to even lower levels, and subnanogram amounts of many elements can now be determined. Many of these developments can be followed in recent reviews.[64,65,349]

Atomic absorption is particularly suitable where a large number of solutions must be analyzed for a single element. There is some loss of convenience in the analysis of solid samples, which must first be taken into solution, and in the sequential determination of many elements in one solution. Approaches to simultaneous multielement analysis by atomic spectroscopy have been reviewed by Busch and Morrison[350] and by Winefordner et al.[351] For fast multielement analysis, systems based on atomic emission or atomic fluorescence appear to be more promising; in other situations, atomic absorption is likely to retain its importance in geochemical trace analysis for many years.

References

1. T. T. Woodson, *Rev. Sci. Instr.*, **10**, 308 (1939).
2. A. E. Ballard and C. D. W. Thornton, *Ind. Eng. Chem. Anal. Ed.*, **13**, 893 (1941).
3. C. W. Zuehlke and A. E. Ballard, *Anal. Chem.*, **22**, 953 (1950).
4. A. Walsh, *Spectrochim. Acta*, **7**, 108 (1955).
5. C. T. J. Alkemade and J. M. W. Milatz, *Appl. Sci. Res.*, **B4**, 289 (1955).
6. C. T. J. Alkemade and J. M. W. Milatz, *J. Opt. Soc. Am.*, **45**, 583 (1955).

7. J. E. Allan, *Analyst*, **83**, 466 (1958).
8. D. J. David, *Analyst*, **83**, 655 (1958).
9. A. Walsh, *Anal. Chem.*, **46**, 698A (1974).
10. E. E. Angino and G. K. Billings, *Atomic Absorption Spectrometry in Geology*, 2nd ed., Elsevier, Amsterdam, 1972.
11. W. T. Elwell and J. A. F. Gidley, *Atomic Absorption Spectrophotometry*, 2nd ed., Pergamon, Oxford, 1966.
12. J. W. Robinson, *Atomic Absorption Spectroscopy*, Arnold, London, 1966.
13. I. Rubeska and B. Moldan, *Atomic Absorption Spectrophotometry*, SNTL, Prague. 1967; English Edition, Iliffe, London, 1969.
14. W. Slavin, *Atomic Absorption Spectroscopy*, Interscience, New York, 1968.
15. J. Ramirez-Munoz, *Atomic Absorption Spectroscopy*, Elsevier, Amsterdam, 1968.
16. J. A. Dean and T. C. Rains (eds.), *Flame Emission and Atomic Absorption Spectrometry*, Vol. 1, Theory, Dekker, New York, 1969.
17. R. J. Reynolds, K. Aldous, and K. C. Thompson, *Atomic Absorption Spectroscopy*, Griffin, London, 1970.
18. B. V. L'vov, *Atomic Absorption Spectrochemical Analysis* (transl. J. H. Dixon), Adam Hilger Ltd., London, 1970.
19. J. B. Willis, in *Analytical Flame Spectroscopy* (ed. R. Mavrodineanu), Macmillan, London, 1970.
20. W. J. Price, *Analytical Atomic Absorption Spectrometry*, Heyden, London, 1972.
21. G. F. Kirkbright and M. Sargent, *Atomic Absorption and Fluorescence Spectroscopy*, Academic Press, London, 1974.
22. H. E. White, *Introduction to Atomic Spectra*, McGraw-Hill, New York, 1934.
23. G. Herzberg, *Atomic Spectra and Atomic Structure*, 2nd ed., Dover, New York, 1944.
24. H. G. Kuhn, *Atomic Spectra*, Longmans Green, London, 1962.
25. C. T. J. Alkemade, *Appl. Opt.*, **7**, 1261 (1968).
26. J. D. Winefordner and T. J. Vickers, *Anal. Chem.*, **36**, 1939 (1964).
27. J. D. Winefordner and T. J. Vickers, *Anal. Chem.*, **36**, 1947 (1964).
28. M. L. Parsons, W. J. McCarthy, and J. D. Winefordner, *Appl. Spectrosc.*, **20**, 223 (1966).
29. P. J. T. Zeegers, R. Smith, and J. D. Winefordner, *Anal. Chem.*, **40** (13), 26A (1968).
30. J. V. Sullivan and A. Walsh, *Spectrochim. Acta*, **21**, 721 (1965).
31. J. B. Dawson and D. J. Ellis, *Spectrochim. Acta*, **23A**, 565 (1967).
32. E. Cordos and H. V. Malmstadt, *Anal. Chem.*, **45**, 27 (1973).
33. R. M. Dagnall, K. C. Thompson, and T. S. West, *Talanta*, **14**, 551 (1967).
34. J. M. Mansfield, M. P. Bratzel, H. O. Norgordon, D. O. Knapp, K. E. Zacha, and J. D. Winefordner, *Spectrochim. Acta*, **23B**, 389 (1968).

35. M. D. Silvester and W. J. McCarthy, *Spectrochim. Acta*, **25B**, 229 (1970).
36. R. F. Browner, B. M. Patel, T. H. Glenn, M. E. Rietta, and J. D. Winefordner, *Spectrosc. Lett.*, **5**, 311 (1972).
37. B. M. Patel, R. F. Browner, and J. D. Winefordner, *Anal. Chem.*, **44**, 2272 (1972).
38. R. F. Browner and J. D. Winefordner, *Spectrochim. Acta*, **28B**, 263 (1973).
39. J. P. S. Haarsma, G. J. de Jong, and J. Agterdenbos, *Spectrochim. Acta*, **29B**, 1 (1974).
40. R. Mavrodineanu and H. Boiteux, *Flame Spectroscopy*, Wiley, New York, 1965.
41. M. D. Amos et al., *Basic Atomic Absorption Spectroscopy*, Varian Techtron Pty. Ltd., Springvale, Australia, 1975.
42. L. R. P. Butler and A. Fulton, *Appl. Opt.*, **7**, 2131 (1968).
43. G. F. Kirkbright, M. K. Peters, M. Sargent, and T. S. West, *Talanta*, **15**, 663 (1968).
44. J. B. Willis, J. O. Rasmuson, R. N. Kniseley, and V. A. Fassel, *Spectrochim. Acta*, **23B**, 725 (1968); **24B**, 155 (1969).
45. H. L. Kahn and J. E. Schallis, *Perkin-Elmer At. Absorpt. Newsl.*, **7**, 5 (1968).
46. P. Johns, *Spectrovision*, **24**, 6 (1970); **26**, 15 (1971).
47. E. F. Dalton and A. J. Malanoski, *Perkin-Elmer At. Absorpt. Newsl.* **10**, 92 (1971).
48. K. C. Thompson and D. R. Thomerson, *Analyst*, **99**, 595 (1974).
49. F. J. Fernandez, *Perkin-Elmer At. Absorpt. Newsl.* **12**, 93 (1973).
50. F. J. Schmidt, J. L. Royer, and S. M. Muir, *Anal. Lett.*, **8**, 123 (1975).
51. A. E. Smith, *Analyst*, **100**, 300 (1975).
52. G. F. Kirkbright and T. S. West, *Appl. Opt.*, **7**, 1305 (1968).
53. G. F. Kirkbright, A. Semb, and T. S. West, *Talanta*, **15**, 441 (1968).
54. J. E. Allan, *Spectrochim. Acta*, **18**, 605 (1962).
55. B. V. L'vov, *Spectrochim. Acta*, **17**, 761 (1961).
56. B. V. L'vov, *Spectrochim. Acta*, **24B**, 53 (1969).
57. M. P. Bratzel, R. M. Dagnall, and J. D. Winefordner, *Anal. Chim. Acta*, **48**, 197 (1969).
58. H. M. Donega and T. E. Burgess, *Anal. Chem.*, **42**, 1521 (1970).
59. T. S. West and X. K. Williams, *Anal. Chim. Acta*, **45**, 27 (1969).
60. M. D. Amos, P. A. Bennett, K. G. Brodie, P. W. Y. Lung, and J. P. Matousek, *Anal. Chem.*, **43**, 211 (1971).
61. C. Fairless and A. J. Bard, *Anal. Chem.*, **45**, 2289 (1973).
62. W. Lund and B. V. Larsen, *Anal. Chim. Acta*, **70**, 299 (1974).
63. G. F. Kirkbright, *Analyst*, **96**, 609 (1971).
64. J. D. Winefordner and T. J. Vickers, *Anal. Chem.*, **46**, 192R (1974).
65. G. M. Hieftje, T. R. Copeland, and D. R. de Olivares, *Anal. Chem.*, **48**, 142R (1976).
66. H. Brandenberger and H. Bader, *Perkin-Elmer At. Absorpt. Newsl.*, **6**, 101 (1967).

REFERENCES

67. M. J. Fishman, *Anal. Chem.*, **42**, 1462 (1970).
68. O. I. Joensuu, *Appl. Spectrosc.*, **25**, 526 (1971).
69. A. M. Ure, *Anal. Chim. Acta*, **76**, 1 (1975).
70. F. J. Langmyhr, J. R. Stubergh, Y. Thomassen, J. E. Hanssen and J. Dolezal, *Anal. Chim. Acta*, **71**, 35 (1974).
71. J. D. Winefordner, *Pure Appl. Chem.*, **23**, 35 (1970).
72. J. W. Robinson, *Anal. Chim. Acta*, **23**, 479 (1960).
73. J. W. Robinson, *Anal. Chem.*, **33**, 1067 (1961).
74. J. E. Allan, *Spectrochim. Acta*, **17**, 467 (1961).
75. R. Lockyer, J. E. Scott, and S. Slade, *Nature*, **189**, 830 (1961).
76. G. Perrault, *Can. Spectrosc.*, **11**, 19 (1966).
77. T. Maruta, M. Suzuki, and T. Takeuchi, *Anal. Chim. Acta*, **51**, 381 (1970).
78. J. B. Willis, *Nature*, **207**, 715 (1965).
79. M. D. Amos and J. B. Willis, *Spectrochim. Acta*, **22**, 1325 (1966).
80. M. D. Amos and J. B. Willis, *Spectrochim. Acta*, **22**, 2128 (1966).
81. J. E. Allan, *Spectrochim. Acta*, **24B**, 13 (1969).
82. D. C. Manning and F. Fernandez, *Perkin-Elmer At. Absorpt. Newsl.*, **7**, 24 (1968).
83. V. A. Fassel, J. O. Rasmuson, and T. G. Cowley, *Spectrochim. Acta*, **23B**, 579 (1968).
84. R. J. Jaworowski and R. P. Weberling, *Perkin-Elmer At. Absorpt. Newsl.*, **5**, 125 (1966).
85. S. R. Koirtyohann and E. E. Pickett, *Anal. Chem.*, **37**, 601 (1965).
86. S. R. Koirtyohann and E. E. Pickett, *Anal. Chem.*, **38**, 585 (1966).
87. B. R. Culver and T. Surles, *Anal. Chem.*, **47**, 920 (1975).
88. M. W. Pritchard and R. D. Reeves, *Anal. Chim. Acta*, **82**, 103 (1976).
89. J. B. Willis, *Methods of Biochemical Analysis*, Vol. 11 (ed. D. Glick), Interscience, New York, 1963.
90. H. L. Kahn, *Perkin-Elmer At. Absorpt. Newsl.*, **7**, 40 (1968).
91. G. K. Billings, *Perkin-Elmer At. Absorpt. Newsl.*, **4**, 357 (1965).
92. K. Fletcher, *Econ. Geol.*, **65**, 588 (1970).
93. C. T. J. Alkemade, *Anal. Chem.*, **38**, 1252 (1966).
94. J. D. Winefordner and T. J. Vickers, *Anal. Chem.*, **44**, 150R (1972).
95. D. J. David, *Analyst*, **84**, 536 (1959).
96. D. J. David, *Analyst*, **85**, 495 (1960).
97. J. B. Willis, *Spectrochim. Acta*, **16**, 259 (1960).
98. J. B. Willis, *Anal. Chem.*, **33**, 556 (1961).
99. D. J. David, *Analyst*, **87**, 576 (1962).
100. W. Slavin, S. Sprague, and D. C. Manning, *Perkin-Elmer At. Absorpt. Newsl.*, **2**, 49 (1963).

101. R. E. Dickson and C. M. Johnson, *Appl. Spectrosc.*, **20**, 214 (1966).
102. A. C. Menzies, *Anal. Chem.*, **32**, 898 (1960).
103. I. Rubeska and B. Moldan, *Anal. Chim. Acta*, **37**, 421 (1967).
104. A. M. Bond and T. A. O'Donnell, *Anal. Chem.*, **40**, 560 (1968).
105. V. S. Sastri, C. L. Chakrabarti, and D. E. Willis, *Can. J. Chem.*, **47**, 587 (1969).
106. V. S. Sastri, C. L. Chakrabarti, and D. E. Willis, *Talanta*, **16**, 1093 (1969).
107. P. E. Thomas and W. F. Pickering, *Talanta*, **18**, 127 (1971).
108. J. T. H. Roos and W. J. Price, *Spectrochim. Acta*, **26B**, 441 (1971).
109. J. Y. Marks and G. G. Welcher, *Anal. Chem.*, **42**, 1033 (1970).
110. S. R. Koirtyohann and E. E. Pickett, *Anal. Chem.*, **40**, 2068 (1968).
111. A. C. West, V. A. Fassel, and R. N. Kniseley, *Anal. Chem.*, **45**, 2420 (1973).
112. J. A. Bowman and J. B. Willis, *Anal. Chem.*, **39**, 1210 (1967).
113. D. C. Manning and L. Capacho-Delgado, *Anal. Chim. Acta*, **36**, 312 (1966).
114. J. E. Allan, *Spectrochim. Acta*, **17**, 459 (1961).
115. J. B. Willis, *Nature*, **191**, 381 (1961).
116. J. B. Willis, *Anal. Chem.*, **34**, 614 (1962).
117. S. Sprague and W. Slavin, *Perkin-Elmer At. Absorpt. Newsl.*, **3**, 37 (1964).
118. R. J. Magee and A. K. M. Rahman, *Talanta*, **12**, 409 (1965).
119. R. E. Mansell and H. W. Emmell, *Perkin-Elmer At. Absorpt. Newsl.*, **4**, 365 (1965).
120. M. C. Greaves, *Nature*, **199**, 552 (1963).
121. D. G. Biechler, *Anal. Chem.*, **37**, 1054 (1965).
122. J. P. Riley and D. Taylor, *Anal. Chim. Acta*, **40**, 479 (1968).
123. U. Westerlund-Helmerson, *Perkin-Elmer At. Absorpt. Newsl.*, **5**, 97 (1966).
124. J. A. Varley and P. Y. Chin, *Analyst*, **95**, 592 (1970).
125. W. S. Zaugg and R. J. Knox, *Anal. Chem.*, **38**, 1759 (1966).
126. T. V. Ramakrishna, J. W. Robinson, and P. W. West, *Anal. Chim. Acta*, **45**, 43 (1969).
127. G. F. Kirkbright, A. M. Smith, T. S. West, and R. Wood, *Analyst*, **94**, 754 (1969).
128. H. N. Johnson, G. F. Kirkbright, and T. S. West, *Proc. Soc. Anal. Chem.*, **9**, 300 (1972).
129. R. Kunkel and S. E. Manahan, *Anal. Chem.*, **45**, 1465 (1973).
130. T. Kumamaru, *Bull. Chem. Soc. Jap.*, **42**, 956 (1969).
131. R. S. Danchik and D. F. Boltz, *Anal. Chim. Acta*, **49**, 567 (1970).
132. R. W. Looyenga and C. O. Huber, *Anal. Chem.*, **43**, 498 (1971).
133. R. W. Looyenga and C. O. Huber, *Anal. Chim. Acta*, **55**, 179 (1971).
134. C. I. Lin and C. O. Huber, *Anal. Chem.*, **44**, 2200 (1972).
135. W. E. Crawford, C. I. Lin, and C. O. Huber, *Anal. Chim. Acta*, **64**, 387 (1973).
136. A. M. Bond and J. B. Willis, *Anal. Chem.*, **40**, 2087 (1968).

137. M. Pinta, *Method. Phys. Anal.*, **6**, 268 (1970).
138. W. M. Barnard and M. J. Fishman, *Perkin-Elmer At. Absorpt. Newsl.*, **12**, 118 (1973).
139. W. M. Edmunds, D. R. Giddings, and M. Morgan-Jones, *Perkin-Elmer At. Absorpt. Newsl.*, **12**, 45 (1973).
140. U. S. Environmental Protection Agency, *Methods of Chemical Analysis of Waters and Wastes*, Cincinnati, Ohio, 1971.
141. R. D. Ediger, *Perkin-Elmer At. Absorpt. Newsl.*, **12**, 151 (1973).
142. E. A. Boettner and F. I. Grunder, in *Trace Inorganics in Water: Advances in Chemistry Series* No. 73, Am. Chem. Soc., Washington, D. C., 1968.
143. S. L. Sachdev and P. W. West, *Environ. Sci. Technol.*, **4**, 749 (1970).
144. T. T. Chao, M. J. Fishman, and J. W. Ball, *Anal. Chim. Acta*, **47**, 189 (1969).
145. T. T. Chao and J. W. Ball, *Anal. Chim. Acta*, **54**, 166 (1971).
146. B. Welz and E. Wiedeking, *Z. Anal. Chem.*, **264**, 110 (1973).
147. D. Y. Hsu and W. O. Pipes, *Environ. Sci. Technol.*, **6**, 645 (1972).
148. M. Fishman, *Perkin-Elmer At. Absorpt. Newsl.*, **11**, 46 (1972).
149. P. D. Goulden and P. Brooksbank, *Anal. Chem.*, **46**, 1431 (1974).
150. A. Zlatkis, W. Bruening, and E. Bayer, *Anal. Chem.*, **41**, 1692 (1969).
151. L. R. P. Butler and D. Brink, *S. Afr. Ind. Chem.*, **17**, 152 (1963).
152. J. A. Platte and V. M. Marcy, *Perkin-Elmer At. Absorpt. Newsl.*, **4**, 289 (1965).
153. A. L. Wilson, *Chem. Ind.*, **36**, 1253 (1969).
154. M. J. Fishman and S. C. Downs, U. S. Geol. Surv. Water Supply Paper 1540-C (1966).
155. J. A. Platte, in *Trace Inorganics in Water: Advances in Chemistry Series*, No. 73, Am. Chem. Soc., Washington, D.C., 1968.
156. J. W. Robinson, D. K. Wolcott, P. J. Slevin, and G. D. Hindman, *Anal. Chim. Acta*, **66**, 13 (1973).
157. S. D. Soman, V. K. Panday, and K. T. Joseph, *Am. Ind. Hyg. Assoc. J.*, **30**, 527 (1969).
158. J. Nix and T. Goodwin, *Perkin-Elmer At. Absorpt. Newsl.*, **9**, 119 (1970).
159. M. J. Fishman and M. R. Midgett, in *Trace Inorganics in Water: Advances in Chemistry Series*, No. 73, Am. Chem. Soc., Washington, D. C., 1968.
160. D. C. Burrell, *Perkin-Elmer At. Absorpt. Newsl.*, **4**, 309 (1965).
161. M. R. Midgett and M. J. Fishman, *Perkin-Elmer At. Absorpt. Newsl.*, **6**, 128 (1967).
162. S. H. Omang, *Anal. Chim. Acta*, **53**, 415 (1971).
163. M. A. Mazurski, A. Chow, and A. D. Gesser, *Anal. Chim. Acta*, **65**, 99 (1973).
164. L. R. P. Butler and P. M. Matthews, *Anal. Chim. Acta*, **36**, 319 (1966).
165. Y. K. Chau and K. Lum-Shue-Chan, *Anal. Chim. Acta*, **48**, 205 (1969).
166. R. B. Baird, S. Pourian, and S. Gabrielian, *Anal. Chem.*, **44**, 1887 (1972).

167. Y. K. Chau and K. Lum-Shue-Chan, *Anal. Chim. Acta*, **50**, 201 (1970).
168. B. P. Fabricand, R. R. Sawyer, S. G. Ungar, and S. Adler, *Geochim. Cosmochim. Acta*, **26**, 1023 (1962).
169. G. K. Billings and R. C. Harriss, *Texas J. Sci.*, **17**, 129 (1965).
170. P. E. Paus, *Z. Anal. Chem.*, **264**, 118 (1973).
171. T. Owa, K. Hiiro, and T. Tanaka, *Bunseki Kagaku*, **21**, 878 (1972).
172. R. R. Brooks, B. J. Presley, and I. R. Kaplan, *Talanta*, **14**, 809 (1967).
173. T. R. Gilbert and A. M. Clay, *Anal. Chim. Acta*, **67**, 289 (1973).
174. G. Tessari and G. Torsi, *Talanta*, **19**, 1059 (1972).
175. K. Kremling and H. Petersen, *Anal. Chim. Acta*, **70**, 35 (1974).
176. R. A. A. Muzzarelli and R. Rocchetti, *Anal. Chim. Acta*, **69**, 35 (1974).
177. T. Joyner and J. S. Finley, *Perkin-Elmer At. Absorpt. Newsl.*, **5**, 4 (1966).
178. D. A. Segar and J. G. Gonzalez, *Anal. Chim. Acta*, **58**, 7 (1972).
179. J. Olafsson, *Anal. Chim. Acta*, **68**, 207 (1974).
180. D. Voyce and H. Zeitlin, *Anal. Chim. Acta*, **69**, 27 (1974).
181. B. P. Fabricand, E. S. Imbibo, M. E. Brey, and J. A. Weston, *J. Geophys. Res.*, **71**, 3917 (1966).
182. D. C. Burrell, *Anal. Chim. Acta*, **38**, 447 (1967).
183. E. E. Angino and G. K. Billings, *Geochim. Cosmochim. Acta*, **30**, 153 (1966).
184. R. A. A. Muzzarelli and R. Rocchetti, *Anal. Chim. Acta*, **64**, 371 (1973).
185. H. Rampon and R. Cuvelier, *Anal. Chim. Acta*, **60**, 226 (1972).
186. R. A. A. Muzzarelli and R. Rocchetti, *Anal. Chim. Acta*, **70**, 283 (1974).
187. H. W. Harvey, *The Chemistry and Fertility of Sea Waters*, Cambridge University Press, London, 1960.
188. O. K. Galle, *Appl. Spectrosc.*, **25**, 664 (1971).
189. B. Delaughter, *Perkin-Elmer At. Absorpt. Newsl.*, **4**, 273 (1965).
190. G. K. Billings and E. E. Angino, *Bull. Can. Petrol. Geol.*, **13**, 529 (1965).
191. H. J. Crump-Wiesner, H. R. Feltz, and W. C. Purdy, *Anal. Chim. Acta*, **55**, 29 (1971).
192. W. Slavin, *Perkin-Elmer At. Absorpt. Newsl.*, **4**, 243 (1965).
193. F. N. Ward, H. M. Nakagawa, T. F. Harms, and G. H. VanSickle, *U. S. Geol. Surv. Bull.* **1289** (1969).
194. S. Slavin, *Perkin-Elmer At. Absorpt. Newsl.*, **11**, 37 (1972); **12**, 9, 77 (1973); **13**, 11, 84 (1974); S. Slavin and D. M. Lawrence, *Perkin-Elmer At. Absorpt. Newsl.*, **14**, 81 (1975).
195. B. S. Rawling, M. D. Amos, and M. C. Greaves, *Nature*, **188**, 137 (1960).
196. F. M. Tindall, *Perkin-Elmer At. Absorpt. Newsl.*, **4**, 399 (1965); **5**, 140 (1966).
197. M. Fixman and L. Boughton, *Perkin-Elmer At. Absorpt. Newsl.*, **5**, 33 (1966).
198. P. E. Moloughney and J. A. Graham, *Talanta*, **18**, 475 (1971).
199. G. Walton, *Analyst*, **98**, 335 (1973).

200. J. G. Sen Gupta, *Anal. Chim. Acta*, **58**, 23 (1972).
201. V. Price and P. C. Ragland, *Southeastern Geol.*, **7**, 93 (1966).
202. C. Huffman, J. D. Mensik, and L. F. Rader, U. S. Geol. Surv. Prof. Paper No. 550-B, B189-B191 (1966).
203. T. T. Chao, J. W. Ball, and H. M. Nakagawa, *Anal. Chim. Acta*, **54**, 77 (1971).
204. M. P. Bratzel, C. L. Chakrabarti, R. E. Sturgeon, M. W. McIntyre, and H. Agemian, *Anal. Chem.*, **44**, 372 (1972).
205. Y. I. Belyaev, A. M. Pchelintsev, and N. F. Zvereva, *Z. Anal. Khim.*, **26**, 1295 (1971).
206. C. Riandey and M. Pinta, *Analusis*, **2**, 179 (1973).
207. I. Rubeska, Z. Sulcek, and B. Moldan, *Anal. Chim. Acta*, **37**, 27 (1967).
208. I. Rubeska, *Anal. Chim. Acta*, **40**, 187 (1968).
209. W. K. Ng, *Anal. Chim. Acta*, **63**, 469 (1973).
210. F. J. Langmyhr and P. E. Paus, *Anal. Chim. Acta*, **43**, 508 (1968).
211. F. J. Langmyhr and P. E. Paus, *Anal. Chim. Acta*, **44**, 445 (1969).
212. P. E. Paus, *Anal. Chim. Acta*, **54**, 164 (1971).
213. H. Heinrichs and J. Lange, *Z. Anal. Chem.*, **265**, 256 (1973).
214. J. C. Van Loon, *Perkin-Elmer At. Absorpt. Newsl.*, **7**, 3 (1968).
215. F. J. Langmyhr and P. E. Paus, *Anal. Chim. Acta*, **43**, 397 (1968).
216. F. J. Langmyhr and P. E. Paus, *Anal. Chim. Acta*, **43**, 506 (1968).
217. F. J. Langmyhr and P. E. Paus, *Anal. Chim. Acta*, **45**, 176 (1969).
218. J. C. Van Loon and C. M. Parissis, *Analyst*, **94**, 1057 (1969).
219. P. L. Boar and L. K. Ingram, *Analyst*, **95**, 124 (1970).
220. S. Pawluk, *Perkin-Elmer At. Absorpt. Newsl.*, **6**, 53 (1967).
221. Y. LaFlamme, *Perkin-Elmer At. Absorpt. Newsl.*, **6**, 70 (1967).
222. O. K. Galle, *Appl. Spectrosc.*, **22**, 404 (1968).
223. J. H. Medlin, N. H. Suhr, and J. B. Bodkin, *Perkin-Elmer At. Absorpt. Newsl.*, **8**, 25 (1969).
224. F. J. Langmyhr and P. E. Paus, *Anal. Chim. Acta*, **50**, 515 (1970).
225. G. I. Spielholtz, G. C. Toralballa, and R. J. Steinberg, *Mikrochim. Acta*, 918 (1971).
226. A. M. Olson, *Perkin-Elmer At. Absorpt. Newsl.*, **4**, 278 (1965).
227. E. C. Simmons, *Perkin-Elmer At. Absorpt. Newsl.*, **4**, 281 (1965).
228. S. L. Law and T. E. Green, *Anal. Chem.*, **41**, 1008 (1969).
229. J. C. Antweiler and J. D. Love, *U. S. Geol. Surv. Circ.* 541 (1967).
230. F. N. Ward and H. M. Nakagawa, U. S. Geol. Surv. Prof. Paper 575-D, 239–241 (1967).
231. J. Husler, *Perkin-Elmer At. Absorpt. Newsl.*, **9**, 31 (1970).
232. L. L. Obermiller and R. W. Freedman, *Fuel*, **44**, 199 (1965).

233. D. J. Trent and W. Slavin, *Perkin-Elmer At. Absorpt. Newsl.*, **3**, 118 (1964).
234. G. K. Billings and J. A. S. Adams, *Perkin-Elmer At. Absorpt. Newsl.*, **3**, 65 (1964).
235. K. Govindaraju, *Appl. Spectrosc.*, **24**, 81 (1970).
236. D. J. Trent and W. Slavin, *Perkin-Elmer At. Absorpt. Newsl.*, **3**, 17 (1964).
237. N. S. Poluektov and R. A. Vitkun, *Z. Anal. Khim.*, **17**, 935 (1962).
238. B. Farrar, *Perkin-Elmer At. Absorpt. Newsl.*, **5**, 62 (1966).
239. Y. I. Belyaev, A. M. Pchelintsev, N. F. Zvereva, and B. I. Kosin, *Z. Anal. Khim.*, **26**, 492 (1971).
240. J. Yamada, C. Iida, and K. Yamasaki, *Bunseki Kagaku*, **19**, 1259 (1970).
241. C. Iida and K. Yamasaki, *Anal. Lett.*, **3**, 251 (1970).
242. A. M. Ure and R. L. Mitchell, *Spectrochim. Acta*, **23B**, 79 (1967).
243. P. T. O'Shaughnessy, *Anal. Chem.*, **45**, 1946 (1973).
244. A. N. Chowdhury and A. K. Das, *Z. Anal. Chem.*, **261**, 126 (1972).
245. L. Beccaluva and G. Venturelli, *Perkin-Elmer At. Absorpt. Newsl.*, **10**, 50 (1971).
246. F. W. E. Strelow, C. L. Liebenberg, and F. S. Toerien, *Anal. Chim. Acta*, **43**, 465 (1968).
247. A. Strasheim, F. W. E. Strelow, and L. R. P. Butler, *J. S. Afr. Chem. Inst.*, **13**, 73 (1960).
248. B. Farrar, *Perkin-Elmer At. Absorpt. Newsl.*, **4**, 325 (1965).
249. C. B. Belt, *Econ. Geol.*, **59**, 240 (1964).
250. G. K. Billings, P. C. Ragland, and J. A. S. Adams, *Nature*, **210**, 829 (1966).
251. G. C. Armitage, R. D. Reeves, and P. Bellwood, *N. Z. J. Sci.*, **15**, 408 (1972).
252. J. E. Allan, *Spectrochim. Acta*, **15**, 800 (1959).
253. E. N. Pollock, *Perkin-Elmer At. Absorpt. Newsl.*, **10**, 77 (1971).
254. G. N. Lypka and A. Chow, *Anal. Chim. Acta*, **60**, 65 (1972).
255. D. H. Anderson, J. H. Evans, J. J. Murphy, and W. W. White, *Anal. Chem.*, **43**, 1511 (1971).
256. B. W. Bailey and F. C. Lo, *Anal. Chem.*, **43**, 1525 (1971).
257. B. W. Bailey and F. C. Lo, *J. Ass. Off. Anal. Chem.*, **54**, 1447 (1971).
258. J. V. O'Gorman, N. H. Suhr, and P. L. Walker, *Appl. Spectrosc.*, **26**, 44 (1972).
259. T. C. Rains and O. Menis, *J. Ass. Off. Anal. Chem.*, **55**, 1339 (1972).
260. W. W. Vaughn and J. H. McCarthy, U. S. Geol. Surv. Prof. Paper 501-D, 123 (1964).
261. W. W. Vaughn, *U. S. Geol. Surv. Circ.* **540** (1967).
262. W. R. Hatch and W. L. Ott, *Anal. Chem.*, **40**, 2085 (1968).
263. S. H. Omang and P. E. Paus, *Anal. Chim. Acta*, **56**, 393 (1971).
264. J. A. Goleb, *Appl. Spectrosc.*, **25**, 522 (1971).
265. B. G. Weissberg, *Econ. Geol.*, **66**, 1042 (1971).
266. R. B. Brand and N. M. Wilkinson, *J. Geochem. Explor.*, **1**, 195 (1972).

267. S. R. Aston and J. P. Riley, *Anal. Chim. Acta*, **59**, 349 (1972).
268. F. D. Deitz, J. L. Sell, and D. Bristol, *J. Ass. Off. Anal. Chem.*, **56**, 378 (1973).
269. A. M. Ure and C. A. Shand, *Anal. Chim. Acta*, **72**, 63 (1974).
270. A. B. Carel, *Anal. Chim. Acta*, **78**, 479 (1975).
271. F. W. E. Strelow, F. S. Toerien, and C. H. S. W. Weinert, *Anal. Chim. Acta*, **50**, 399 (1970).
272. R. Ohrdorf, *Geochim. Cosmochim. Acta*, **32**, 191 (1968).
273. M. Stone and S. E. Chesher, *Analyst*, **94**, 1063 (1969).
274. W. Leithe and A. Hofer, *Mikrochim. Acta*, **2**, 268 (1961).
275. R. W. Nesbitt, *Anal. Chim. Acta*, **35**, 413 (1966).
276. T. R. Williams, B. Wilkinson, G. A. Wadsworth, D. H. Barter, and W. J. Beer, *J. Sci. Food Agr.*, **17**, 344 (1966).
277. L. R. Hossner and L. W. Ferrara, *Perkin-Elmer At. Absorpt. Newsl.*, **6**, 71 (1967).
278. P. D. Rao, *Perkin-Elmer At. Absorpt. Newsl.*, **10**, 118 (1971).
279. D. Hutchinson, *Analyst*, **97**, 118 (1972).
280. C. L. McIsaac, *Eng. Mining J.*, **170**, 55 (1969).
281. J. C. Van Loon, *Perkin-Elmer At. Absorpt. Newsl.*, **11**, 60 (1972).
282. I. Rubeska, B. Moldan, and Z. Valny, *Anal. Chim. Acta*, **29**, 206 (1963).
283. J. Husler, *Talanta*, **19**, 863 (1972).
284. J. Kinnunen and L. Lindsjo, *Chemist-Analyst*, **56**, 25, 76 (1967).
285. A. S. Bazhov, *Z. Anal. Khim.*, **23**, 1640 (1968).
286. C. Iida, T. Tanaka, and K. Yamasaki, *Bunseki Kagaku*, **15**, 1100 (1966).
287. B. Moldan, I. Rubeska, and M. Miksovsky, *Anal. Chim. Acta*, **50**, 342 (1970).
288. R. T. Swider, *Perkin-Elmer At. Absorpt. Newsl.*, **7**, 111 (1968).
289. A. Simonsen, *Anal. Chim. Acta*, **49**, 368 (1970).
290. J. C. Van Loon, J. H. Galbraith, and H. M. Aarden, *Analyst*, **96**, 47 (1971).
291. M. Vosters and S. Deutsch, *Earth Planet. Sci. Lett.*, **2**, 449 (1967).
292. I. Roelandts, *Perkin-Elmer At. Absorpt. Newsl.*, **11**, 48 (1972).
293. J. Stupar, *Z. Anal. Chem.*, **203**, 401 (1964).
294. E. V. Elliott, K. R. Stever, and H. H. Heady, *Perkin-Elmer At. Absorpt. Newsl.*, **13**, 113 (1974).
295. M. M. Schnepfe and F. S. Grimaldi, *Talanta*, **16**, 1461 (1969).
296. A. Hofer, *Z. Anal. Chem.*, **253**, 206 (1972).
297. W. K. Ng, *Anal. Chim. Acta*, **64**, 292 (1973).
298. D. J. Nicolas, *Anal. Chim. Acta*, **55**, 59 (1971).
299. V. Stresko and E. Martiny, *Perkin-Elmer At. Absorpt. Newsl.*, **11**, 4 (1972).
300. E. P. Welsch and T. T. Chao, *Anal. Chim. Acta*, **76**, 65 (1975).
301. M. Miksovsky and I. Rubeska, *Chem. Listy*, **68**, 299 (1974).
302. C. S. Rann and A. N. Hambly, *Anal. Chim. Acta*, **32**, 346 (1965).

303. B. C. Severne and R. R. Brooks, *Talanta*, **19**, 1467 (1972).
304. B. C. Severne and R. R. Brooks, *Anal. Chim. Acta*, **58**, 216 (1972).
305. B. J. Heffernan, R. O. Archbold, and T. J. Vickers, *Proc. Austr. Inst. Min. Met.*, **223**, 65 (1967).
306. J. A. Bowman, *Anal. Chim. Acta*, **42**, 285 (1968).
307. J. D. Mensik and H. J. Seidemann, *Perkin-Elmer At. Absorpt. Newsl.*, **13**, 8 (1974).
308. B. Moldan, I. Rubeska, M. Miksovsky, and M. Huka, *Anal. Chim. Acta*, **52**, 91 (1970).
309. C. B. Belcher and K. A. Brooks, *Anal. Chim. Acta*, **29**, 202 (1963).
310. A. Hofer, *Z. Anal. Chem.*, **249**, 115 (1970).
311. C. Huffman and J. D. Mensik, *Appl. Spectrosc.*, **21**, 125 (1967).
312. R. F. Goecke, *Perkin-Elmer At. Absorpt. Newsl.*, **8**, 106 (1969).
313. B. Bernas, *Anal. Chem.*, **40**, 1682 (1968).
314. M. Fratta, *Can. J. Spectrosc.*, **19**, 33 (1974).
315. R. F. Goecke, *Talanta*, **15**, 871 (1968).
316. D. C. G. Pearton, J. D. Taylor, P. K. Faure, and T. W. Steele, *Anal. Chim. Acta*, **44**, 353 (1969).
317. B. F. Quin and R. R. Brooks, *Anal. Chim. Acta*, **65**, 206 (1973).
318. E. Keller and M. L. Parsons, *Perkin-Elmer At. Absorpt. Newsl.*, **9**, 92 (1970).
319. P. D. Rao, *Perkin-Elmer At. Absorpt. Newsl.*, **9**, 131 (1970).
320. J. E. Allan, *Analyst*, **86**, 530 (1961).
321. F. J. Langmyhr and S. Sveen, *Anal. Chim. Acta*, **32**, 1 (1965).
322. C. B. Belt, *Anal. Chem.*, **39**, 676 (1967).
323. F. J. Langmyhr, *Anal. Chim. Acta*, **39**, 516 (1967).
324. A. M. Bond and D. R. Canterford, *Anal. Chem.*, **43**, 134 (1971).
325. N. H. Suhr and C. O. Ingamells, *Anal. Chem.*, **38**, 730 (1966).
326. C. O. Ingamells, *Anal. Chem.*, **38**, 1228 (1966).
327. O. A. Ohlweiler, J. O. Meditsch, and C. M. S. Piatnicki, *Anal. Chim. Acta*, **67**, 283 (1973).
328. S. Abbey, Geol. Surv. Can. Pap. No. 70-23 (1970).
329. S. Abbey, N. J. Lee, and J. L. Bouvier, Geol. Surv. Can. Pap. No. 74-19 (1974).
330. A. M. Hartstein, R. W. Freedman, and D. W. Platter, *Anal. Chem.*, **45**, 611 (1973).
331. J. Anderson, *Perkin-Elmer At. Absorpt. Newsl.*, **13**, 31 (1974).
332. C. A. Black et al. (eds.), *Methods of Soil Analysis*, Part 2, American Society of Agronomy, Madison, Wisc. (1965).
333. W. S. G. McPhee and D. F. Ball, *J. Sci. Food Agr.*, **18**, 376 (1967).
334. Y. LaFlamme, *Perkin-Elmer At. Absorpt. Newsl.*, **7**, 101 (1968).
335. L. C. Blakemore, *N. Z. J. Agr. Res.*, **11**, 515 (1968).

336. A. J. McLean, R. L. Halstead, and B. J. Finn, *Can. J. Soil Sci.*, **49**, 327 (1969).
337. M. Nadirshaw and A. H. Cornfield, *Analyst*, **93**, 475 (1968).
338. A. M. Ure and M. L. Berrow, *Anal. Chim. Acta*, **52**, 247 (1970).
339. S. T. Holding and P. H. D. Matthews, *Analyst*, **97**, 189 (1972).
340. J. F. Alder and T. S. West, *Anal. Chim. Acta*, **58**, 331 (1972).
341. R. C. Barras and J. D. Helwig, *Proc. Amer. Petrol. Inst. Sec. III*, **43**, 223 (1963).
342. E. J. Moore, O. I. Milner, and J. R. Glass, *Microchem. J.*, **10**, 148 (1966).
343. D. J. Trent and W. Slavin, *Perkin-Elmer At. Absorpt. Newsl.*, **3**, 131 (1964).
344. J. D. Kerber, *Appl. Spectrosc.*, **20**, 212 (1966).
345. S. H. Omang, *Anal. Chim. Acta*, **56**, 470 (1971).
346. J. F. Alder and T. S. West, *Anal. Chim. Acta*, **61**, 132 (1972).
347. L. Capacho-Delgado and D. C. Manning, *Perkin-Elmer At. Absorpt. Newsl.*, **5**, 1 (1966).
348. C. L. Chakrabarti and G. Hall, *Spectrosc. Lett.*, **6**, 385 (1973).
349. C. Woodward (ed.), *Annual Reports on Analytical Atomic Spectroscopy 1974*, Society for Analytical Chemistry, London, 1975.
350. K. W. Busch and G. H. Morrison, *Anal. Chem.*, **45**, 712A (1973).
351. J. D. Winefordner, J. J. Fitzgerald, and N. Omenetto, *Appl. Spectrosc.*, **29**, 369 (1975).

CHAPTER

11

FLAME EMISSION AND ATOMIC FLUORESCENCE SPECTROMETRY

11.1 GENERAL INTRODUCTION

The possibility of using the characteristic radiation emitted by atoms excited in flames for qualitative analysis of certain elements arose directly from the work of Kirchhoff and Bunsen[1,2] in the early 1860s. Shortly afterward, an instrument was developed for the quantitative analysis of sodium in plant ash,[3] using a Bunsen flame and visual comparison of the emission intensity with that from a reference flame. During the next 60 years, interest in flame spectrochemical analysis was maintained by several groups of workers. However, the widespread use of flame emission for quantitative work awaited the development of reproducible methods of introducing samples into flames and the use of convenient objective methods of measuring the intensity of the emitted radiation.

Particularly significant advances were made in the 1920s by Lundegårdh.[4] In Lundegårdh's apparatus the sample, in solution form, was sprayed into a nebulizer, and the resulting aerosol was carried into an air/acetylene flame. The emission was dispersed by a quartz prism spectrograph and recorded photographically. Precision of analysis was typically 5% to 10%.

Greater convenience resulted from the introduction of simple colored filters for wavelength selection, together with photocell/galvanometer combinations for measuring intensities directly. Simple and inexpensive instruments of this type, using air/coal-gas or air/acetylene flames, became widely used from the late 1940s for the determination of sodium, potassium, lithium, and calcium.

Work on many other elements became possible with the use of grating spectrometers equipped with photomultiplier detectors. Commercial instruments included flame emission attachments for absorption spectrophotometers. Extensive use was made of total consumption burners with air/acetylene, oxygen/acetylene, and oxygen/hydrogen flames.

With the development of atomic absorption in the 1960s, the use of flame emission has become very restricted. In many laboratories only the alkali metals are determined this way, using either a simple filter photometer or an atomic absorption apparatus with minor modification (p. 219). This neglect of flame emission is largely unjustified, as many developments that have assisted the practice of atomic absorption (such as long-path burners for premixed flames, and the use of the nitrous oxide/acetylene flame) are also beneficial for flame emission. Pickett and Koirtyohann[5] emphasize that flame emission should now be regarded as complementary to atomic absorption. With comparable apparatus, better detection limits can be obtained by flame atomic emission than by atomic absorption for more than 20 elements, including the alkali and alkaline earth metals, aluminum, gallium, indium, rhodium, tungsten, and most of the lanthanides.

11.2 THEORETICAL AND EXPERIMENTAL CONSIDERATIONS

When a solution containing an analyte element M is aspirated into a flame, a sequence of processes (vaporization of solvent, vaporization of salt molecules, decomposition of molecular species into atoms) takes place, as discussed in more detail in Section 10.3. Neutral atoms of M, and molecules such as MO and MOH, may undergo excitation to higher electronic energy levels as a result of collisions with energetic species produced in the flame combustion reaction. Some ionization of M atoms may also occur. The excited atoms and molecules can revert to the ground state, losing energy either by collision with other molecules or by the spontaneous emission of radiation of characteristic wavelengths.

For any given temperature, the equilibrium ratio of the number of atoms in an excited state to the number in the ground state can be calculated as shown by equation 10.2 and Table 10.1 (p. 161). However, the actual concentrations of both ground- and excited-state atoms are dependent on flame chemistry as well as temperature, being largely governed by equilibria involving species such as MO, MOH, and M^+. Furthermore, although the Boltzmann equilibrium expression (equation 10.2) is applicable above the reaction zone in many flames, there are cases where the excited-state population exceeds that calculated on the basis of simple collisional (thermal) excitation. This situation (suprathermal chemiluminescence) arises when part of the chemical energy released in a reaction is used directly for atomic excitation, if the reactant concentrations exceed their equilibrium values.[6] The latter condition may occur in some flame environments, and suprathermal chemiluminescence may result from reactions such as

$$H + OH + M \longrightarrow H_2O + M^*$$

and

$$CH + O + M \longrightarrow CHO + M^*$$

If only equilibrium excitation is involved and the excited metal atoms emit radiation at wavelength λ_0, then the intensity of the radiation emitted at this wavelength is[6]

$$I = \frac{hc}{4\pi\lambda_0} lA_{1\to 0}[\text{M}^*] \tag{11.1}$$

where I is the flux of radiant energy per unit solid angle per unit surface area, integrated over the width of the emission line, l is the thickness of the flame section along the axis of observation, and $A_{1\to 0}$ is the transition probability (in s^{-1}) for the line in question. This can be combined with equation 10.2 to give I in terms of the ground-state metal atom concentration [M] (assumed to be uniform) in the region of the flame being observed:

$$I = \frac{hc}{4\pi\lambda_0} lA_{1\to 0}[\text{M}] \frac{g_1}{g_0} \exp\frac{-E_1}{kT} \tag{11.2}$$

where g_1, g_0, and E_1 are as defined on p. 139. Because of transmission losses in the optical system, only a portion of this flux falls on the detector surface.

Several important points can be made in connection with equation 11.2:

1. Because of the exponential dependence on $(-E_1/kT)$, the emission intensity will be greatest from lines of lowest excitation energy, other things being equal. At 3000 K the exponential term is over 2000 times greater for a line with $E_1 = 2.0$ eV than for a line with $E_1 = 4.0$ eV. Atomic lines with wavelengths below approximately 300 nm, corresponding to excitation energies above approximately 4.1 eV, are therefore seldom emitted strongly from flame sources.

2. A further consequence of the exponential dependence of emission intensity on $(-E_1/kT)$ is that a rather small relative variation in temperature leads to a large relative variation in intensity. Since $E_1 = hc/\lambda_0$, differentiation of 11.2 leads to

$$\frac{dI}{I} = \frac{hc}{\lambda_0 kT} \frac{dT}{T} \tag{11.3}$$

Hence a variation of 25 K in a flame at 2500 K leads to intensity variations of about 8% for a line at 700 nm and 14% for a line at 400 nm. Maintenance of constant flame temperature is therefore much more critical in flame emission than in atomic absorption spectrometry.

3. Provided virtually all the emitted photons leave the flame and are not reabsorbed by ground-state atoms, the emission intensity is proportional to [M], the concentration of ground-state atoms in the flame. Under carefully controlled aspiration conditions, it is possible to ensure that [M] is proportional to the concentration of M in the original solution. It is important

to note that any type of matrix effect that alters the fraction of free atoms of M produced in the flame, acts as an interference in flame emission in the same way as in atomic absorption. Such effects, which include changes in aspiration efficiency, ionization effects, incomplete decomposition of molecular species, and formation of stable compounds, are discussed in more detail in Section 10.3.

4. It is popularly supposed that atomic absorption is inherently more sensitive than flame emission, simply because the population of ground-state atoms (capable of absorbing) is very much greater than the population of excited-state atoms (capable of emitting). This fallacy has been exposed by Alkemade[7] who compared the theoretical bases of the two methods. It was shown that atomic absorption can only be more sensitive if the brightness of the atomic absorption light-source at the analyte wavelength exceeds that of a black body at flame temperature at this wavelength. For some nonthermal sources, such as hollow cathode lamps, this appears to be the case, particularly at short wavelengths. For flame emission to be superior at short wavelengths, flame temperatures much higher than the usual 2000 to 3000 K would be required. Detailed discussion of factors affecting the limits of detection by both techniques has been given by Winefordner et al.[8,9]

A great deal of basic information on both theoretical and practical aspects of flame atomic emission, together with details of the spectra emitted by salted and unsalted flames, can be found in the volume by Mavrodineanu and Boiteux.[10] An authoritative account of flame spectra is that of Gaydon.[11] Alkemade and Zeegers[12] have given a detailed discussion of excitation processes in flames.

11.3 INSTRUMENTATION

A wide variety of instrumental components has been used at various stages in the development of flame-emission spectrometry. Much useful work, principally on the alkali metals and calcium, has been carried out with simple and inexpensive filter photometers. However, where a wide range of elements is to be investigated, selection of a very narrow band of wavelengths with a prism or grating monochromator becomes essential.

11.3.1 Filter Spectrophotometers

A simple type of filter spectrophotometer is shown schematically in Fig. 11.1. A Meker-type burner with a fuel such as coal gas, natural gas, or propane burning in air gives a flame suitable for the excitation of the alkali metals. Higher temperatures are achieved with air/acetylene flames from other types of premix burners. An increase in the measured radiant flux is obtained by

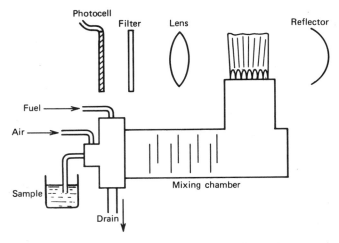

Fig. 11.1. Schematic diagram of a simple filter flame photometer.

placing behind the flame a concave reflector with its center of curvature in the flame. An appropriate 30 to 50 nm band of wavelengths is selected by using one of several interchangeable colored absorption filters. These are commonly used in the determination of sodium, potassium, and lithium. Instruments equipped with suitable interference filters isolating a narrower bandwidth (10 to 20 nm) can be used to determine all of the alkali and alkaline earth metals. Detection systems consist of a selenium barrier-layer photocell with a direct-reading meter, or a phototube or photomultiplier with amplifier and meter.

11.3.2 Prism and Grating Spectrophotometers

For the best flame-emission work a prism or grating spectrophotometer should be used, preferably a grating instrument capable of giving a spectral bandpass of about 0.05 nm in the first order. This makes it possible to eliminate most atomic emission lines near the line of interest and to decrease the contribution from molecular emission bands and flame background. With a suitable choice of flame it is possible to use flame emission to determine most elements (see Table 11.1, p. 222).

Diffusion flames from hydrogen or acetylene burning in air or oxygen were formerly used extensively, but more recently greater use has been made of slot burners for premixed air/acetylene and nitrous oxide/acetylene flames, as in atomic absorption spectrophotometry. Detection can be carried out with a photomultiplier and d.c. amplifier, as shown schematically in Fig. 11.2.

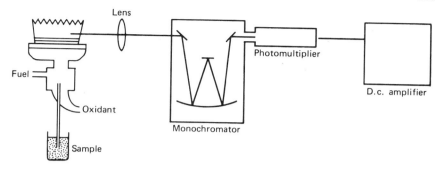

Fig. 11.2. Schematic diagram of a flame emission spectrophotometer.

Alternatively, an a.c. amplifier may be used, as in atomic absorption instruments. In the latter case the signal emitted from the flame must be modulated at the appropriate frequency, usually by placing a mechanical chopper between the flame and the entrance slit of the monochromator.

When a slot burner is used, self-absorption (and consequent curvature of working curves) becomes a problem at lower concentrations than is the case with more compact flames of circular cross section. However, in measuring trace concentrations the greater flame length from the slot burner is an advantage. As in atomic absorption the burner can be rotated to reduce the sensitivity if necessary.

A useful discussion of optical systems for flame spectroscopy, together with a summary of the essential features of commercial instruments, has been given by Müller-Herget.[13] Further details of instrumentation can be found in standard works.[10,14,15]

11.4 PRACTICE AND APPLICATIONS

Most flame emission work is carried out by comparing the net emission intensity (line minus background) of the sample solution at the analyte wavelength with the intensities of a set of standard solutions. Stock solutions similar to those used for atomic absorption (Table 10.4) are generally suitable for the preparation of standards. When air/acetylene or nitrous oxide/acetylene flames are being used for the determination of readily ionizable elements (i.e., when the ionization potential is less than about 7.5 eV), it is advisable to suppress ionization of the analyte by adding to the sample and standards an excess (up to 1000 μg/ml) of another easily ionized element, such as potassium or cesium. The use of cesium as an ionization suppressor in the nitrous oxide/acetylene flame has been studied by Kornblum and de Galan.[16]

Where such additions are made, it is necessary to ensure that traces of the analyte are absent from the added solution.

Over a limited concentration range, linear analytical curves are obtained. Where a large range is to be covered (e.g., two to three orders of magnitude) the analytical curve may deviate from linearity, as a result of self-absorption, for example. In this case it is convenient to plot a curve of the logarithm of the emission intensity against the logarithm of the concentration. Where bending of the analytical curve is serious, it is advisable to reduce the length of the flame by rotating the slot burner or to dilute the solutions.

Standard Additions

The possibility that constituents of the sample matrix are influencing the emission intensity can be checked by the method of standard additions. A multiple addition technique similar to that described for atomic absorption (p. 187) can be used, provided the analytical curve is linear over the whole concentration range involved. Alternatively, a single addition method can be used,[14] in which the apparent concentrations of the original solution (c'_A) and the solution after standard addition (c'_B) are measured with reference to the standard calibration curve. If the known concentration added is c_S, then it can be shown that the true concentration c_X of the unknown is

$$c_X = c_S \frac{c'_A}{c'_B - c'_A} \qquad (11.4)$$

These methods assume that any interference occurs to the same degree to the standard addition and to the original amount of analyte.

Internal Standards

The effects of variable efficiency of aspiration, nebulization and to some extent flame fluctuations can be minimized by an internal standard technique. A fixed concentration of a selected element is added to all samples and standard solutions. The emission intensity of the analyte and standard element are measured simultaneously or sequentially, using dual detectors or by scanning a range of wavelengths for both emission lines. The two lines should have similar wavelengths and excitation potentials, and the ionization potentials of the two elements should be similar. The internal standard element must be available in high purity and should be absent from the sample, or its concentration in the sample should be known independently. Examples of this technique include the use of lithium in the determination of potassium and sodium[17] and the use of silver in the determination of copper.[18] The use of the internal standard method, however, tends to be limited by the rather stringent requirements of the identity of behavior of the analyte and standard elements.

Interferences

Flame emission is subject to the same kinds of interference as atomic absorption, although the relative importance of different interferences may differ for the two techniques. Chemical interferences (p. 179) alter the fraction of analyte existing as free atoms, both in the ground state and in excited states. Flame emission and atomic absorption are equally affected. Many interferences of the type described in Section 10.3.3 were first discovered in the course of flame emission work. Remedies such as the use of releasing agents (e.g., lanthanum,[19] strontium[20]) or chelating agents (e.g., EDTA,[21] 8-hydroxyquinoline[22]) for interferences such as that of phosphate on calcium were also introduced by analysts using flame emission.

Some particular cases of interference from overlap of two atomic spectral lines have been noted in Chapter 10 (p. 178). Care in checking spectral line interference is particularly important in flame emission, as the wavelength range over which the signal is measured is determined by the monochromator slit width and grating characteristics, rather than by the narrow line width of the light source, as is normally the case in atomic absorption.

Background emission from the flame can also be a greater problem in flame emission than in atomic absorption. Emission from the analyte is superimposed upon emission band spectra from various molecular species produced by the flame reaction. These species vary from flame to flame, but include molecules such as OH, C_2, CH, and NH. Additional emission may come from band spectra of molecules formed from sample matrix components. For example, MO and MOH bands appear when solutions containing appreciable amounts of the alkaline earth elements are aspirated into various flames. However, even if the background is high at the analyte wavelength, provided it is also stable, emission measurements can be made by scanning over a narrow wavelength range (up to 1 nm), using a small spectral bandpass (0.05 to 0.1 nm).

Organic Solvents

It was noted in Chapter 10 that modest improvements in atomic absorption sensitivity can be obtained by aspirating solutions containing the analyte in an organic solvent. In most cases the improvement is due mainly to the increased efficiency with which the sample is introduced into the flame, although some modifications of flame temperature and chemistry also occur.

In flame emission, however, the intensities of some lines show exceptional enhancement (e.g., 100- to 1000-fold) in the presence of organic solvents. A

TABLE 11.1 Detection Limits by Flame Emission[a]

Element	Wavelength (nm)	Detection limit (μg/ml)	Flame[b]
Ag	328.07	0.02	N
Al	396.15	0.005[c]	N
Au	267.60	0.5	N
Ba	553.55	0.001[c]	N
Be	470.86[d]	0.2	N
Ca	422.67	0.0001[c]	N
Cd	326.11	2	N
Co	345.35	0.05	N
Cr	425.44	0.005[c]	N
Cu	327.40	0.01	N
Dy	404.60	0.07	N
Er	400.80	0.04	N
Eu	459.40	0.0006	N
Fe	371.99	0.05	N
Ga	403.30	0.01[c]	N
Gd	440.19	2	N
Ge	265.12	0.5	N
Ho	405.39	0.02	N
In	451.13	0.002[c]	N
Ir	380.01	30	N
K	766.49	0.0005	A
La	441.82[d]	0.1	N
Li	670.78	0.00003[c]	N
Lu	451.86	1	N
Mg	285.21	0.005	N
Mn	403.08	0.005	N
Mo	390.30	0.1	N
Na	589.00	0.0005	A
Nb	405.89	1	N
Nd	292.45	0.2	N
Ni	341.48	0.03	N
Pb	405.78	0.2	N
Pd	363.47	0.05	N
Pr	495.14	1	N
Pt	265.95	2	N
Rb	780.02	0.001	A
Re	346.05	0.2	N
Rh	369.24	0.02	N
Ru	372.80	0.02	N
Sc	402.04	0.03	N
Sm	476.03	0.2	N
Sn	284.00	0.5	N
Sr	460.73	0.0001[c]	N
Ta	481.28	5	N

Table 11.1 (*continued*)

Element	Wavelength (nm)	Detection limit (μg/ml)	Flame[b]
Tb	431.89	0.4	N
Ti	399.86	0.2	N
Tl	535.05	0.02	N
Tm	371.79	0.02	N
V	437.92	0.01	N
W	400.88	0.5	N
Y	362.09	0.4	N
Yb	398.80	0.002	N
Zr	360.12	3	N

Source: After Pickett and Koirtyohann.[5]
[a] Aqueous solutions, nitrous oxide/acetylene or air/acetylene premixed flames, 5- or 10-cm slot burners.
[b] N: nitrous oxide/acetylene; A: air/acetylene.
[c] In presence of ionization suppressant such as KCl.
[d] Molecular band emission from monoxide.

notable example is that of tin[23,24] introduced into an air/hydrogen flame in the presence of propan-2-ol. The enhancement has been shown not to be caused by an increase of flame temperature, but has been attributed to suprathermal chemiluminescence (p. 215). Enhancement for many other elements has been reported by Buell,[25] although only a few of the lines have analytical significance. The role of chemiluminescence in flame spectrometry has recently been reviewed by Glover.[26]

Detection Limits

Detection limits determined for aqueous solutions aspirated into nitrous oxide/acetylene or air/acetylene premixed flames on 5- or 10-cm slot burners have been given in several reports.[5,27,28] Some of these data are shown in Table 11.1. Detection limits quoted for some of the lanthanides may be further improved when an ionization suppressant is used. Optimization of conditions for determination of many elements by emission in the nitrous oxide/acetylene flame has been studied by Christian and Feldman.[29] Lanthanide emission spectra have also been examined extensively in the oxyacetylene flame[30] and in the separated nitrous oxide/acetylene flame.[31]

In a few cases better detection limits than those of Table 11.1 have been achieved when other flames have been used or in the presence of organic solvents.[32] The detection limits for 67 elements in premixed oxyacetylene flames have been tabulated by Fassel and Golightly.[33] Very low detection limits were observed for some of the alkali metals (Cs 0.008 μg/ml at 852.11 nm, Li 0.000003 μg/ml at 670.78 nm, Na 0.0001 μg/ml at 589.00 nm) when red- or yellow-sensitive photomultipliers were used. Detection limits for many elements in the argon/hydrogen/entrained air flame were reported by Zacha and Winefordner.[34]

Detection limits are generally poor for atomic spectral lines of As, B, Bi, Hg, Sb, Te, and Zn. Values for these elements in various flames can be found elsewhere.[32,33,35]

A very comprehensive review of the flame spectroscopy of nonmetals has been presented by Gilbert.[36] Elements discussed and species whose spectra are of some significance for qualitative or quantitative analysis include boron (BO_2, B), phosphorus (HPO), sulfur (S_2), and the halogens (copper and indium monohalides).

Very low absolute detection limits (10^{-10} to 10^{-12} g) for many elements by emission have resulted from the use of pulsed evaporation from a graphite microprobe into a nitrous oxide/acetylene flame.[37]

Applications

In the analysis of geological materials most applications of flame emission have been to the determination of Na, K, Ca, Mg, Sr, Li, Rb, Cs, and occasionally a few other elements. For both trace and higher concentrations of alkali metals, flame emission has been widely adopted because of its precision (better than 1% in favorable cases) and sensitivity. For other elements flame emission generally has been superseded by atomic absorption, not always with justification.[5]

Solid geological samples are usually dissolved in an acid medium (Chapters 4 and 10). Alkali fusions are avoided unless it can be shown that the fusion reagents are free from significant traces of the analyte. For the elements that commonly occur in high concentrations in rocks and soils (e.g., Na, K, Ca, Mg), separation and concentration procedures are rarely necessary. In other cases, ion-exchange or solvent extraction may be applicable. Appropriate additions can be made to solutions to overcome any problems from ionization or chemical interferences.

Detailed procedures for many elements have been given by Dean[14] and Pinta,[38,39] and bibliographies of work prior to 1968 have been prepared by Mavrodineanu.[40,41] It should be noted, however, that much of this work predates the use of the premixed nitrous oxide/acetylene flame, which has greatly increased the potential of flame emission.

Examples of more recent work include determinations of the alkali metals in geological materials using the nitrous oxide/acetylene flame,[42,43] cesium in seawater using an oxy-hydrogen flame,[44] rubidium and cesium in ores,[45] and rubidium, potassium, and strontium in rocks.[46] Sulfide and silicate minerals and ores have been analyzed for aluminum,[47] and calcium has been determined in rainwater.[48] Gallium has been determined in rocks after an extraction step.[49] Aluminum, calcium, iron, and titanium have been determined simultaneously in standard rocks with a nitrous oxide/acetylene flame, using a silicon-diode vidicon tube as detector.[50] A combination of atomic absorption and flame emission has been used for determining the lanthanides in rocks.[51] Additional recent work has been summarized in Dean and Rains (Ref. 15, Vol. 3) and elsewhere.[52,53]

It is apparent from the detection limits of Table 11.1 that rather greater use of flame emission could be made in trace-element work than is the case at present. Although instrumental parameters (e.g., gas mixture, position of optical path in the flame, choice of wavelength and slit width) tend to be more critical in flame-emission work than in atomic absorption, there are advantages in not needing hollow cathode lamps, and in the greater sensitivity that can be obtained for more than 20 elements.

11.5 ATOMIC FLUORESCENCE SPECTROMETRY

In atomic fluorescence spectrometry, atoms produced by a flame or other atomizing device are excited by an external source of radiation containing the appropriate wavelengths. Some of the atoms excited in this way undergo a radiational transition to the ground state or other lower energy state, the radiation being described as atomic fluorescence. Although fluorescence of atomic vapors was first studied early in this century[54] and work with flame atom sources dates from the 1920s,[55,56] it was not until 1962 that the possibility of its being used for analysis was noted.[57] The first papers on analytical atomic fluorescence appeared in 1964.[58,59]

Atomic fluorescence spectrometry offers the possibility of better detection limits than atomic absorption for many elements, especially those with resonance lines below about 320 nm. However, largely because of a number of instrumental problems, no commercial atomic fluorescence spectrometer has remained in production, and analysts wishing to use atomic fluorescence have had to assemble their own apparatus.

A notable difficulty has been the provision of suitably intense and stable sources of the appropriate wavelengths. This problem has to some extent been overcome as a result of recent work on thermostatically controlled electrodeless discharge lamps[60,61] and on intermittently pulsed hollow cathode lamps.[62] Some problems have also been presented by the superimposed flame emission and by light scattering. Scattering is generally small, however, when narrow-line sources are used in conjunction with premixed air/acetylene or nitrous oxide/acetylene flames. Emission problems have been minimized by the use of low-background separated flames.

Theoretical expressions have been derived[63-65] that relate the radiance of the fluorescence to the atomic concentration, the spectral radiance of the source, various atomic parameters, and parameters determined by the geometry of the source/sample cell/detector arrangement. For both continuous and line sources, the fluorescence radiance is proportional to the radiance of the source at the peak absorption wavelength and proportional to the ground-state concentration of atoms in the vapor, provided the vapor is sufficiently dilute. As in the case of molecular fluorescence, the greatest analytical utility therefore lies in the field of trace analysis.

Instrumentation

Although continuum sources such as xenon arc lamps have been investigated extensively, most recent work has made use of microwave-excited electrodeless discharge lamps (EDLs), pulsed hollow cathode lamps or pulsed laser sources.

Both flames and electrothermal devices have been used as atomizers. The wide variety of flames studied includes premixed air/acetylene, nitrous oxide/acetylene, and hydrogen/air flames. Burners of circular cross section are commonly used. The desirability of combining efficient atomization with a low flame-emission background has led to considerable use of nitrous oxide/acetylene or air/acetylene flames in which the secondary reaction zone is separated with a concentric stream of nitrogen or argon.[66,67] Because atomic fluorescence is much more effectively quenched by polyatomic species (e.g., CO_2, N_2, O_2, H_2O) than by monatomic gases, environments containing significant proportions of an inert gas such as argon are often favored.

The features of electrothermal atomization noted in connection with atomic absorption (p. 171) apply also to atomic fluorescence. Detection limits for a number of elements in atomic fluorescence with electrothermal atomizers have been given by Patel et al.[68] and Browner.[69] A comparison of detection limits for flame atomic emission, absorption, and fluorescence[69] shows atomic fluorescence to be superior to the other two techniques for 14 elements (Ag, Au, Bi, Cd, Ce, Co, Cu, Ge, Hg, Mn, Sc, Te, Tl, Zn) and comparable to atomic absorption (and better than emission) for an additional six elements (As, Fe, Ni, Pb, Sb, Se).

A wide variety of optical systems has been used, some of which have been illustrated in the comprehensive review by Browner.[69] A typical atomic fluorescence arrangement, using a thermostatically controlled EDL and dispersive optics (grating monochromator) is shown in Fig. 11.3. Because a high degree of wavelength selection is implicit in the absorption-fluorescence

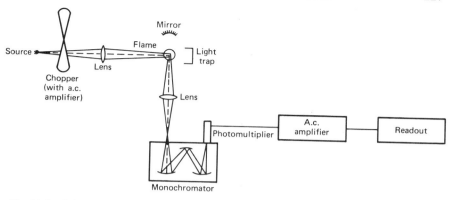

Fig. 11.3. Schematic diagram of a dispersive system for atomic fluorescence spectrometry. Mechanical chopping is required if an unmodulated source (e.g., EDL) is used together with an a.c. amplifier. See Ref. 69 for alternative optical systems, and Ref. 70 for a nondispersive system.

combination of processes, it has also been possible to construct simple and inexpensive nondispersive atomic fluorescence systems.[70,71] The advantages and disadvantages of both monochromator and nondispersive systems have been summarized.[69] Further discussions of atomic fluorescence theory and instrumentation have been given elsewhere.[72-74]

Applications

Up to the present time, much more work has been devoted to the development and optimization of apparatus than to the atomic fluorescence analysis of real samples. The limited range of applications to geological materials includes the determination of mercury in soil and water by a flameless fluorescence version of the reduction-aeration technique[75-78] originally developed for atomic absorption measurements. Detection limits of 0.5 to 3 ng have been quoted. More recently,[79] amounts of mercury down to 5 pg have been detected by reducing the dead volume of the vapor-generation system.

Gold has been determined in mine-water[80] with a detection limit of 0.015 μg/ml, using a separated air/acetylene flame. Soil extracts were analyzed for Ca, Cu, Mg, Mn, and Zn,[81] all with detection limits in the range 0.4 to 1 μg/ml using a multichannel atomic fluorescence spectrometer with an argon-shielded air/acetylene flame. A similar apparatus with a nitrogen-separated air/acetylene flame was used[82] in an automated solvent extraction-atomic fluorescence system to determine Co, Cr, Cu, and Zn in seawater with detection limits in the range 0.0002 μg/ml (Zn) to 0.003 μg/ml (Cr). Selenium has also been determined in seawater[83] using the hydride generation technique and an argon/hydrogen diffusion flame.

Apart from the possibility of obtaining better sensitivity for some elements (by comparison with atomic absorption), atomic fluorescence continues to be investigated because it lends itself to simultaneous multielement analysis more readily than atomic absorption does. Approaches to multielement atomic fluorescence have included the use of rapid sequential pulsing of as many as six hollow cathode lamps with coincident use of rotating filters to isolate the required wavelengths.[84] The use of pulsed hollow cathode lamps[85,86] or multielement EDLs[87] together with a monochromator capable of undergoing a wavelength-scanning program has also been described. A seven-channel nondispersive atomic fluorescence spectrometer was described by Walsh.[88] Johnson et al.[89] have developed a scanning spectrometer for use with a continuum (xenon arc) source.

For comprehensive surveys of these and other possibilities in multielement atomic spectroscopy the reviews of Busch and Morrison[90] and of Winefordner et al.[91] should be consulted.

References

1. G. R. Kirchhoff and R. Bunsen, *Pogg. Ann.*, **110**, 161 (1860); **113**, 337 (1861).
2. G. R. Kirchhoff, *Phil. Mag.*, **105**, 250 (1862).
3. P. Champion, H. Pellet, and M. Grenier, *Compt. Rend.*, **76**, 707 (1873).
4. H. Lundegårdh, *Die quantitative Spektralanalyse der Elemente*, Fischer, Jena, Vol. 1, 1929; Vol. 2, 1934.
5. E. E. Pickett and S. R. Koirtyohann, *Anal. Chem.*, **41** (14), 28A (1969).
6. C. T. J. Alkemade, in *Analytical Flame Spectroscopy* (ed. R. Mavrodineanu), Macmillan, London, 1970, p. 41.
7. C. T. J. Alkemade, *Appl. Opt.*, **7**, 1261 (1968).
8. J. D. Winefordner and T. J. Vickers, *Anal. Chem.*, **36**, 1939, 1947 (1964).
9. J. D. Winefordner and C. Veillon, *Anal. Chem.*, **37**, 416 (1965).
10. R. Mavrodineanu and H. Boiteux, *Flame Spectroscopy*, Wiley, New York, 1965.
11. A. G. Gaydon, *The Spectroscopy of Flames*, 2nd ed., Halsted, New York, 1974.
12. C. T. J. Alkemade and P. J. T. Zeegers, in *Spectrochemical Methods of Analysis* (ed. J. D. Winefordner), Wiley-Interscience, New York, 1971.
13. W. Müller-Herget, in *Analytical Flame Spectroscopy* (ed. R. Mavrodineanu), Macmillan, London, 1970.
14. J. A. Dean, *Flame Photometry*, McGraw-Hill, 1960.
15. J. A. Dean and T. C. Rains (eds.), *Flame Emission and Atomic Absorption Spectrometry*, Dekker, New York; Vol. 1, *Theory*, 1969; Vol. 2, *Components and Techniques*, 1971; Vol. 3, *Elements and Matrices*, 1975.

16. G. R. Kornblum and L. de Galan, *Spectrochim. Acta*, **28B**, 139 (1973).
17. C. L. Fox, E. B. Freeman, and S. E. Lasker, *ASTM Spec. Tech. Publ. No. 116* (1951).
18. J. A. Dean, *Anal. Chem.*, **27**, 1224 (1955).
19. J. Yofe and R. Finkelstein, *Anal. Chim. Acta*, **19**, 166 (1958).
20. J. I. Dinnin, *Anal. Chem.*, **32**, 1475 (1960).
21. J. Wirtschafter, *Science*, **125**, 603 (1957).
22. J. Debras-Guédon and I. Voinovitch, *Compt. Rend.*, **248**, 3421 (1959).
23. P. T. Gilbert, in *Proc. Xth Colloquium Spectroscopicum Internationale* (eds. E. R. Lippincott and M. Margoshes), Spartan Press, Washington, D. C., 1963, pp. 171–215.
24. J. H. Gibson, W. E. L. Grossman, and W. D. Cooke, *Anal. Chem.*, **35**, 266 (1963).
25. B. E. Buell, *Anal. Chem.*, **35**, 372 (1963).
26. J. H. Glover, *Analyst*, **100**, 449 (1975).
27. G. D. Christian and F. J. Feldman, *Appl. Spectrosc.*, **25**, 660 (1971).
28. R. N. Kniseley, C. C. Butler, and V. A. Fassel, *Anal. Chem.*, **41**, 1494 (1969).
29. G. D. Christian and F. J. Feldman, *Anal. Chem.*, **43**, 611 (1971).
30. R. N. Kniseley, V. A. Fassel, and C. C. Butler, in *Analytical Flame Spectroscopy* (ed. R. Mavrodineanu), Macmillan, London, 1970.
31. D. N. Hingle, G. F. Kirkbright, and T. S. West, *Analyst*, **94**, 864 (1969).
32. O. Menis and T. C. Rains, in *Analytical Flame Spectroscopy* (ed. R. Mavrodineanu), Macmillan, London, 1970, pp. 49–51.
33. V. A. Fassel and D. W. Golightly, *Anal. Chem.*, **39**, 466 (1967).
34. K. Zacha and J. D. Winefordner, *Anal. Chem.*, **38**, 1537 (1966).
35. J. A. Dean and W. J. Carnes, *Analyst*, **87**, 743 (1962).
36. P. T. Gilbert, in *Analytical Flame Spectroscopy* (ed. R. Mavrodineanu), Macmillan, London, 1970, pp. 181–377.
37. E. D. Prudnikov, *Z. Anal. Khim.*, **27**, 2327 (1972).
38. M. Pinta, *Detection and Determination of Trace Elements*, Ann Arbor Science Publishers, Ann Arbor, Mich., 1966, pp. 255–290.
39. M. Pinta, in *Analytical Flame Spectrometry* (ed. R. Mavrodineanu), Macmillan, London, 1970.
40. R. Mavrodineanu, *Bibliography on Flame Spectroscopy, Analytical Applications, 1800–1966*, Natl. Bur. Standards Misc. Publ. 281, U. S. Government Printing Office, Washington, D. C., 1967.
41. R. Mavrodineanu, in *Analytical Flame Spectrometry*, (ed. R. Mavrodineanu), Macmillan, London, 1970, pp. 651–715.
42. M. A. Hildon and W. J. F. Allen, *Analyst*, **96**, 480 (1971).
43. W. J. F. Allen, *Anal. Chim. Acta*, **59**, 111 (1972).
44. T. R. Folsom, N. Hansen, G. J. Parks, and W. E. Weitz, *Appl. Spectrosc.*, **28**, 345 (1974).

45. I. Rubeska and M. Miksovsky, *Chem. Listy*, **67**, 1197 (1973).
46. K. Cammann, H. D. Huckenholz, K. Kohler, and D. Müller-Sohnius, *Fortschr. Mineral.*, **50**, 18, 55 (1973).
47. R. J. Guest and D. R. MacPherson, *Anal. Chim. Acta*, **78**, 299 (1975).
48. P. L. Searle and G. Kennedy, *Analyst*, **97**, 457 (1972).
49. M. S. Cresser and J. Torrent-Castellet, *Talanta*, **19**, 1478 (1972).
50. K. W. Busch, N. G. Howell, and G. H. Morrison, *Anal. Chem.*, **46**, 575 (1974).
51. J. G. Sen Gupta, *Talanta*, **23**, 343 (1976).
52. J. D. Winefordner and T. J. Vickers, *Anal. Chem.*, **46**, 192R (1974).
53. G. M. Hieftje, T. R. Copeland, and D. R. de Olivares, *Anal. Chem.*, **48**, 142R (1976).
54. R. W. Wood, *Phil. Mag.*, **10**, 513 (1905).
55. E. L. Nichols and H. L. Howes, *Phys. Rev.*, **23**, 472 (1924).
56. R. M. Badger, *Z. Phys.*, **55**, 56 (1929).
57. C. T. J. Alkemade, in *Proc. Xth Colloquium Spectroscopicum Internationale* (eds. E. R. Lippincott and M. Margoshes), Spartan Press, Washington, D. C., 1963.
58. J. D. Winefordner and T. J. Vickers, *Anal. Chem.*, **36**, 161 (1964).
59. J. D. Winefordner and R. A. Staab, *Anal. Chem.*, **36**, 165 (1964).
60. R. F. Browner, B. M. Patel, T. H. Glenn, M. E. Rietta, and J. D. Winefordner, *Spectrosc. Lett.*, **5**, 311 (1972).
61. R. F. Browner and J. D. Winefordner, *Spectrochim. Acta*, **28B**, 263 (1973).
62. E. Cordos and H. V. Malmstadt, *Anal. Chem.*, **45**, 27 (1973).
63. H. P. Hooymayers, *Spectrochim. Acta*, **23B**, 567 (1968).
64. P. J. T. Zeegers, R. Smith, and J. D. Winefordner, *Anal. Chem.*, **40** (13), 26A (1968).
65. P. J. T. Zeegers and J. D. Winefordner, *Spectrochim. Acta*, **26B**, 161 (1971).
66. G. F. Kirkbright and T. S. West, *Appl. Opt.*, **7**, 1305 (1968).
67. R. S. Hobbs, G. F. Kirkbright, M. Sargent, and T. S. West, *Talanta*, **15**, 997 (1968).
68. B. M. Patel, R. D. Reeves, R. F. Browner, C. J. Molnar, and J. D. Winefordner, *Appl. Spectrosc.*, **27**, 171 (1973).
69. R. F. Browner, *Analyst*, **99**, 617 (1974).
70. P. L. Larkins, *Spectrochim. Acta*, **26B**, 477 (1971).
71. P. L. Larkins and J. B. Willis, *Spectrochim. Acta*, **26B**, 491 (1971).
72. J. D. Winefordner, S. G. Schulman, and T. C. O'Haver, *Luminescence Spectrometry in Analytical Chemistry*, Wiley-Interscience, New York, 1972.
73. G. F. Kirkbright and M. Sargent, *Atomic Absorption and Fluorescence Spectroscopy*, Academic Press, London, 1974.
74. V. Sychra, V. Svoboda, and I. Rubeska, *Atomic Fluorescence Spectroscopy*, Van Nostrand, Princeton, N.J., 1975.

REFERENCES

75. V. I. Muscat and T. J. Vickers, *Anal. Chim. Acta*, **57**, 23 (1971).
76. V. I. Muscat, T. J. Vickers, and A. Andren, *Anal. Chem.*, **44**, 218 (1972).
77. K. C. Thompson and G. D. Reynolds, *Analyst*, **96**, 771 (1971).
78. F. L. Corcoran, *Am. Lab.*, **6**(3), 69 (1974).
79. J. E. Hawley and J. D. Ingle, *Anal. Chim. Acta*, **77**, 71 (1975).
80. J. Matousek and V. Sychra, *Anal. Chim. Acta*, **49**, 175 (1970).
81. R. M. Dagnall, G. F. Kirkbright, T. S. West, and R. Wood, *Anal. Chem.*, **43**, 1765 (1971).
82. M. Jones, G. F. Kirkbright, L. Ranson, and T. S. West, *Anal. Chim. Acta*, **63**, 210 (1973).
83. K. C. Thompson, *Analyst*, **100**, 307 (1975).
84. D. G. Mitchell and A. Johansson, *Spectrochim. Acta*, **25B**, 175 (1970); **26B**, 677 (1971).
85. H. V. Malmstadt and E. Cordos, *Am. Lab.*, **4** (8), 35 (1972).
86. E. Cordos and H. V. Malmstadt, *Anal. Chem.*, **44**, 2277, 2407 (1972); **45**, 425 (1973).
87. J. D. Norris and T. S. West, *Anal. Chem.*, **45**, 226 (1973).
88. A. Walsh, *Pure Appl. Chem.*, **23**, 1 (1970).
89. D. J. Johnson, F. W. Plankey, and J. D. Winefordner, *Anal. Chem.*, **47**, 1739 (1975).
90. K. W. Busch and G. H. Morrison, *Anal. Chem.*, **46**, 712A (1973).
91. J. D. Winefordner, J. J. Fitzgerald, and N. Omenetto, *Appl. Spectrosc.*, **29**, 369 (1975).

CHAPTER

12

X-RAY EMISSION SPECTROSCOPY

12.1 INTRODUCTION

Electrons in a beam impinging on a target material are slowed by multiple interactions with electrons of the target. The energy is lost in the form of a continuum of electromagnetic radiation, *X radiation*, with a minimum wavelength λ_{min}, corresponding to the maximum energy of the electrons in the beam. If the bombarding electrons are sufficiently energetic, inner-orbital electrons may be completely removed from the target atoms. The vacancy can then be filled by an electron from an orbital of higher energy, a process that is accompanied by emission of an X-ray photon with a wavelength characteristic of the energy levels involved.

An inner orbital electron can also be lost when a γ ray or high-energy X ray interacts with an atom A:

$$h\nu_1 + A \longrightarrow A^{+*} + e^- \text{(photoelectron)} \qquad (12.1)$$

If the exciting radiation is monoenergetic, then the photons ejected from a given electron shell of atom A will also be approximately monoenergetic, the photoelectron energy E_e being given by

$$E_e = h\nu_1 - E_b \qquad (12.2)$$

where E_b is the binding energy of the electron ejected. It is therefore possible to characterize a material by studying the energies of the photoelectrons emitted as a result of X-ray bombardment. This constitutes the field of X-ray photoelectron spectroscopy,[1] which is developing rapidly as a technique for examining the major-element composition of solid surfaces, but which has been little explored for its potential in trace analysis.[2,3]

Relaxation of the excited ion A^{+*} can take place (*a*) by the Auger effect, represented by

$$A^{+*} \longrightarrow A^{++} + e^- \qquad (12.3)$$

in which a second electron is ejected from the atom, or (*b*) by X-ray emission:

$$A^{+*} \longrightarrow A^+ + h\nu_2 \qquad (12.4)$$

Because the primary excitation (12.1) is frequently carried out with X-ray photons, the secondary X-ray emission (12.4) is often described as X-ray fluorescence. The wavelengths of the X-ray emission are characteristic of the atoms in the bombarded target.

For many elements, some of the transitions between the inner (K, L, M) electron shells involve energies in the range 2.5 to 40 keV, corresponding to X-ray wavelengths from approximately 0.5 to 0.03 nm. Measurement of the energy and intensity of X-ray fluorescence in this range has become a convenient method of carrying out qualitative and quantitative analysis on materials of many kinds.

Although X-rays were discovered by Röntgen[4] in 1895, it was not until 1948 that the first commercially feasible instrument for X-ray fluorescence was developed.[5] Originally X-ray fluorescence was not considered to be particularly sensitive, and its use was restricted to the determination of elements at levels down to the parts per thousand range. However, advances in instrumentation have resulted in better sensitivities, and the method is now a valuable tool for the analyst concerned with trace analysis of geological materials.

12.2 X-RAY SPECTRA

12.2.1 Primary X Rays

Primary X rays are produced when a beam of electrons is accelerated across an evacuated tube containing a target of tungsten or some other metal. If V is the accelerating potential, e is the electronic charge, c is the speed of light, and h is Planck's constant, then the minimum wavelength of X rays emitted from the tube is

$$\lambda_{min} = \frac{hc}{eV} \qquad (12.5)$$

A typical X-ray emission spectrum of a tungsten target is illustrated in Fig. 12.1. This shows the continuum, produced by deceleration of the electrons when they strike the target, and the superimposed line spectrum resulting from inner-electron ejection followed by descent of another electron from a higher energy level.

Electronic transitions leading to X-ray emission are shown schematically in Fig. 12.2. Spectral lines are indicated by K, L, M, ... according to the shell from which an electron has been ejected. For example, a Kα line is one resulting from the filling of a K-shell vacancy by an L-shell electron. Designations such as Kα_1, Kα_2 are used to distinguish the particular L-shell orbital involved.

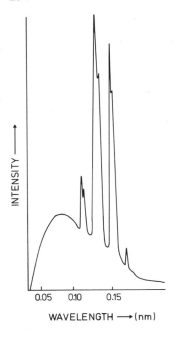

Fig. 12.1. Characteristic spectrum of tungsten superimposed on the continuous spectrum. *Source*: Shalgosky.[6]

Moseley[7] discovered that the wavelength of a given emission line (e.g., Kα) is related to the atomic number Z of the radiating element, as follows:

$$\left(\frac{1}{\lambda}\right)^{1/2} = k(Z - \sigma)$$

This equation is commonly given in the form

$$\frac{1}{\lambda} = A(Z - \sigma)^2 \tag{12.6}$$

where A is a constant and σ is a constant that indicates the screening action of other electrons in effectively reducing the nuclear charge. Equation 12.6 is not exact but is useful for many purposes. It is possible, for example, to calculate (to within 1%) the wavelengths of the Kα lines of the elements with Z between 11 and 46, using equation 12.6 with $A = 8.3 \times 10^6$ m^{-1} and $\sigma = 1.0$. Similarly, the wavelengths of the Lα lines for elements with $Z > 30$ can be found by putting $A = 1.48 \times 10^6$ m^{-1} and $\sigma = 6.5$. The relationship between wavelength and atomic number for the K and L spectra is shown in Fig. 12.3. For the study of X-ray emission it is usually convenient to make use of the wavelength range from 0.03 to 0.5 nm. Use is therefore made of the

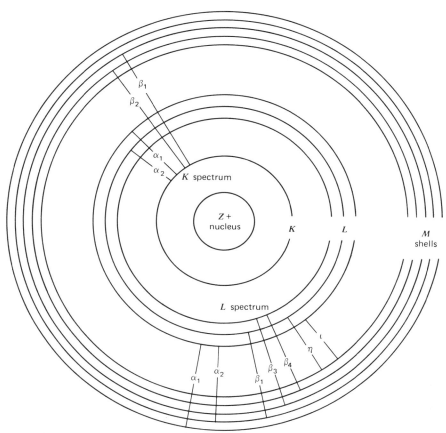

Fig. 12.2. Diagrammatic representation of electronic transitions producing X-ray emission lines. *Source*: Shalgosky.[6]

L spectra of heavy elements ($Z > 46$) and the K spectra of lighter elements. The lightest elements ($Z < 10$) cannot be readily investigated because even their K spectra occur at wavelengths too long to be dispersed by the analyzing crystals currently available.

12.2.2 X-Ray Absorption

A beam of X-rays, passing through a thin absorber, may undergo scattering (with or without change of wavelength), or, if the X-rays are energetic enough, may eject photoelectrons from the absorber. The ability of a pure element to

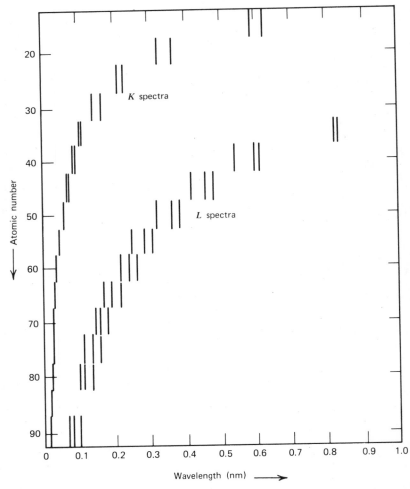

Fig. 12.3. The relationship between atomic numbers and wavelengths of X-ray emission lines. *Source*: Shalgosky.[6]

absorb X-rays varies with the X-ray energy or wavelength, as shown in Fig. 12.4. As the wavelength increases, a sharp discontinuity ("absorption edge") occurs where the X rays no longer have sufficient energy to remove K electrons from the absorber, and are then transmitted more readily. Further discontinuities occur at the maximum wavelengths capable of removing an electron from other successive subshells.

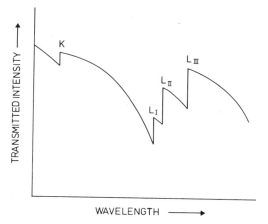

Fig. 12.4. The intensity of a beam of X rays after transmission through an absorber, plotted as a function of wavelength. *Source*: Shalgosky.[6]

The absorption of monochromatic X rays by an element follows a Beer's law type of relationship:

$$P = P_0 \exp(-\mu x) \qquad (12.7)$$

where P is the power of the X-ray beam of initial power P_0, after transmission through an absorber of thickness x, and μ is the linear absorption coefficient. More often, use is made of the mass absorption coefficient μ_m defined by

$$\mu_m = \frac{\mu}{\rho} \qquad (12.8)$$

where ρ is the density of the absorber. The mass absorption coefficient of an element varies with wavelength in a manner opposite to that of the transmitted beam (Fig. 12.4). Between absorption edges the value of μ_m for an element of atomic number Z varies with wavelength according to the empirical formula

$$\mu_m = \frac{CZ^4 \lambda^n}{A} \qquad (12.9)$$

where A is the atomic weight of the element, n is an exponent between 2.5 and 3.0, and C is a constant which changes value at each absorption edge.

12.2.3 Fluorescence Yield

X-ray fluorescence spectroscopy involves some rather inefficient processes. The low efficiency of conversion of electron energy into primary

X radiation in the X-ray tube is further compounded by the poor efficiency of the production of secondary X rays. The term *fluorescence yield*[8] is used to indicate the number of quanta emitted for every ionization occurring. Consider a single atomic species losing electrons from the K shell. When a steady state is reached, N_K atoms per unit time will be returning to the ground state, of which N_1 atoms yield the $K\alpha_1$ line, N_2 the $K\alpha_2$ line etc. The fluorescence yield W_K is given by the expression

$$W_K = \frac{(N_1 + N_2 + N_3 \ldots)}{N_K} = \frac{\sum N_f}{N_K} \qquad (12.10)$$

where $\sum N_f$ is the number of transitions for which K quanta emerge from the atoms. W_K is always less than unity because of the existence of radiationless processes such as the Auger effect. Radiationless deactivation is related to atomic number and increases appreciably as atomic numbers decrease. The problems of low fluorescence yields and unsuitable wavelengths therefore combine to make investigation of the lighter elements very difficult.

12.3 INSTRUMENTATION

12.3.1 Dispersive X-Ray Spectrometers

General Design

The design of a dispersive X-ray spectrometer is illustrated in Fig. 12.5. The sample, usually in the form of a compressed powder, is placed near the X-ray tube and the fluorescent X rays are collimated and diffracted by a suitable crystal according to the well-known Bragg equation:

$$n\lambda = 2d \sin \theta \qquad (12.11)$$

where n is the order of diffraction (1, 2, 3, ...), λ is the wavelength of the radiation, d is the distance between diffracting planes within the crystal, and θ is the angle of incidence of the radiation upon the crystal. The diffracted energy is collected by a suitable detector that was formerly a Geiger counter but is now usually a proportional or scintillation counter. The sample is scanned by rotating the crystal through an angle θ and the detector through an angle 2θ. The output from the detector can appear as a recorder tracing on paper graduated in values of 2θ. The position of a peak gives the identity of the element and its area or height is proportional to the concentration of the same element.

For hard X rays (<0.2 nm) being collected by a scintillation counter, an airpath is permissible in the various parts of the optical system. For inter-

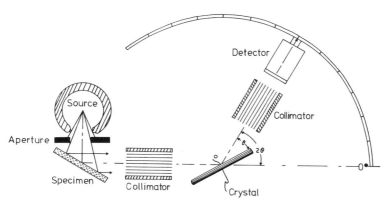

Fig. 12.5. Basic design of a dispersive X-ray emission spectrometer. *Source*: after Adler and Rose.[9]

mediate X rays (0.2–0.5 nm) the apparatus may be flushed with helium, whereas for soft X rays (>0.5 nm) a vacuum is essential. As a general rule only those elements being determined using a flow proportional counter ($Z < 26$) require a vacuum path.

Crystals

The crystals available commercially for use in dispersive instruments have values of $2d$ ranging from 0.16 nm (quartz) to 2.64 nm (potassium hydrogen phthalate). The optimum choice of crystal depends on the atomic number of the element(s) concerned and the series to which the lines belong. Other factors involve optimum resolution and/or intensity and the degree of interference due to crystal fluorescence.

Detectors

Although early spectrometers used Geiger counters for counting purposes, such counters suffer from the disadvantage that they have a limited ability to resolve rapidly occurring pulses (dead-time loss). This is not usually a problem in trace analysis because counting rates are small. Nevertheless, X-ray emission spectrometers are seldom designed exclusively for trace-element work and there is therefore a trend toward the use of proportional and scintillation counters. These two types of counter are energy-sensitive, whereas the Geiger counter with its higher applied voltage gives counts of equal size, independent of the energy of the ionizing pulse. The reader is referred to Chapter 14 and to various texts[10–14] for further information on counter design and operation.

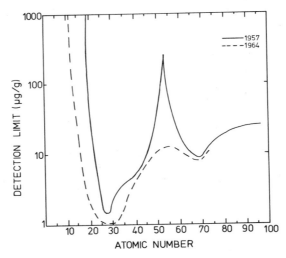

Fig. 12.6. Relationship between detection limit and atomic number for dispersive X-ray emission spectrometers. *Source*: Jenkins.[15] Copyright by N. V. Philips' Gloeilampenfabriken, Eindhoven, The Netherlands.

Detection Limits

Recent instrumentation has greatly improved the sensitivity of X-ray emission analysis. Figure 12.6 shows approximate detection limits for elements determined by dispersive instruments. The lower limit of detection is given as a concentration yielding a count rate equal to three times the standard deviation of the background count rate (for 99.7% confidence). It may be noted that, in general, only elements 18 to 40 can be determined with a sensitivity comparable to that obtained by emission and atomic absorption spectroscopy.

12.3.2 Nondispersive X-Ray Spectrometers

Some of the greatest advances in instrument design have been in the development of nondispersive X-ray spectrometers. Since the energy of X rays is inversely related to wavelength, any device that will sort X rays according to energy will obviate the need for a diffracting crystal and also avoid the problem of multiple-order diffraction. Furthermore, crystal optics are very inefficient and can result in a thousandfold loss of intensities of the fluorescent X rays. On the other hand, the resolution of nondispersive systems is usually poorer than for crystals. The poorer resolution of these systems, although greatly improved by the development of lithium-drifted silicon detectors, tends to cancel out some of the advantages of greater efficiency, at least as far as trace analysis is concerned.

Although it is doubtful at this stage whether nondispersive systems are the more satisfactory for trace analysis in terms of quantitative accuracy and precision, the continuing improvements in these systems will make them increasingly attractive for determination of trace constituents in geological materials.

12.3.3 Gamma-Source X-Ray Spectrometers

Gamma-source X-ray spectrometers[16] (Fig. 12.7) have been available commercially for the past decade. They are essentially very simple instruments. The X-ray tube is replaced by a radioactive gamma emitter. The gamma radiation irradiates the sample and induces X-ray fluorescence. The system is nondispersive as the crystal is replaced by balanced filters that are used as a matched pair in which the X-ray energy of the analyte element is bracketed

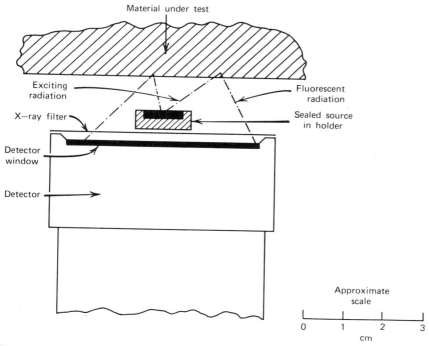

Fig. 12.7. Diagrammatic representation of a γ-source X-ray fluorescence analyzer. *Source*: Bowie.[16]

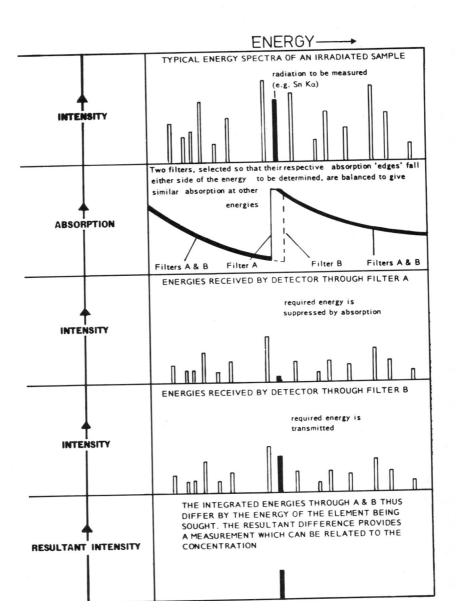

Fig. 12.8. Mechanism of operation of balanced filters for γ-source X-ray fluorescence analyzers. *Source*: after Hilger and Watts.[17] Reproduced by kind permission of Rank Hilger, Westwood, Margate, England.

TABLE 12.1 Sources, Filters, and Detection Limits for Gamma-Source X-Ray Emission Analysis of Rocks and Ores

Element	Source	Filter pair	Detection limit (μg/g)
Arsenic	^{238}Pu	Ge/Ga	1000
Chromium	^{238}Pu	V/Ti	1000
Cobalt	^{238}Pu	Mn/Fe	1000
Copper	^{238}Pu	Ni/Co	400
Gold	^{57}Co	W/Ta	400
Iron	^{238}Pu	Mn/Cr	2000
Lead	^{57}Co	Ir/Re	300
Manganese	^{238}Pu	V/Cr	1000
Molybdenum	^{109}Cd	Zr/Y	50
Nickel	^{238}Pu	Co/Fe	600
Niobium	^{109}Cd	Y/Sr	100
Silver	^{147}Pm	Ru/Rh	300
Tin	^{147}Pm	Ag/Pd	200
Titanium	^{55}Fe	Ti/Sc	1000
Tungsten	^{238}Pu	Cu/Ni	2000
Zinc	^{238}Pu	Cu/Ni	200

Source: Ekco.[18]

by the absorption edges of the filters. This is illustrated in Fig. 12.8. Two filters are selected with thicknesses that result in identical absorption except in the narrow window bracketing the analyte energy. Readings are taken first with one filter in place and then with the other. When the readings are subtracted, the residual is a measure of the intensity of the analyte line.

Gamma-source X-ray spectrometry is necessarily restrictive because only a limited number of elements can be measured with a given source. However, by use of a range of sources and balanced filters, 15–20 elements can readily be measured. Table 12.1 gives data on sources, filters and detection limits for 16 elements.

Because detection limits are not particularly good, the spectrometers are seldom used for trace analysis. At present the instruments are used mainly for the analysis of ores and as portable units suitable for semiquantitative field work. Improvements in instrumentation are resulting in lower detection limits, and it is probable that gamma-X units will have an increasing application to trace analysis in the future.

An alternative to gamma sources is provided by gamma-X spectrometry.[19] In this case a target cone is irradiated with gamma rays to produce monoenergetic primary X rays, which in turn excite X-ray fluorescence from the sample. In this case only a single filter is needed and a wide range of target cones is available to ensure selective excitation of the specific element being determined.

12.4 X-RAY SPECTROCHEMICAL ANALYSIS

The principles of qualitative and quantitative analysis have been discussed in several standard works.[10-14] Because of the intrinsic simplicity of X-ray spectra, qualitative analysis is very rapid. A scan of a typical rock specimen will give about 100 lines (compared with several thousand for atomic emission spectra), and the operation can be carried out in about 30 min with a dispersive instrument and in a much shorter time with a nondispersive spectrometer.

Quantitative analysis is considerably more involved than qualitative analysis because the emission intensity of an X-ray line is influenced by several factors other than the concentration of the analyte element. Two of these factors are the bulk composition of the sample and the thickness of that sample. Matrix effects are a problem in most instrumental methods of analysis but are particularly severe in X-ray emission analysis. A significant proportion of secondary (and primary) X-rays is absorbed by major constituents of the matrix and, unless standards are close in composition to the samples, accuracy will be poor.

The intensities of X-ray lines are a function of the thickness of the sample. As the thickness increases, so does the signal until a limiting thickness (critical thickness) is reached. Beyond this point the signal remains constant. It is essential that the size of the particles in the powder be appreciably less than the critical thickness, and that where possible "infinitely thick" samples (with respect to the wavelength of interest) be presented for analysis.

The homogeneity of the sample is also very important because the incident X-rays do not penetrate far into the sample. Rock samples should be ground to at least -200 mesh. Fine grinding also assists in overcoming particle-size effects, since the effective penetration depth for elements of $Z < 14$ in rocks and soils is only approximately 10 to 50 μm. The bulk density of the sample should also be kept approximately constant.

Some of the above problems are overcome by pressing the powdered rock sample (-200 mesh) into a briquette with a hydraulic press. It is usual to surround the sample with a cup made from compressed borax or salts of other light elements that do not produce an interfering spectrum. This treatment gives a homogeneous sample of approximately constant bulk density and smooth surface finish but does not take care of the matrix effect. Some workers prefer to fuse silicate samples into a glass with lithium tetraborate or borax.[20,21] This treatment reduces the matrix effect and ensures constant bulk density and homogeneity but also dilutes the sample and reduces sensitivity. The method is very useful for major element analysis, however, where count rates are relatively high.

Liquid samples may also be used in X-ray emission analysis, particularly where homogeneity of the sample in undissolved form is a problem, but there is a disadvantage in the high background intensity due to scattering of the primary radiation.

Since the interaction of X rays with the sample, and the fluorescent emission, are random processes, recorded count rates have a Gaussian distribution about a mean value. If time t is needed to collect N counts, the standard deviation of t, σ_t, is given by

$$\sigma_t = \frac{t}{N^{1/2}} \qquad (12.12)$$

As in arc emission spectroscopy, a background intensity must be subtracted from each line count, and since background is also subject to random error, the standard deviation for the net intensity of the line is

$$\sigma = \frac{(R/N)^{1/2}(R + 1)^{1/2}}{R - 1} \qquad (12.13)$$

where R is the ratio of the times taken to record N counts from background and sample, respectively.

Because of matrix effects, the method of additions is sometimes used in determination of individual elements over limited concentration ranges. Only two additions are required if the plots are linear. The original concentration is obtained by extrapolation in a manner analogous to that shown in Fig. 10.6 (p. 190).

When several elements are to be determined, internal standards may be used, as in optical emission spectroscopy. Alternatively, use may be made of previously analyzed standards of similar composition to the sample. A working curve can be constructed from these standards, minimizing the matrix problem.

An alternative way of dealing with the problem of variable matrices consists of correcting the measured analyte line intensity for the effects of other elements present. Many of the correction techniques use empirical equations to express the interelement effects and employ computer facilities to carry out the rather complex calculations. The empirical equations take a form such as

$$R_i = \frac{C_i}{\sum K_{ij} C_j} \qquad (12.14)$$

where C is a weight fraction, K_{ij} is the correction coefficient for the effect of element j on the analyte i, and R_i is the intensity of the analyte line in the multicomponent sample relative to that in pure element i. The summation is over all j, including $j = i$. Measurements with standards containing the elements of interest are used to generate the coefficients K_{ij}. Many equations of this kind have been developed, and their use has been reviewed by Rasberry and Heinrich.[22] These authors favor the use of an equation in which coefficients A_{ij} describing the depression caused by X-ray absorption are separated from coefficients B_{ij} describing the enhancement due to secondary fluorescence from the matrix:

$$\frac{C_i}{R_i} = 1 + \sum_{j \neq i} A_{ij} C_j + \sum_{j \neq i} \frac{B_{ij} C_j}{1 + C_i} \qquad (12.15)$$

The development of fully automated instrumental systems and computerized calculation facilities now allows the simultaneous determination of 15 or 20 elements in about a minute in a sample mounted in a vacuum instrument. Such instruments are analogous to direct-reading emission spectrometers. They have the disadvantage of employing a nondispersive system with lithium-drifted silicon detectors with resolution no better than about 130 eV at present.

Some of the advantages and disadvantages of X-ray fluorescence compared with atomic emission and atomic absorption spectroscopy have already been discussed (see Table 9.3), and the reader is referred to standard works for further details. It should be noted here, however, that X-ray fluorescence possesses the advantages of being nondestructive (at least in the sense that the same sample preparation may be analyzed repeatedly), that solids may be analyzed directly, and that it is useful for determinations (e.g., of some nonmetals such as the halogens[23,24]) that are difficult or impossible by other atomic spectroscopic methods. The nondestructive feature has made X-ray fluorescence valuable in the analysis of lunar material.[25]

Recent advances in X-ray spectrometric analysis can be followed in a number of reviews,[26-32] and books,[10-13,33] and in an abstracting service catering specifically for the field of X-ray emission.[34]

12.5 OTHER X-RAY EMISSION TECHNIQUES

12.5.1 The Electron Microprobe

In the electron microprobe[35] a beam of electrons about 0.25 μm in diameter from an electron gun is focused electromagnetically on to a small part of the sample. The optical arrangement is shown in Fig. 12.9. The sample is viewed through a microscope and the emitted X rays are examined by a dispersive instrument. The microprobe is capable of supplying *in situ* information on elemental composition by way of fixed bombardment or

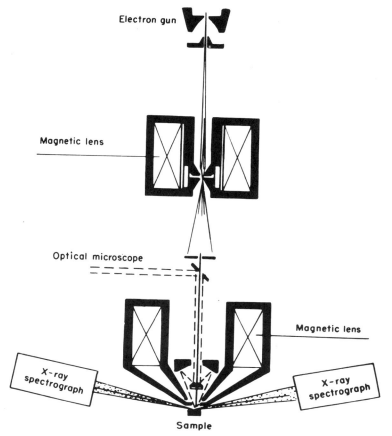

Fig. 12.9. Schematic diagram of electron microprobe. *Source*: Adler and Rose.[9]

X, Y scan on very small samples. The volume examined can be as small as 1 μm^3 (10^{-12} cm^3). The information can be displayed on a screen and photographed if needed, introducing a spatial dimension into chemical analysis. The probe is particularly useful in the examination of minerals. Although the absolute detection limits for many elements are very low because of the small sample size, this technique is for the investigation of major and minor elements and not for trace (<1000 μg/g) constituents.

12.5.2 Proton-Induced X-Ray Emission

Since the early 1970s, X-ray emission analysis has also been carried out using charged particles from accelerators to excite the X-ray emission.[36–39]

This procedure has the advantages over normal X-ray emission analysis of being able to use smaller samples (approximately 20 mg), and of having lower background levels.

The advantage of the small sample size is partially negated by the fact that protons have a low penetrating power and therefore extreme homogeneity of sample is needed. A further advantage of proton-induced emission is that matrix effects are much less pronounced than for other excitation procedures. In spite of these apparent advantages, proton-induced emission is being used mainly for the analysis of small samples (e.g., blood). The limited availability of the source and the necessity for use of poor-resolution nondispersive systems will probably have a limiting effect on the use of this technique for future trace analysis of geological material. Recently X rays from laser-induced plasmas have been studied,[40] but the full potential of the method may not become evident for some time.

References

1. K. Siegbahn et al., *ESCA Atomic Molecular and Solid State Studies by Means of Electron Spectroscopy*, Almquist and Wiksells, Uppsala, Sweden, 1967.
2. D. M. Hercules, L. E. Cox, S. Onisick, G. D. Nichols, and J. C. Carver, *Anal. Chem.*, **45**, 1973 (1973).
3. M. Czuha and W. M. Riggs, *Anal. Chem.*, **47**, 1836 (1975).
4. W. C. Röntgen, *Ann. Phys. Chem.*, **64**, 1 (1898).
5. H. Friedman and L. S. Birks, *Rev. Sci. Instrum.*, **19**, 323 (1948).
6. H. I. Shalgosky, in *Methods in Geochemistry* (eds. A. A. Smales and L. R. Wager), Interscience, New York, 1960, p. 111.
7. H. G. J. Moseley, *Phil. Mag.*, **26**, 1024 (1913).
8. A. H. Compton and S. K. Allison, *X-Rays in Theory and Experiment*, Van Nostrand, New York, 1951.
9. I. Adler and H. J. Rose, in *Trace Analysis—Physical Methods* (ed. G. H. Morrison), Interscience, New York, 1965, p. 271.
10. L. V. Azaroff (ed.), *X-Ray Spectroscopy*, McGraw-Hill, New York, 1974.
11. E. P. Bertin, *Principles of X-Ray Spectrochemical Analysis*, Plenum, New York, 1970.
12. L. S. Birks, *X-Ray Spectrochemical Analysis*, 2nd ed., Interscience, New York, 1969.
13. R. Jenkins and J. L. De Vries, *Practical X-Ray Spectrometry*, 2nd ed., Macmillan, London, 1970.
14. H. A. Liebhafsky, H. G. Pfeiffer, E. H. Winslow, and P. D. Zemany, *X-Ray Absorption and Emission in Analytical Chemistry*, Wiley, New York, 1960.
15. R. Jenkins, in *Proceedings of the 4th Conference on X-Ray Analytical Methods* (Sheffield), Philips, Eindhoven, 1964, p. 49.

16. S. H. O. Bowie, *Min. Mag.*, **118**, 1 (1968).
17. Hilger and Watts, *Catalogue CH-444*, London, 1968.
18. EKCO Instruments Ltd., *Catalogue L3182–12–69*, London, 1969.
19. J. S. Watt, *Int. J. Appl. Rad. Isot.*, **18**, 383 (1967).
20. K. Norrish and B. W. Chappell, in *Physical Methods in Determinative Mineralogy* (ed. J. Zussman), Academic Press, New York, 1967.
21. K. Norrish and J. Hutton, *Geochim. Cosmochim. Acta*, **33**, 431 (1969).
22. S. D. Rasberry and K. F. J. Heinrich, *Anal. Chem.*, **46**, 81 (1974).
23. B. P. Fabbi and L. F. Espos, *Appl. Spectrosc.*, **26**, 293 (1972).
24. P. J. Dunton, *Appl. Spectrosc.*, **22**, 99 (1968).
25. W. Gose (ed.), *Geochim. Cosmochim. Acta, Suppl.*, **4**, 1 (1973); **5**, 1 (1974).
26. R. Jenkins and J. L. De Vries, *Analyst*, **94**, 447 (1969).
27. L. S. Birks, *Appl. Spectrosc.*, **23**, 303 (1969).
28. K. G. Carr-Brion and K. W. Payne, *Analyst*, **95**, 977 (1970).
29. W. J. Campbell and J. V. Gilfrich, *Anal. Chem.*, **42**, 248R (1970).
30. L. S. Birks, *Anal. Chem.*, **44**, 557R (1972); **46**, 361R (1974); L. S. Birks and J. V. Gilfrich, *Anal. Chem.*, **48**, 273R (1976).
31. D. S. Urch, *Quart. Rev. Chem. Soc.*, **25**, 343 (1971).
32. R. Jenkins, in *M.T.P. International Review of Science*, Physical Chemistry Series One, Vol. 13: Analytical Chemistry, Part 2 (ed. T. S. West), Butterworth, London, 1973.
33. L. S. Birks, C. S. Barrett, J. B. Newkirk, and C. O. Ruud (eds.), *Advances in X-Ray Analysis*, Vol. 16, Plenum, New York, 1973.
34. P. R. Masek, I. Sutherland, and S. Grivell (eds.), *X-Ray Fluorescence Spectrometry Abstracts*, Sci. Technol. Agency, London, 1973.
35. S. J. B. Reed, *Electron Microprobe Analysis*, Cambridge University Press, Cambridge, England, 1975.
36. R. Akselsson, T. B. Johansson, and S. A. E. Johansson, *Norsk. Hyg. Tidskr.*, **53**, 11 (1972).
37. T. A. Cahill, *Bull. Am. Phys. Soc.*, **17**, 505 (1972).
38. T. B. Johansson, R. Akselsson, and S. A. E. Johansson, *Nucl. Phys. Rep.*, **LUNP-7109**, 24 (1971).
39. N. E. Whitehead, Inst. Nucl. Sci. (N. Z.) Rep., INS-R-136, 1 (1973).
40. D. J. Nagel et al., *Proc. 6th Eu. Conf. Fusion Plasma Phys.*, 1973, p. 447.

CHAPTER

13

RADIOMETRIC AND RADIOACTIVATION METHODS

13.1 RADIOACTIVITY AND ITS MEASUREMENT

13.1.1 Natural and Artificial Radioactivity

Several techniques important in the investigation of geological materials are based on the measurement of radiation emitted as a result of changes in atomic nuclei. Such changes occur spontaneously in naturally occurring unstable nuclei, but can also be induced in many others after activation of the nuclei by bombardment with particles such as neutrons or protons or with high-energy photons.

Measurements of natural radioactivity find particular use in prospecting for minerals containing uranium and thorium, and in the investigation of radioelements occurring in natural waters. Studies of the concentration of uranium, thorium, and other radioelements, are also important in the testing of theories of the origin and geological history of various rock bodies.

The artificial induction of radioactivity has been developed as a general analytical technique, applicable in some form to almost every element. For many elements the nuclear activation methods are among the most sensitive available, are capable of good precision at trace levels, and are applicable to a wide variety of materials whether they be solids, liquids, or gases. Activation by neutron bombardment is of particular importance, being used for trace analysis in high-purity metals and semiconductor materials as well as for many materials of biological and geological origin.

13.1.2 Radioactive Decay

Nuclear instability is characteristic of heavy nuclei and of nuclei with an unfavorable neutron:proton ratio. All nuclei heavier than ^{209}Bi undergo radioactive decay at measurable rates, and those with neutron:proton ratios falling outside the narrow range indicated in Fig. 13.1 are also unstable. The radioisotopes that survive in natural terrestrial matter are obviously those with half-lives that are not small compared to the age of the earth

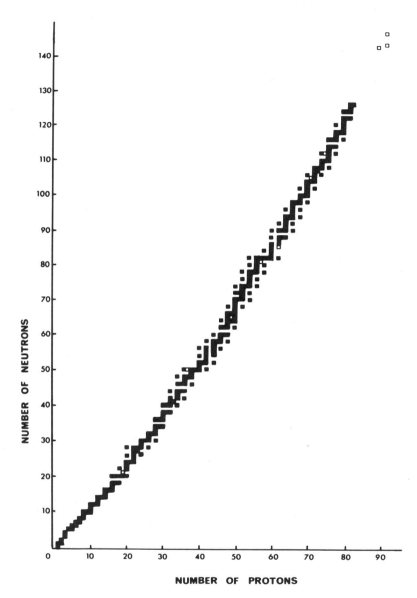

Fig. 13.1. Numbers of protons and neutrons in stable nuclei. ■ Stable nuclei (and those with $T_{1/2} > 2 \times 10^{11}$ years). □ Naturally occurring unstable nuclei with $T_{1/2}$ between 2×10^8 and 2×10^{11} years.

TABLE 13.1 Data on Naturally Occurring Radioisotopes[a]

Isotope	Isotopic abundance (%)	Half-life (years)	Mode of decay[b]	Decay products
$^{40}_{19}K$	0.0118	1.26×10^9	β^-; EC, γ (89%) (11%)	$^{40}_{20}Ca$; $^{40}_{18}Ar$
$^{87}_{37}Rb$	27.85	4.9×10^{10}	β^-	$^{87}_{38}Sr$
$^{138}_{57}La$	0.089	1.12×10^{11}	β^-, γ; EC, γ (30%) (70%)	$^{138}_{58}Ce$; $^{138}_{56}Ba$
$^{147}_{62}Sm$	14.97	1.05×10^{11}	α	$^{143}_{60}Nd$
$^{176}_{71}Lu$	2.59	2.6×10^{10}	β^-, γ	$^{176}_{72}Hf$
$^{187}_{75}Re$	62.93	4.3×10^{10}	β^-	$^{187}_{76}Os$

[a] Data mainly obtained from Lederer et al.[1]
[b] EC: electron capture.

(i.e., greater than about 10^9 years), or those derived from such long-lived isotopes. Natural radioisotopes include ^{40}K, ^{87}Rb, and the members of the radioactive decay series derived from ^{238}U, ^{235}U, and ^{232}Th. Some data on the lighter natural radioisotopes are given in Table 13.1, and the three naturally occurring radioactive decay series are given in Table 13.2.

Unstable nuclei decay by emitting an α particle (^{4}He nucleus), β particle (electron or positron), or γ radiation. In some cases decay consists of the capture by the nucleus of an inner-orbital electron, the process being described as electron-capture. Various secondary processes that occur as a consequence of the initial event may also be observed.

The rate of decay of any radioisotope is described by the equation

$$\frac{dN}{dt} = -\lambda N \tag{13.1}$$

where N is the number of radioactive atoms present at time t, and the proportionality constant λ is known as the decay constant. If N_0 is the number of radioactive atoms present at $t = 0$, then equation 13.1 can be integrated to give

$$N = N_0 \exp(-\lambda t) \tag{13.2}$$

The half-life of the radioisotope, $T_{1/2}$, is the time taken for N to decrease to $\frac{1}{2}N_0$.

$$T_{1/2} = \frac{\ln 2}{\lambda} = \frac{0.693}{\lambda} \tag{13.3}$$

TABLE 13.2 Natural Radioactive Decay Series[a]

Isotope	Half-life	Radiation emitted, and major energies[b] (MeV)	
^{238}U	4.51×10^9 yr	α; 4.20, 4.15	γ; 0.048
^{234}Th	24.1 day	β; 0.19	γ; 0.093, 0.063, 0.030
^{234m}Pa	1.17 min	β; 2.29	γ; 0.044
^{234}U	2.17×10^5 yr	α; 4.77, 4.72	γ; 0.053
^{230}Th	8.0×10^4 yr	α; 4.68, 4.62	γ; 0.068
^{226}Ra	1.60×10^3 yr	α; 4.78, 4.60	γ; 0.186
^{222}Rn	3.823 day	α; 5.49	
^{218}Po[c]	3.05 min	α; 6.00	
^{214}Pb	26.8 min	β; 1.03, 0.73, 0.67	γ; 0.352, 0.295, 0.242
^{214}Bi[c]	19.7 min	β; 3.26, 1.51, 1.0	γ; 0.609, 1.76, 1.12
^{214}Po	1.64×10^{-4} s	α; 7.69	
^{210}Pb	20.4 yr	β; 0.061, 0.015	γ; 0.047
^{210}Bi[c]	5.01 day	β; 1.16	
^{210}Po	138.4 day	β; 5.31	
^{206}Pb	stable		
^{232}Th	1.41×10^{10} yr	α; 4.01, 3.95	γ; 0.059
^{228}Ra	6.7 yr	β; 0.048	
^{228}Ac	6.13 hr	β; 2.18, 1.11, 0.45	γ; 0.91, 0.97, 0.338
^{228}Th	1.91 yr	α; 5.43, 5.34	γ; 0.084
^{224}Ra	3.64 day	α; 5.68, 5.45	γ; 0.241
^{220}Rn	55 s	α; 6.29	
^{216}Po[c]	0.15 s	α; 6.78	
^{212}Pb	10.64 hr	β; 0.35, 0.59, 0.17	γ; 0.239, 0.300
^{212}Bi	60.6 min	β; 2.27, 1.55	γ; 0.727, 1.62, 0.785
		α; 6.05, 6.09	γ; 0.040
^{212}Po[d]	3.04×10^{-7} s	α; 8.79	
^{208}Tl[d]	3.10 min	β; 1.79, 1.29, 1.52	γ; 2.615, 0.583, 0.511, 0.860
^{208}Pb	stable		
^{235}U	7.1×10^8 yr	α; 4.40, 4.37	γ; 0.185, 0.143
^{231}Th	25.5 hr	β; 0.30	γ; 0.084
^{231}Pa	3.25×10^4 yr	α; 5.01, 5.02, 4.94, 4.73, 5.06	γ; 0.027, 0.097, 0.299, 0.329
^{227}Ac[c]	21.6 yr	β; 0.046	γ; 0.009, 0.015, 0.025
^{227}Th	18.2 day	α; 6.04, 5.98, 5.76, 5.72	γ; 0.236, 0.050
^{223}Ra	11.4 day	α; 5.71, 5.61, 5.75, 5.54	γ; 0.270, 0.149
^{219}Rn	4.0 s	α; 6.82, 6.55, 6.42	γ; 0.271, 0.401
^{215}Po[c]	1.78×10^{-3} s	α; 7.38	
^{211}Pb	36.1 min	β; 1.36, 0.53	γ; 0.832, 0.405, 0.427
^{211}Bi[c]	2.15 min	α; 6.62, 6.28	γ; 0.351
^{207}Tl	4.79 min	β; 1.44	γ; 0.897
^{207}Pb	stable		

[a] Data mainly obtained from Lederer et al.[1]
[b] For β^- particles, E_{max} is quoted.
[c] Minor amount of chain branching occurs at this point.
[d] ^{212}Po from β^- decay of ^{212}Bi (64%); ^{208}Tl from α decay of ^{212}Bi (36%).

In the case where the daughter isotope is itself radioactive, as in the scheme

$$A \xrightarrow{\lambda_A} B \xrightarrow{\lambda_B} C$$

the rate of change of N_B, the number of atoms of B, is given by

$$\frac{dN_B}{dt} = \lambda_A N_A - \lambda_B N_B \qquad (13.4)$$

If $(N_A)_0$ atoms of isotope A alone are present at time $t = 0$, the variation of N_B with time is given by

$$N_B = \frac{\lambda_A (N_A)_0}{\lambda_B - \lambda_A} [\exp(-\lambda_A t) - \exp(-\lambda_B t)] \qquad (13.5)$$

Where $\lambda_A \ll \lambda_B$ (i.e., where the parent isotope A has a much longer half-life)

$$N_B \doteq \frac{\lambda_A (N_A)_0 \exp(-\lambda_A t)}{\lambda_B} \{1 - \exp[-(\lambda_B - \lambda_A)t]\}$$

that is

$$\lambda_B N_B \doteq \lambda_A (N_A)_0 \exp(-\lambda_A t)[1 - \exp(-\lambda_B t)]$$

After about five or six half-lives of B have elapsed, $\exp(-\lambda_B t) \ll 1$ and

$$\lambda_B N_B \doteq \lambda_A (N_A)_0 \exp(-\lambda_A t) \doteq \lambda_A N_A \qquad (13.6)$$

Where λ_A is sufficiently small that N_A can be considered constant over the period during which the system is observed, equation 13.6 describes the condition known as secular radioactive equilibrium.

In the cases of the radioactive decay series based on ^{238}U, ^{235}U, and ^{232}Th, the parent isotope half-lives are all several orders of magnitude greater than those of any other member of the series. After a sufficiently long period the secular equilibrium is reached, with

$$\lambda_A N_A = \lambda_B N_B = \lambda_C N_C = \cdots$$

The actual number of atoms of any isotope in the series is therefore inversely proportional to its decay constant, or directly proportional to its half-life. Inspection of the data of Table 13.2 shows that the thorium series attains equilibrium in about 40 years, but in the ^{238}U series equilibrium is reached only after about 1.5×10^6 years.

The relative amounts of the isotopes of a decay series occurring in a sample of a natural geological material depend both on the time elapsed since the parent isotope was last separated and on whether or not preferential loss of any member of the series has since occurred.

Amounts of radioisotopes are often expressed in terms of the curie (Ci), this being the quantity of the isotope undergoing 3.7×10^{10} disintegrations per second. This is approximately the activity of 1 g of ^{226}Ra or of any of its radioactive decay products in equilibrium with it.

13.1.3 Radiations Emitted and Their Interaction With Matter

α Decay

Many unstable heavy nuclei decay by α emission, the particles from a given source being of the same energy. An α particle loses energy in passing through matter by causing ionization and excitation of atoms and molecules in its path. Most α particles are emitted with an energy in the range 4 to 9 MeV, but the energy loss is so rapid that the particles travel only a few centimeters in air and considerably less in solid absorbers.

β^- Decay

Where the neutron:proton ratio is too high for the nucleus to be stable, decay can occur by a transformation represented by

$$^A_Z X \longrightarrow \ ^A_{Z+1} Y + \beta^- + \nu$$

The nucleus becomes that of the element with the next higher atomic number, and a β^- particle (electron) and an antineutrino (ν) are emitted. The β^- particles from a given source are emitted with a continuous range of energies from zero to a characteristic maximum energy, E_{max}, the decay energy being distributed between the β^- particle and the antineutrino. The greatest number of β^- particles is emitted with energies in the vicinity of $0.3 E_{max}$.

A β^- particle loses kinetic energy by causing ionization of the medium through which it passes. In addition, β^- particles of sufficient energy (>0.1 MeV) may interact with the electromagnetic field of an atomic nucleus and lose energy in the form of X rays with a continuous distribution of energies from zero up to E_{max}. The probability of occurrence of this type of emission (bremsstrahlung) is proportional to the β^- particle energy and to the square of the atomic number of the nucleus with which it interacts.

It is found that in many cases the logarithm of the β^- activity measured by a detector decreases approximately linearly with increases in the thickness of an absorbing material placed between the source and the detector. The detected β^- activity ceases at an absorber thickness, or range, related to E_{max} for the β^- particles concerned. The range in aluminum is approximately 0.7 mm for 0.6 MeV β^- and about 3.5 mm for 2.0 MeV β^- particles.

β^+ Decay

Where the nucleus has a neutron:proton ratio that is too low for stability, one possible mode of decay is by positron (β^+) and neutrino emission, represented by

$$_Z^A X \longrightarrow {}_{Z-1}^A Y + \beta^+ + \nu$$

The new atom also loses an orbital electron to preserve electrical neutrality. Positrons are generally short-lived: in the presence of an electron, mutual destruction occurs:

$$\beta^+ + \beta^- \longrightarrow 2\gamma$$

Mass-energy and momentum are conserved by the emission of two γ-ray photons, each of energy ≥ 0.51 MeV. Such γ-rays are described as annihilation radiation.

Electron Capture

Consideration of the atomic mass balance shows that positron emission occurs only when the energy of the daughter atom is at least 1.02 MeV lower than that of the parent atom. In other cases, a nucleus with a neutron deficiency can reach a stable state by capturing an orbital electron, frequently a K-shell electron, the process being described as electron capture (EC) or K-capture. Rearrangement of the orbital electrons then gives rise to X-ray emission characteristic of the daughter element.

γ Decay

Gamma rays consist of electromagnetic radiation, usually of higher energy than X rays, emitted as a result of transitions between nuclear energy levels. Decay by α or β emission may leave the daughter nucleus in an excited state, from which the ground state is reached almost immediately (in $< 10^{-9}$ s) by γ emission of one or several well-defined energies. In some cases the transition from the lowest excited nuclear energy level to the ground state is highly forbidden, giving rise to an identifiable metastable nuclear isomer. Examples include ^{110m}Ag($T_{1/2} = 253$ day), ^{46m}Sc($T_{1/2} = 18.7$ s) and ^{196m}Au($T_{1/2} = 14$ hr), which decay by γ emission to ^{110}Ag, ^{46}Sc, and ^{196}Au, respectively. Such transitions are described as isomeric transitions.

A γ ray emitted from an excited nucleus may interact with an orbital electron of the same atom, ejecting the electron with an energy equal to the original γ energy minus the binding energy of the electron. This process, internal conversion, leads to the emission of electrons that are monoenergetic, in contrast to β^- emission from the nucleus. The loss of an inner-orbital electron by internal conversion can be followed by X-ray emission as the orbital electron structure rearranges.

Gamma-radiation may interact with matter in several different ways, leading to complexities in observed γ-ray spectra. The most important interactions are the following.

1. *The photoelectric effect*, in which a low-energy γ-ray photon liberates an orbital electron from an atom of the absorbing material, with an energy equal to the photon energy minus the electron binding energy. This process is followed by X-ray emission.

2. *The Compton effect*, in which collision between a γ-ray photon and an outer-orbital electron of an atom causes transfer of only part of the γ energy. A continuous energy distribution between the deflected γ rays and the Compton electrons can be observed, the latter having energies up to a maximum value given by

$$(E_e)_{max} = E_\gamma \left(1 + \frac{m_0 c^2}{2E_\gamma}\right)^{-1} \tag{13.7}$$

where m_0 is the rest mass of the electron, c is the speed of light, and E_γ is the energy of the incident γ ray. The probability of both the photoelectric and the Compton effects decreases with increasing γ energy.

3. *Pair production*, in which a γ ray of energy greater than 1.02 MeV interacts with the intense electromagnetic field near an atomic nucleus to give rise to an electron-positron pair. The γ energy thus appears in the form of the rest mass of the two particles (equivalent to 1.02 MeV) and as kinetic energy of the particles. The positron, in turn, can undergo annihilation by interacting with an electron in its path, generating two γ rays of combined energy ≥ 1.02 MeV. The probability of pair production increases with increasing γ energy above 1.02 MeV.

Although the interactions of γ rays with matter take these various forms, with a relative importance which is energy-dependent, a plot of the logarithm of the observed activity of a simple γ emitter against absorber thickness is often linear. This plot can be used to obtain a half-thickness (i.e., thickness that reduces the observed activity by one-half), which is characteristic of the γ energy and of the absorbing material. Half-thicknesses in lead are approximately 3 g cm^{-2} for 0.4 MeV γ rays and approximately 15.6 g cm^{-2} for 2.0 MeV γ rays.

13.1.4 Methods of Measuring Nuclear Radiation

Most devices for measuring nuclear radiation use ionization or excitation processes that occur when the radiation interacts with the detector material.

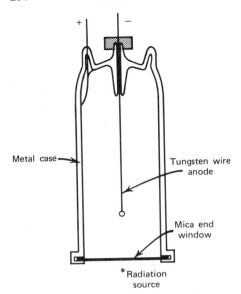

Fig. 13.2. Schematic diagram of a Geiger–Müller counter.

Ionization of gases, or the liberation of electron-hole pairs in semiconductors, can be followed by collection of electrical charge, leading to the generation of an output signal. Electronic excitation of various solid materials or solutes in solution can lead to fluorescent or phosphorescent emission (scintillation), which can be measured with a photomultiplier or similar detector. Several detection and measurement devices are discussed briefly below.

Gas-Ionization Counters

The most important device of this type is the *Geiger-Müller* (G-M) *counter* (Fig. 13.2), widely used as a β^- counter. It is seldom used for α particles because of their difficulty in penetrating the counting tube; γ radiation is measured more effectively with scintillation or semiconductor counters. The G-M tube consists of a central wire anode insulated from a surrounding cylindrical metal cathode. The tube contains argon or argon/neon as the counting gas at a pressure of a few centimeters of mercury, together with a small proportion of a polyatomic quenching gas, such as bromine or ethanol.

Ionizing radiation enters the tube through the thin end window, and causes the formation of a primary shower of electrons and positively charged ions of the inert gas. With a sufficiently high voltage applied between the electrodes, the primary electrons are accelerated toward the anode, causing further ionization. In the intense electric field around the anode, an avalanche of electrons is produced along the length of the wire and is collected as a current pulse, which operates the counting equipment. Under these conditions the magnitude of the pulse is independent of the extent of the primary ionization; amplification of the primary effect is of the order of 10^7 to 10^8. The speed of operation of the G-M tube is limited by the time taken ($\sim 100\ \mu s$) for the positive-ion cloud to travel outward toward the cathode to be neutralized. The quenching gas aids the dispersal of the energy of the positive ions and prevents them from liberating further electrons from the cathode wall. Because the tube is effectively inoperative for about $100\ \mu s$ after registering a pulse, it is desirable that count rates should not exceed about 10,000 counts per minute.

In the *proportional counter* the tube geometry, applied voltage, and gas pressure are such that the secondary electron shower is not propagated along the whole length of the anode wire, but is much more localized. Methane or argon/methane mixtures at about atmospheric pressure are used. The localized nature of the discharge ensures that a second pulse can be collected elsewhere on the wire within a few microseconds, and high count rates can be recorded. The gas amplification factor is of the order of 10^4 to 10^5, and external amplification of the pulses is needed to operate a scaler. If the energy of the radiation is completely absorbed in ionizing the counting gas, the output pulse is proportional to the energy of the radiation entering the counter. Under these conditions it is possible to use pulse height analysis to count particles of different energy and ionizing power.

Scintillation Detectors

It has long been known that when a particle or γ ray of sufficient energy strikes the surface of certain crystals, an excitation process occurs, followed by emission of a flash of light. Early studies of radioactive materials involved an observer counting scintillations from a zinc sulfide screen.

In modern apparatus, light of visible or ultraviolet wavelengths (often 410–450 nm) emitted by organic or inorganic "phosphors" is detected with a photomultiplier tube and amplifier. Organic phosphors include aromatic hydrocarbons such as anthracene, *p*-terphenyl, and stilbene; commonly used inorganic phosphors include crystals of sodium iodide activated with about 0.1% thallium impurity and zinc sulfide activated with silver.

Organic phosphors are used particularly for the counting of β^- particles. Where the β^- energy is low (e.g., ^{3}H, ^{14}C, ^{35}S) a liquid scintillation medium can be used, with the sample incorporated in the solution. Examples include *p*-terphenyl or 2,5-diphenyloxazole (PPO) in toluene as a solvent. In the latter case the ultraviolet emission can be translated into visible radiation, if desired, by adding 1,4-*bis*-[2-(5-phenyloxazolyl)]benzene (POPOP), which undergoes a secondary excitation and fluorescence.

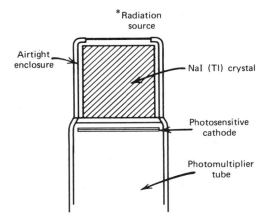

Fig. 13.3. Schematic diagram of a NaI(Tl) crystal scintillation counter.

For γ radiation the most widely used scintillation detector is NaI(Tl); ZnS(Ag) detectors are used for α particles. A typical arrangement of a sodium iodide detector is shown in Fig. 13.3. Various other types of sample-scintillator-detector geometry allow a larger proportion of the emitted γ-radiation to be detected.

Scintillation counting is of the proportional type: the amplitude of the output pulse is proportional to the energy of the radiation absorbed by the scintillator. Pulse-height analysis can therefore be carried out, and a multichannel analyzer can be used to give a complete energy spectrum of the emitted radiation. This is of particular value when the emitted radiation is of a limited number of discrete energies, as is the case with α and γ radiation. Figure 13.4a shows the γ-ray spectrum of ^{137}Cs obtained with a NaI(Tl) detector; the half-width (i.e., full width at half maximum height) of the 0.662 MeV photopeak is approximately 0.05 MeV or 7% of the γ-ray energy.

Because of the variety of interactions between γ radiation and matter, even isotopes emitting γ rays of a single energy can give rather complex γ-ray spectra. Where there is total absorption of the γ energy within the detector material (by the photoelectric effect and Compton effect, and by pair-production if the γ energy exceeds 1.02 MeV), this gives rise to the main high-energy photopeak. If, however, there is escape from the detector of (a) the X ray resulting from the photoelectric effect in the detector, (b) the secondary γ ray from the Compton effect, or (c) one or both γ rays from annihilation radiation following pair production, then pulses are recorded at energies below that of the main photopeak. The main features of γ- and β-spectra observed when using a NaI(Tl) detector are summarized in Table

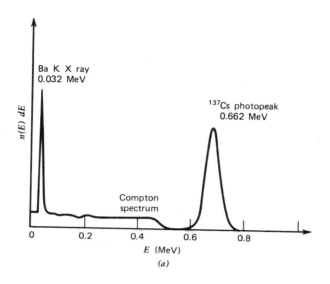
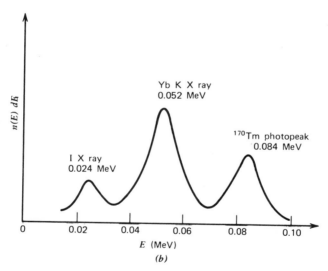

Fig. 13.4. Spectra of some γ emitters obtained with a NaI(Tl) detector. Number of counts per second in energy range between E, $E + dE$ plotted against E. (a) ^{137}Cs, showing main photopeak, Compton continuum, and X ray from Ba daughter. (b) ^{170}Tm, showing main photopeak, and X rays from Yb daughter and I in detector.

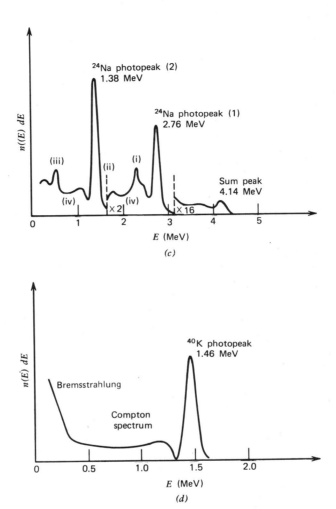

Fig. 13.4. Spectra of some γ emitters obtained with a NaI(Tl) detector. Numbers of counts per second in energy range between E, $E + dE$ plotted against E. (c) ^{24}Na, showing two main photopeaks (1) at 2.76 MeV, (2) at 1.38 MeV, and a sum peak at 4.14 MeV; a single escape peak at 2.25 MeV (i), a double escape peak at 1.74 MeV (ii), external annihilation peak at 0.51 MeV (iii), and Compton continua (iv). (d) ^{40}K, showing main photopeak, Compton continuum and bremsstrahlung. See text and Table 13.3 for additional details. *Source*: spectra (a), (b), and (c) after De Soete et al.[2]

RADIOACTIVITY AND ITS MEASUREMENT   263

TABLE 13.3   Features of γ- and β-Absorption Spectra with a NaI(Tl) Detector

Radiation reaching detector	Mechanism of energy loss	Radiation escaping from detector	Resulting energy detected
γ ray from source	Photoelectric effect, Compton effect, pair production (if $E_\gamma > 1.02$ MeV)	Nil	Photopeak corresponding to energy E_γ
γ ray from source	Photoelectric effect	X ray from detector (0.028 MeV for I)	Peak at E_γ minus 0.028 MeV; not detected if E_γ is large
	Compton effect	Scattered γ ray	Continuum below energy of Compton edge (equation 13.7)
	Pair production (if $E_\gamma > 1.02$ MeV)	One or both γ rays from annihilation	Peaks at E_γ minus 0.51 MeV; E_γ minus 1.02 MeV
γ ray from positron annihilation in materials surrounding detector	Photoelectric effect Compton effect	Nil	Peak at 0.51 MeV
β^- particle from source	Bremsstrahlung	Nil	Continuum increasing with decreasing β^- energy, below E_{max} of β^- particle
X ray from daughter isotope, following γ emission and internal conversion	Photoelectric effect, Compton effect	Nil	Peak corresponding to X-ray energy
		X ray from detector (0.028 MeV for I)	Peak at X-ray energy minus 0.028 MeV

13.3 and can be identified in the spectra, shown in Fig. 13.4, of the β^- and γ emitters ^{137}Cs, ^{170}Tm, ^{24}Na, and ^{40}K. In addition to the processes listed in Table 13.3, there is the possibility of observing the simultaneous absorption of two γ rays, which gives rise to sum peaks such as that shown in the ^{24}Na spectrum.

Semiconductor Detectors

Crystals of semiconductor materials can be prepared in which the absorption of radiation liberates electrons and positive holes. The movement of these electrons and holes toward opposite electrodes under an applied potential difference leads to the output of a current pulse that is proportional to the energy of the radiation absorbed. Semiconductor detectors include the

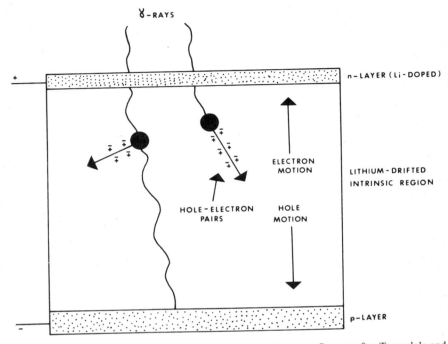

Fig. 13.5. Schematic diagram of Ge(Li) semiconductor detector. *Source*: after Tavendale and Ewan.[3]

surface-barrier and the PIN type (such as the lithium-drifted germanium detector), the latter being shown schematically in Fig. 13.5.

The silicon barrier detector, useful in the measurement of α particles, consists of a p-type semiconductor wafer (i.e., one containing an excess of holes from doping with an electron acceptor) with a thin metal layer deposited on each side. The radiation-sensitive region is a depletion layer approximately 0.1 to 1 mm thick below one of the electrodes.

The lithium-drifted germanium detector is made from a p-type germanium crystal. Diffusion and electrical drifting of lithium through much of the crystal, supplies electrons and returns the germanium to a so-called intrinsic state. Ionization is caused within this sensitive volume (up to about 50 cm^3) by the passage of γ rays. The electrons and positive holes migrate under the influence of the applied electric field, and a current pulse is recorded. The production of an electron-hole pair in germanium requires only 2.9 eV (3.6 eV in silicon), compared to the approximately 300 eV required to generate the photoelectron that leads to an output pulse of a NaI(Tl) scintillator. As a result, the

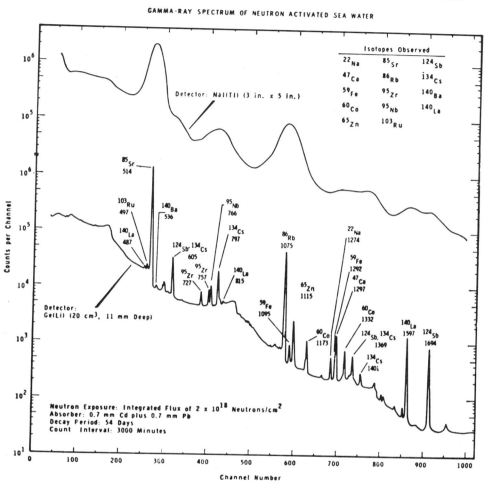

Fig. 13.6. Spectra of neutron-activated sea-water, taken with NaI(Tl) and Ge(Li) detectors, showing selectivity obtained by the higher resolution of the Ge(Li) detector. *Source*: Nielsen.[4]

statistical distribution of the conversion of radiant energy into an electrical signal is much less broad in semiconductor detectors. The half-width of the 0.662 MeV γ-ray of ^{137}Cs is approximately 2 to 5 keV with a Ge(Li) detector, a resolution about 10 to 25 times better than that achieved with a NaI(Tl) scintillator. To take advantage of the high resolution, a multichannel analyzer with a large number of channels (e.g., 1024–4096) is generally used.

The Ge(Li) detector requires cooling to liquid-nitrogen temperature to reduce thermal noise, and its efficiency is less than that for NaI(Tl). Silicon-based semiconductor detectors, which operate satisfactorily at dry-ice temperature, give good resolution and sensitivity for the measurement of X rays and low-energy γ rays.

The resolutions obtained with NaI(Tl) and Ge(Li) detectors are illustrated in Fig. 13.6, which shows the γ-ray spectrum of a neutron-activated seawater sample.

Further discussion of measuring devices for nuclear radiation, and on the techniques of γ-ray spectrometry, can be found elsewhere.[2,4-9] Data on scintillation γ-ray spectrometry have been assembled by Heath.[10]

13.2 TRACE ANALYSIS USING NATURAL RADIOACTIVITY

The long half-lives of the surviving natural radioisotopes make trace analysis by simple radiometric methods relatively unattractive in most cases. One milligram of potassium, for example, contains 0.0118 % of ^{40}K, sufficient to yield only 1.656 disintegrations per minute (dpm) by β^- decay and 0.204 dpm by electron capture and γ emission. Even with high counting efficiencies and elaborate precautions to minimize the recording of background radiation, radiometric determination of traces of potassium is not feasible. Radiometric analysis is, however, possible for potassium at higher concentrations (e.g., $>0.1\%$).[11-14] It must also be borne in mind that the decay of ^{40}K from the potassium concentrations of 0.5 % to 10 % commonly present in geological materials may make a significant contribution to the total radioactivity of samples being analyzed for other radioelements.

Because of the short range of α particles, and because the combined equilibrium α activity of ^{232}Th and ^{228}Th is only about 0.48 dpm per microgram, direct radiometric measurement of thorium at trace levels requires chemical separation and concentration of the α emitters (p. 270). An alternative method of analysis for thorium involves counting the 2.62 MeV γ radiation emitted by ^{208}Tl, assuming this to be in radioactive equilibrium with the parent thorium. In this case, adequate activity can be obtained even from samples containing part-per-million levels of thorium by using a large sample (e.g., 10–1000 g) and a large crystal NaI(Tl) scintillator. (If only submicrogram amounts of thorium are present in the amount of sample available, greater sensitivity is needed, and can be obtained by using neutron activation analysis.)

Direct determination of uranium by counting the α particles emitted by traces of ^{238}U, ^{235}U, and ^{234}U is also of limited applicability unless chemical separations are employed. However, use can be made of measurements of the 1.76 MeV γ-radiation emitted by ^{214}Bi (in the ^{238}U series). For samples enriched in ^{235}U, the 0.185 MeV γ ray of ^{235}U itself can be measured. Again, for small samples containing low levels of uranium, neutron activation analysis provides the necessary sensitivity (p. 289).

Radiometric measurements have found particular application in exploration for uranium and thorium, and apparatus has been developed both for laboratory assays and for studies in the field. Simple radiometric equipment such as the G-M counter gives an indication only of the total radioactivity of the sample to which it is exposed. Much more comprehensive information is provided by γ-ray spectrometry,[11-13,15-20] which allows simultaneous determinations to be made of thorium (via the 2.62 MeV γ-ray of ^{208}Tl), uranium (via the 1.76 MeV γ ray of ^{214}Bi), and potassium (1.46 MeV γ ray), as shown in Fig. 13.7.

Derivation of thorium and uranium concentrations from the γ-ray spectrum is possible if two assumptions can be made: (a) that radioactive equilibrium exists throughout the ^{232}Th and ^{238}U series, and (b) that the isotopic ratio ^{238}U/^{235}U is constant (approximately 138) in all terrestrial material. Although there is no evidence for fractionation of ^{238}U and ^{235}U in nature, the assumption of radioactive equilibrium is not valid in some circumstances. Soils, recent volcanic materials and sediments, and recently formed secondary minerals, may be found without equilibrium levels of their decay products. Alternatively, in some materials preferential loss of one or more members of the series (such as radium or radon) may have occurred; thorium and uranium are then best determined by a direct method.

The use of γ-ray spectrometry with NaI(Tl) or CsI(Tl) detectors in prospecting has been discussed by Adams et al.,[18] Lövborg et al.,[19] and Rybach.[20] For laboratory measurements standard procedures can be followed with respect to sample preparation (e.g., fine grinding to a uniform particle size), counting geometry, and counter shielding from natural background radiation. Spectrometers have also been developed that are suitable for use on foot or for carborne or airborne surveys.

In the field, the validity of the measurements is largely dependent on maintenance of known or constant sample-detector geometry. Most of the γ radiation recorded by a detector 2 m above the ground comes from a circular area approximately 30 m in diameter; that at 150 m is an average over a circle approximately 600 m in diameter. The observed count rate is sensitive to altitude as a result of the gradual attenuation of γ rays on passing through air and the change of effective solid angle subtended at the detector. Measurements of γ activity can usefully be made at altitudes up to about 300 m. Large areas can be surveyed rapidly by helicopter-borne detectors at lower altitudes, and linear anomalies with uranium and thorium levels approximately 15 to 20 μg/g above those of the surrounding rocks can be detected if the anomaly exceeds about 3 m in width.[18] It should be noted that, as more

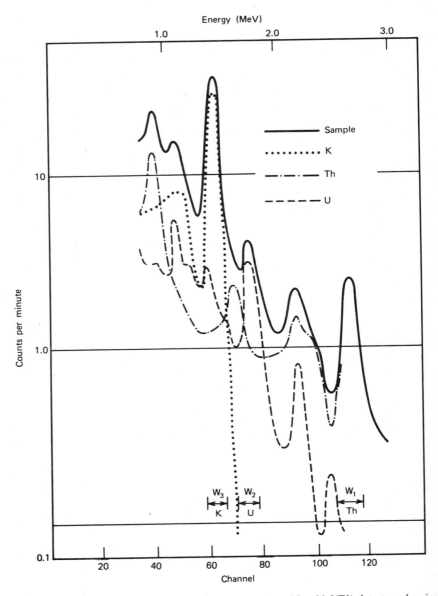

Fig. 13.7. Gamma-ray spectrum of a rock sample, taken with a NaI(Tl) detector, showing total spectrum and individual contributions from U(^{214}Bi), Th(^{208}Tl) and K. "Windows" W1, W2, and W3 indicate the range of channel numbers (or energies) used for obtaining activities. *Source*: After Rybach.[20]

than 90% of the observed radiation comes from the topmost 20 cm of rock material,[20] this type of measurement is unsuitable for detecting deeply buried ore bodies.

It is apparent from a spectrum such as that shown in Fig. 13.7 that Compton scattering of the 2.62 MeV γ ray from ^{208}Tl makes a contribution to the counts recorded in the channels corresponding to the 1.76 MeV γ ray from ^{214}Bi, necessitating a correction to the measured intensity of the latter peak. If potassium is also being determined, correction for the combined effects of Compton scattering of both the 2.62 MeV and 1.76 MeV γ rays must be made. Correction formulae can be derived that are valid for a given sample type and sample-detector geometry, calibrations being based on spectra obtained from standards (e.g., uraninite for uranium, monazite for thorium), which have been analyzed accurately by other methods. Details of calculation procedures have been described.[12,18,20]

Airborne γ-ray spectrometers can be calibrated by recording the γ-ray spectrum over several homogeneous rock areas of known radioelement content. Errors in aerial prospecting are introduced by the presence of soil zones, where there may be some departure from secular equilibrium in the decay series, and in areas of very variable topography, where it may be impossible to ensure that the detector is recording γ radiation from a constant mass of rock. Allowance for this mass effect must therefore be made when aerial prospecting is carried out over mountainous terrain and in narrow valleys.

Under laboratory conditions, precision of 4–10% (based on the standard deviation of the counting statistics) can be attained for thorium levels down to about 2 μg/g and for uranium levels down to about 1 μg/g.[11] If poorer precision is acceptable, γ spectrometry can be used down to about 0.2 μg/g Th and 0.1 μg/g U.[16,20] With stationary counters in the field, abundances of uranium and thorium in an average granite can be obtained with a precision of about 15% with 5-min counting[20]; precision of about 20% can be achieved in airborne surveys.[18]

Gamma-ray spectrometry with Ge(Li) detectors has recently been used for the laboratory study of naturally occurring γ emitters. The high resolution of peaks at low energies (e.g., 0.03 to 0.38 MeV) has made it possible to obtain direct measurements of the concentrations of various γ-emitting members of the decay series and hence also to investigate departures from radioactive equilibrium.[21,22] However, the low efficiency, high cost, and instrumental complexity appear to make the use of Ge(Li) detectors unsuitable for routine prospecting or ore analysis at the present time.

Many investigations have been made of the α-particle activity of materials of geological and environmental importance. In some cases measurements have been confined to total α activity. More often, chemical separations or specialized counting techniques have been used to relate the observed α activity to the concentrations of particular α emitters. Thorium, for example, has been estimated from the frequency of emission of pairs of α particles, about 0.2 s apart, by ^{220}Rn and ^{216}Po in the ^{232}Th decay series.[23,24] Uranium can then be determined from the total α activity by difference.

The chemical separation of α emitters is often a rather complex process, which may involve coprecipitation, solvent extraction, or ion exchange, followed by preparation of a thin source suitable for α counting. For elements such as uranium, thorium, plutonium, and americium, electrodeposition of hydrous oxides on a disk of platinum or stainless steel has been used.[25–29] The concentration of radium in solution can be determined conveniently by removing the ^{222}Rn ($T_{1/2}$ = 3.823 d) formed by decay of ^{226}Ra. About 1.2 μg of radon is produced per day from 1 g of radium. The equilibrium ratio is about 6.5 μg of radon (0.66 cm^3 at standard temperature and pressure) per gram of radium. After equilibrium has been established, or after a measured period, the radon can be bubbled out of solution with a stream of inert gas and transferred to a gas-ionization or scintillation counter.[30–32] The low background encountered in α counting enables radium to be determined at levels down to 0.1–1 pCi, corresponding to 10^{-12}–10^{-13} g Ra. Radium in solution has also been determined directly by coprecipitation with barium sulfate followed by proportional or scintillation counting.[30,33]

The investigation of natural and man-made α emitters has been facilitated by the recent development of α particle spectrometers incorporating silicon surface-barrier semiconductor detectors and multichannel analyzers. Applications have included the determination of thorium and uranium isotopes in sediments,[34] plutonium in seawater,[26] radium, actinium, and thorium isotopes in wastes and other environmental samples,[35] and a large number of α emitters, from radium to californium, in soil.[29]

A great deal of information about the occurrence and measurement of naturally occurring radioisotopes can be found elsewhere.[36] Detailed analytical procedures and calculation methods for the determination of radioisotopes of H, P, Sr, Ru, I, Cs, Po, Rn, Ra, Th, U, and Pu and of fission products such as ^{144}Ce, ^{140}Ba, ^{95}Zr, and ^{95}Nb have also been published.[30]

13.3 NEUTRON ACTIVATION ANALYSIS

13.3.1 Introduction

Activation analysis is based on the production of radioactive nuclei from stable nuclei subjected to irradiation by a flux of particles such as neutrons, protons, or deuterons, or by a flux of high-energy radiation (γ rays or energetic X rays). Some analytical work was performed by charged-particle activation[37] in the 1930s. However, since the development in the 1940s and 1950s of nuclear reactors with high fluxes of thermal neutrons, neutron activation has become the most commonly used form of activation analysis.

Most elements have at least one isotope that can be transformed by absorption of a thermal neutron into the isotope with mass number one greater. In many cases the new isotope is radioactive, as in the following examples.

$$^{75}_{33}\text{As} + ^{1}_{0}n \longrightarrow {}^{76}_{33}\text{As} + \gamma \qquad {}^{76}_{33}\text{As} \xrightarrow[26.4 \text{ hr}]{\beta^-, \gamma} {}^{76}_{34}\text{Se}(\text{stable})$$

$$^{133}_{55}\text{Cs} + ^{1}_{0}n \longrightarrow {}^{134m}_{55}\text{Cs} + \gamma \qquad {}^{134m}_{55}\text{Cs} \xrightarrow[2.9 \text{ hr}]{\gamma} {}^{134}_{55}\text{Cs} \xrightarrow[2.05 \text{ yr}]{\beta^-, \gamma}$$
$$^{134}_{56}\text{Ba}(\text{stable})$$

The above neutron activation processes are usually written more briefly as $^{75}\text{As}(n, \gamma)^{76}\text{As}$, and $^{133}\text{Cs}(n, \gamma)^{134m}\text{Cs}$.

By characterizing and measuring the radioactivity produced in a sample, both qualitative and quantitative analysis can be performed. There are many instances where the sensitivity of the method permits the determination of amounts of an element in the range 10^{-7}–10^{-11} g. A wide range of sample sizes can be used, from the microgram level up to 10–100 g.

Although it can be used to determine macroconstituents, neutron activation has been particularly powerful as a means of trace and ultratrace analysis, not only of geological materials, but also of samples of archaeological, biological, clinical, forensic, and metallurgical origin, and of semiconductors, polymers, and other industrial materials.

13.3.2 Growth and Decay of Induced Radioactivity

Consider a sample containing N_A nuclei of the stable isotope A, which is bombarded with a neutron flux ϕ, producing a nucleus B that is radioactive and has decay constant λ_B:

$$A \xrightarrow{n, \gamma} B \xrightarrow{\lambda_B} C \text{ (stable)}$$

The rate of formation of B nuclei is given by

$$\frac{dN_B}{dt} = \sigma_{\text{act}} N_A \phi \tag{13.8}$$

where σ_{act} is a constant, characteristic of this particular nuclear reaction for neutrons of a specified energy. This constant has dimensions of area and is known as the neutron activation cross section. When the rate of formation is expressed as a number of nuclei formed per second, and the neutron flux is expressed as a number of neutrons traversing each square centimeter of sample per second, the cross section σ_{act} has units of square centimeters. Values of σ_{act} are often quoted in barns (b); $1\ b = 10^{-24}\ cm^2$.

The radioisotope B decays according to

$$\frac{dN_B}{dt} = -\lambda_B N_B \qquad (13.9)$$

During the time of bombardment, the net rate of change of the number of radioactive nuclei is therefore

$$\frac{dN_B}{dt} = \sigma_{act} N_A \phi - \lambda_B N_B \qquad (13.10)$$

In the case where $N_B = 0$ at $t = 0$, and N_A remains virtually unchanged during the time of bombardment, equation 13.10 can be integrated to give

$$N_B = \frac{\sigma_{act} N_A \phi}{\lambda_B}[1 - \exp(-\lambda_B t)] \qquad (13.11)$$

The activity of B at any time during bombardment is therefore

$$a_B = \lambda_B N_B = \sigma_{act} N_A \phi[1 - \exp(-\lambda_B t)] \qquad (13.12)$$

As t becomes large, $\exp(-\lambda_B t)$ approaches zero. After a bombardment time greater than about six half-lives of B, the activity of B reaches the limiting value $\sigma_{act} N_A \phi$, as shown in Fig. 13.8. Where bombardment can be carried out for only a small fraction of the half-life of B, the activity developed is approximately

$$a_B = \lambda_B N_B \doteq \sigma_{act} N_A \phi \lambda_B t = \frac{0.693 \sigma_{act} N_A \phi t}{T_{1/2}} \qquad (13.13)$$

If the bombardment is stopped after a time t_b, the activity at this time is

$$(a_B)_{t_b} = \sigma_{act} N_A \phi[1 - \exp(-\lambda_B t_b)] \qquad (13.14)$$

and if a further time t_d (delay time) elapses before the activity of the sample is measured, the activity has decayed to

$$(a_B)_{t_b + t_d} = \sigma_{act} N_A \phi[1 - \exp(-\lambda_B t_b)]\exp(-\lambda_B t_d) \qquad (13.15)$$

In some cases the nucleus C formed by decay of B is itself radioactive:

$$A \xrightarrow{(n,\gamma)} B \xrightarrow{\lambda_B} C \xrightarrow{\lambda_C} D\ (stable)$$

It may be more convenient to measure the activity of C rather than that of B, especially if the half-life of B is very short, or if chemical separation of C is simpler, or if the radiation emitted by C is more easily measured. Provided

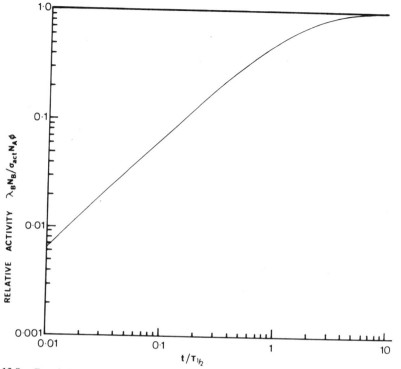

Fig. 13.8. Development of radioactivity in an isotope undergoing neutron activation, shown as a function of the number of product half-lives, $t/T_{1/2}$, for which activation is carried out.

the decay of B leads only to the formation of C, the net rate of formation of C at any time during bombardment is given by

$$\frac{dN_C}{dt} = \lambda_B N_B - \lambda_C N_C \qquad (13.16)$$

from which it can be shown[2] that the activity of C after bombardment for time t_b is

$$(a_C)_{t_b} = \frac{\sigma_{act} N_A \phi}{\lambda_C - \lambda_B} \{\lambda_C[1 - \exp(-\lambda_B t_b)] - \lambda_B[1 - \exp(-\lambda_C t_b)]\} \qquad (13.17)$$

and the activity of C after a delay time t_d is

$$(a_C)_{t_b + t_d} = \frac{\sigma_{act} N_A \phi}{\lambda_C - \lambda_B} \{\lambda_C[1 - \exp(-\lambda_B t_b)]\exp(-\lambda_B t_d) \\ - \lambda_B[1 - \exp(-\lambda_C t_b)]\exp(-\lambda_C t_d)\} \qquad (13.18)$$

In the case where B is very short-lived compared to C, where the sample is irradiated to the limiting B activity, and where the delay time allows B to decay completely, i.e., $\lambda_B \gg \lambda_C$, and $\exp(-\lambda_B t_b)$ and $\exp(-\lambda_B t_d)$ are each approximately zero, then

$$(a_C)_{t_b + t_d} = \sigma_{act} N_A \phi [1 - \exp(-\lambda_C t_b)] \exp(-\lambda_C t_d) \qquad (13.19)$$

More complex cases, including branching activation and second-order reactions (e.g., activation of A followed by a second activation of B or C), are discussed by De Soete et al.[2] General methods of solving differential equations encountered in radioisotope formation and decay schemes are given by Rubinson.[38]

If the sample irradiated contains a mass w of the element to be determined and if the fractional abundance of its isotope A is θ, then the number of atoms of A in the sample is

$$N_A = 6.02 \times 10^{23} \frac{\theta w}{M}$$

where M is the molar mass of the element. Substitution for N_A in the appropriate equation, such as equation 13.15, enables the relation between activity and mass to be obtained:

$$(a_B)_{t_b + t_d} = 6.02 \times 10^{23} \theta \frac{w}{M} \sigma_{act} \phi [1 - \exp(-\lambda_B t_b)] \exp(-\lambda_B t_d) \qquad (13.20)$$

The measured activity a'_B (in counts per unit time) is actually only a proportion of the activity a_B (in disintegrations per unit time), this proportion being determined by the efficiency with which the radiation is detected and counted, and on the extent of losses occurring during any chemical treatment of the sample after irradiation.

In principle, equations such as equation 13.20 allow w to be determined without reference to standards. However, because of difficulties in guaranteeing the constancy and homogeneity of the neutron flux, and some uncertainty in the value of σ_{act}, much better precision can be achieved by a comparative procedure. The sample and a standard containing a known mass w_s, of the element being determined, are irradiated together, taken through similar postirradiation steps, and their activities measured. It is then possible to calculate w from

$$\frac{w}{w_s} = \frac{a'_B}{(a'_B)_s} \qquad (13.21)$$

Equations such as equation 13.20 are still important, in that they enable predictions of activity to be made for given values of w, t_b, and t_d. The feasibility of an analysis by radioactivation can therefore be calculated before the analysis is attempted.

13.3.3 Neutron Capture and Neutron Sources

Neutron Capture

The capture of a neutron by a nucleus results in the formation of a compound nucleus; this process may be followed immediately by emission of a photon [(n, γ) reaction], emission of another particle such as a proton or α-particle [(n, p) or (n, α) reaction], or, in some cases, fission of the compound nucleus [(n, f) reaction]. Activation analysis makes use of processes in which the product of the nuclear reaction is radioactive.

Each reaction $A(n, x)B$ has a cross section, $\sigma(n, x)$, which is a measure of the probability of neutron capture by nucleus A, multiplied by the relative probability of emission of the photon or particle x. The magnitude of $\sigma(n, x)$ is characteristic of each isotope A for a particular x and for a given neutron energy. Each $\sigma(n, x)$, however, varies with neutron energy.

For most nuclei irradiated with thermal neutrons (i.e., those in thermal equilibrium with their surroundings) which have an average energy of about 0.025 eV and a most probable velocity of about 2200 m s^{-1} at 20°C, the (n, γ) process is the most important. For irradiations with fast neutrons with energies above 0.1 MeV, the (n, γ) processes are generally much less important than reactions such as (n, p), (n, α), or $(n, 2n)$. Further information on cross sections for (n, γ) and other nuclear reactions, and on the dependence of cross sections on neutron energy, can be found elsewhere.[2,39-41] Values of isotopic neutron capture cross sections for thermal neutrons are included in the compilation of Lederer et al.[1] Where thermal neutrons are used for irradiation, the activation cross section, σ_{act} (equation 13.8), can generally be taken as the capture cross section for the (n, γ) reaction for those isotopes that give a radioactive product.

Nuclear Reactors

The greatest amount of trace-element activation analysis has made use of reactor thermal neutrons. In a nuclear reactor, capture of a neutron by a ^{235}U nucleus leads to fission of the compound nucleus into two unequal fragments, generally with atomic masses in the ranges 80 to 103 and 131 to 147, and two or three very energetic neutrons. The fission neutrons have energy up to about 25 MeV and a most probable energy of about 1 MeV. They are slowed by elastic collisions with atoms of the moderator, and eventually reach thermal energies. In the region surrounding the reactor core where samples are placed for thermal neutron activation, the flux is typically in the range 10^{12} to 6×10^{13} thermal neutrons cm^{-2} s^{-1}. Significant fluxes of fast neutrons and partially slowed epithermal neutrons are also present.

Most elements have at least one isotope that gives rise to β^-, γ activity after irradiation with thermal neutrons. In a few cases (Si, P, S, Cr, Fe, Y, Bi, and the first eight elements of the periodic table), irradiation of microgram amounts of the element with thermal neutrons gives little or no γ activity, although useful β^- activity is given by yttrium. Sensitivities for activation by fast neutrons are better than those for thermal neutrons in some instances (O, Si, P, Fe, Y, Pb). Also, irradiation with epithermal neutrons can provide a more selective activation for some elements (U, Rb, Cs, Sb, Sr, Ta, Sm, Tb, Hf, Th, Ba), which is particularly useful when the analysis is being completed by purely instrumental means.[42-44] Potentially interfering thermal neutron reactions are prevented by enclosing the sample in a container of cadmium or boron, which absorbs thermal neutrons strongly.

There are a few cases in thermal neutron activation analysis where the fast neutron flux leads to interfering (n, α) or (n, p) reactions.[2,45] For example, in the determination of vanadium in a chromium-rich matrix, ^{52}V is formed both by (n, γ) activation of ^{51}V and by (n, p) activation of ^{52}Cr. Trace levels of aluminum in a silicon-rich matrix can be determined by neutron activation of ^{27}Al, but ^{28}Al is also produced from ^{28}Si by a (n, p) reaction and from ^{31}P by a (n, α) reaction if much phosphorus is also present. In such cases it may be possible to correct for the interfering fast-neutron activation processes by carrying out activations both with and without a thermal-neutron absorber.

14-MeV Neutron Generators

The bombardment of tritium with deuterons from a low-energy deuteron accelerator leads to the reaction ^{3}H$(d, n)^4$He, producing monoenergetic 14 MeV neutrons. These can be used to irradiate stable isotopes, in many cases yielding radioactive products as a result of processes such as (n, p) and $(n, 2n)$. Fluxes of the order of 10^8 to 10^9 neutrons cm^{-2} s^{-1} can be produced by bombardment of the tritiated target (usually a stable metal hydride formed by saturating a metal surface with tritium gas). The neutron flux is relatively inhomogeneous. The tritiated targets eventually become depleted, making this type of source most satisfactory for applications requiring only short irradiation times. Because the detection limits for many elements lie in the range 10^{-3} to 10^{-6} g, activation analysis with 14 MeV neutrons has been used more extensively for major and minor constituents of geological samples, rather than for trace elements. The reader is referred elsewhere for further discussion of 14 MeV neutron sources and activation analysis techniques.[2,45-50]

Isotopic Neutron Sources

Relatively small neutron fluxes can also be obtained from (α, n) or (γ, n) reactions involving certain light nuclei, the necessary α or γ radiation being provided by radioactive decay. Suitable α particles can be provided by the decay of isotopes such as ^{226}Ra or ^{241}Am. The α emitter is mixed with beryllium, which acts as the target material for the reaction ^{9}Be$(\alpha, n)^{12}$C. Because most of the neutrons produced have energies of 2 to 4 MeV, moderation is required if thermal neutrons are to be used for activation.

Alternatively, an emitter of γ radiation of energy above 1.67 MeV, such as ^{124}Sb, can produce neutrons from a beryllium target by the process ^{9}Be$(\gamma, n)2^4$He. Most of the neutrons from this reaction are of 25 to 30 keV, and can be moderated with water or paraffin wax.

Small neutron sources can be provided by the spontaneous fission of some transuranium isotopes. A 10-μg mass of ^{252}Cf ($T_{1/2}$ = 2.6 yr), for example, emits 3×10^7 neutrons s^{-1}, with a mean energy of 1.5 MeV. Largely because the neutron fluxes from isotopic sources are about 10^7 to 10^9 cm^{-2} s^{-1} (i.e., about four orders of magnitude lower than those of nuclear reactors), their use for trace analysis has been limited.

13.3.4 Activation Analysis Procedures

There are two basic approaches to the neutron activation analysis procedure. In one the sample after irradiation is taken through a sequence of chemical processes designed to isolate the induced radioactivity of interest. In the other the irradiated sample is exposed directly to a discriminatory counting device such as a γ-ray spectrometer. Most geochemical analyses by neutron activation reported before the late 1960s used the former approach. The development of Ge(Li) γ-ray spectrometer systems, however, has increased the use being made of the nondestructive, purely instrumental technique, which is obviously simpler and less time-consuming, and is suitable if the desired γ-ray energies are well resolved from one another. In some cases where many radioactive species are produced and are to be measured, it is possible to carry out rapid chemical separations into several groups of elements, before subjecting these groups to γ-ray spectrometry.

Sample Preparation

Solid geochemical samples are wrapped directly in aluminum foil and placed in aluminum cans, or are powdered and placed in ampoules of aluminum, silica, or polyethylene. It is only occasionally necessary or desirable to carry out concentration or separation steps prior to activation. Such a need may arise if the sample volume is very large, if activation of the sample matrix would lead to a total activity which is too great to handle conveniently, or if the desired activity is too short-lived to allow separations after activation. Particular precautions, including reagent blanks for chemical steps, need to be taken to guard against contamination of the sample during any process prior to activation. Inactive trace contamination occurring after irradiation is relatively unimportant, as the amount added will be negligible compared to that of the added carrier.

It is also possible to irradiate liquid samples in ampoules of quartz or polyethylene, although precautions must be taken to ensure that the samples and containers can withstand safely the temperatures of 100°C or more to which they are likely to be subjected. The problem of the possible loss of activity through adsorption from solution on to container walls is discussed by De Soete et al. (Ref. 2, pp. 253–255).

Irradiation

Details of sample transfer and irradiation facilities in various reactors are given by De Soete et al.,[2] Guinn,[45] and Fite et al.[47] The time for which irradiation is carried out is determined largely by the half-life of the radionuclide being produced (see equation 13.12 and Fig. 13.8). For short half-lives, irradiation to the saturation level (at least five or six half-lives) is possible, otherwise it is necessary to compromise, using a shorter irradiation time and obtaining a poorer sensitivity.

Samples and standards should be subjected to the same neutron flux. Standards should be thermally and radiolytically stable, but the choice of standard material is less critical than in analytical techniques such as emission spectrography and X-ray fluorescence, as matrix effects in irradiation are usually not large.

Chemical Separation

Chemical treatment after irradiation usually involves the following steps.

1. The sample is dissolved, using suitable acids or an alkali fusion mixture, with the addition of a known amount of the analyte element in nonradioactive form (inactive carrier). The solution is treated to ensure that there is complete chemical exchange between the element in its active and inactive forms. This may involve ensuring that all of the element is converted into a particular chemical species.

2. Chemical separations are carried out to isolate the element from others with interfering radioactivity. The irradiated standard is put through the same sequence of operations.

TABLE 13.4 Trace Elements Determined in Geological Materials by Neutron Activation with Chemical Separations

Element	Ref.	Element	Ref.	Element	Ref.
Ag	51–53	I	100	Sc	83, 111, 115, 133–136
As	54, 55	In	53, 76, 101–105	Se	107, 137, 138
Au	56–63	Ir	61, 106–108	Sn	119, 139, 140
Ba	64, 65	La[a]	76, 83, 109–118	Sr	141, 142
Bi	65–69	Mo	119	Ta	119–121, 143–146
Br	70–72	Nb	120, 121	Te	100, 137
Cd	68, 73–76	Ni	58, 79, 80, 122	Th	44, 147–155
Cl	77, 78	Os	108, 123–125	Ti	156
Co	58, 79–82	Pb	65	Tl	52, 65, 66, 68
Cr	81, 83, 84	Pd	58, 126, 127	U	42, 44, 64, 65, 100, 149–151, 157–161
Cs	85–87	Pt	61, 108, 126	V	134, 162
Cu	54, 58, 79, 80, 82, 88, 89	Rb	58, 85–87	W	82, 119, 143, 163
Fe	82	Re	57, 124, 125, 128–131	Y	76, 111, 115
Ga	82, 89–91	Rh	53	Zn	82, 89, 164–167
Hf	92–97	Ru	108, 123	Zr	92–97, 168
Hg	65, 66, 68, 98, 99	Sb	54, 55, 132		

[a] Indicates one or more of the lanthanides.

3. The specimens of the compound isolated from both the standard and the sample are weighed, and counted under identical conditions. The mass of compound obtained, in conjunction with the known mass of added carrier, is used to determine the yield of the chemical separation procedure.

4. Tests on the radiochemical purity of the separated radioactivity can be carried out by investigating both the half-life and the energy of the emitted radiation.

Widely used separation methods include precipitation, ion exchange, and solvent extraction. An element-by-element discussion cannot be given here, but the references listed in Table 13.4 can be be consulted for experimental details on neutron activation analysis of geological materials for single elements or small numbers of elements.

The chemical separation steps need not involve quantitative extraction of the element being determined. Suppose that w is the mass of the element in the sample, irradiated to an activity a' counts/s, w_c is the mass added as carrier, and m is the mass separated for counting. Then, provided $w_c \gg w$, the activity actually measured (a'' counts/s) is given by

$$a'' = \frac{m}{w + w_c} a' \doteq \frac{m}{w_c} a' \qquad (13.22)$$

If similar quantities are defined for the standard, denoted by the addition of the subscript s, then

$$a_s'' \doteq \frac{m_s}{w_{cs}} a_s' \qquad (13.23)$$

Combination of equations 13.21, 13.22, and 13.23 gives

$$w = w_s \frac{(w_c/m)a''}{(w_{cs}/m_s)a_s''} \qquad (13.24)$$

The measurement of m, m_s, provides a determination of the yield of the chemical separation process. This may be done by weighing the precipitates directly, or by other standard analytical techniques such as molecular or atomic absorption spectrophotometry. Details of methods for chemical yield determination for many elements have been summarized (Ref. 2, Appendix 7).

Reactivation of the sample provides an alternative method of yield determination.[55,155,169,170] The separated sample is counted, and its activity is then allowed to decay to a negligible level. It is then reactivated and counted again. In effect, the second measurement is used to calculate the yield, and the first measurement can then be used to calculate w. If the measured activities of the separated amounts of sample and standard after reactivation are a''', a_s''' counts/s, respectively, then

$$\frac{m}{m_s} = \frac{a'''}{a_s'''}$$

and

$$w = w_s \left(\frac{w_c}{w_{cs}}\right)\left(\frac{a_s'''}{a'''}\right)\left(\frac{a''}{a_s''}\right) \qquad (13.25)$$

Activation analysis can be performed without a yield determination if equal (but unmeasured) amounts of the element are extracted from both the sample and the standard. If the same amount of carrier is added to sample and standard, and if w_c, $w_{cs} \gg w$, w_s, any separation performed so that $m = m_s$ causes equation 13.24 to reduce to

$$w = w_s \frac{a''}{a_s''} \qquad (13.26)$$

Equal masses of the element can be separated by ensuring that the separation is carried out with a substoichiometric amount of reagent.[171,172] For example, in the substoichiometric extraction of silver with dithizone into carbon tetrachloride[173] the amount of dithizone used is insufficient to react with all the added silver carrier. Equal masses of silver are extracted, controlled by the equal amounts of dithizone used.

The use of substoichiometry in various types of radiochemical analysis is dealt with in a book[174] and several reviews.[175–177] Substoichiometric methods have been used in several of the applications listed in Table 13.4.[62,67,88,131,136]

Multielement Analysis with Chemical Separations

Although the development of γ spectrometry with the Ge(Li) detector has allowed activation analysis to be performed without chemical destruction of the sample, the purely instrumental technique does not always possess sufficient sensitivity, precision, or freedom from interference, and some degree of chemical separation may be required. Recent multielement work has involved the post-irradiation separation of up to 40 or 50 elements into a number of groups, which can be subjected to γ spectrometry without serious interference. Where NaI(Tl) detectors alone have been used, separation schemes have usually been designed to isolate a purified form of each element individually. A comprehensive scheme was developed by Schutz and Turekian,[178] for example, for determining nearly 20 trace elements in seawater. In the analysis of granite and diabase, Alian and Shabana[179] used multiple solvent extractions to separate seven elements prior to counting with a single-channel NaI(Tl) detector.

The use of Ge(Li) detectors has allowed the development of schemes that keep the chemical separations to a minimum. Sample dissolution (e.g., by alkali fusion with sodium peroxide or peroxide-hydroxide mixtures) is followed by group separations based on precipitation, volatilization, solvent extraction, or ion exchange. The immediate postirradiation activity of many geological materials is dominated by isotopes such as ^{24}Na ($T_{1/2} = 15.0$ hr) and ^{56}Mn ($T_{1/2} = 2.58$ hr) derived from precursors with high abundances and large neutron-capture cross sections. If it is not possible to wait for these activities to decay, it may be desirable to separate them from the matrix, even when high-resolution γ spectrometry is being used for the activity measurement. Sodium can conveniently be removed by adsorption on hydrated antimony pentoxide (HAP),[180–183] which also removes any tantalum activity.

An example of a separation scheme suitable for the determination of a large number of trace elements in small geological samples is that of Allen et al.,[184] shown in Fig. 13.9. Other schemes can be found in the work summarized in Table 13.5.

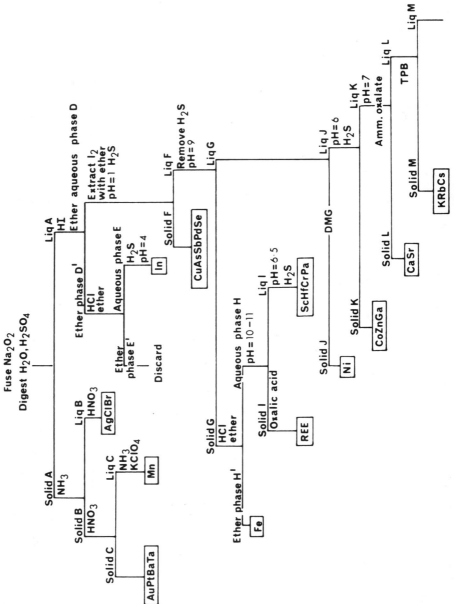

Fig. 13.9. Flow scheme for chemical separations in multielement neutron activation analysis. *Source:* Alian et al.[184]

TABLE 13.5 Multielement Analysis with Chemical Separations and γ Spectrometry

Elements	Sample	Notes	Ref.
Lanthanides	Rocks, minerals	Alkali fusion, group separation; lanthanides separated as hydroxides for Ge(Li) γ spectrometry	185
Al, As, Au, Ba, Ca, Cd, Co, Cr, Cs, Cu, Fe, Ga, Hf, Hg, In, K, Mo, Na, Ni, Np, Pa, Rb, Re, Sb, Sc, Sr, Ta, W, Zn, Zr, and 12 lanthanides	Rocks	Eight elements determined instrumentally before separation; others by acid dissolution and separation into six groups of 2–20 elements for Ge(Li) γ spectrometry	181
Ag, As, Au, Ba, Br, Cl, Co, Cr, Cs, Cu, Fe, Ga, Hf, In, K, Ni, Mn, Rb, Sb, Sc, Se, Sr, Ta, Zn, and 12 lanthanides	Lunar fines, basalt	Alkali fusion; separation into 12 groups; Ge(Li) and NaI(Tl) γ spectrometry	184
As, Au, Co, Cs, Ga, Ge, Hg, Mo, Os, Re, Sb, Sc, Se, Tl, Zn	Rocks, chondrites	Separation into groups for Ge(Li) γ spectrometry, using solvent extraction, ion exchange, distillation	186
Au, Cu, Ga, La, Mn, Sb	Glass	Acid dissolution, Ge(Li) γ spectrometry after separation of Na, Ta on HAP	182
Cr, Cs, P, Rb	Silicates	Alkali fusion; hydroxides precipitated; Rb, Cs, Cr by Ge(Li) γ spectrometry on filtrate	187
Lanthanides (12)	Lunar samples	Separated as hydroxides, converted into oxalates for Ge(Li) γ spectrometry	188
Ba, Cs, Rb, Sr, and 8 lanthanides	Ultramafic rocks	Alkali fusion; Rb, Cs precipitated as tetraphenylborates; ion exchange and solvent extraction gives remaining elements in three fractions	189
Au, Ir, Os, Pd, Pt, Ru	Geological materials	Alkali fusion; noble metals extracted on to chelating resin for Ge(Li) γ spectrometry	190
Ag, Au, Bi, Br, Cd, Co, Cs, Ga, Ge, In, Ir, Ni, Rb, Re, Sb, Se, Te, Tl, U, Zn	Meteorites, lunar and terrestrial samples	Alkali fusion; extensive chemical separations for NaI(Tl) γ counting; some β and X-ray counting also	191

Instrumental Neutron Activation Analysis

In some cases it is possible to carry out trace analysis on geological samples without chemical destruction of the sample and separation of the elements being sought. Activation of the sample is followed by direct measurement of the induced γ or X radiation or positron-annihilation activity, using a γ-ray spectrometer.

Both NaI(Tl) and Ge(Li) detectors have been used, in conjunction with multichannel pulse-height analyzers. The ouput from a NaI(Tl) scintillation detector is often recorded in 128 to 512 channels, but the higher resolution of the Ge(Li) semiconductor detector warrants the use of 1024 to 4096 channels. The relatively poor resolution of the NaI(Tl) detector restricts its use to matrices containing only a small number of γ emitters, as illustrated in the work of Brunfelt and Steinnes[192] and of Stueber and Goles.[193] As the bombardment of most geological materials gives rise to a great variety of γ emitters, the Ge(Li) detector is much more generally useful for multielement neutron activation analysis by the purely instrumental technique.

Treatment of the large amount of data acquired by a Ge(Li) detector-spectrometer system is facilitated by computer programs. These can aid such tasks as identifying all statistically significant photopeaks in the spectrum, calculating net photopeak areas above the Compton continuum, finding a weighted combination of pure radionuclide reference spectra that gives the best fit to the observed spectrum, and calculating concentrations of each element detected. Discussions of automated neutron activation analysis systems and of computer programs for the analysis of γ-ray spectra are given elsewhere.[2,47,194,195]

Instrumental neutron activation with Ge(Li) detectors has been used in the investigation of ores of silver[196] and the platinum metals,[197] and in the study of trace and minor elements in sulfide ores.[198] Many different rock types, including the U. S. Geological Survey standards, have been analyzed for a wide variety of elements by Cobb,[199] Gordon et al.,[200] and several other groups.[201–204] Filby et al.[183] have determined 32 elements in seven standard rocks, using a chemical separation of sodium on HAP prior to measurement of nine elements, the remainder being determined without chemical attack of the sample. Many elements have been measured instrumentally in atmospheric aerosols.[205,206] Lunar soil and rock samples have been investigated by instrumental methods,[207–210] radiochemical methods,[211,212] and a combination of the two.[213]

Several general reviews of developments in activation analysis have been published,[195,214–217] and other articles have dealt with applications to the analysis of ores and minerals,[218] to geochemical, meteoritic, and lunar studies,[219] and to environmental research.[215,220] The proceedings of an international colloquium[221] contain many papers of geochemical relevance.

13.3.5 Analysis by Neutron Activation and Fission Track Measurement

Neutron bombardment of samples containing uranium or other elements with fissionable isotopes, leads to the formation of fission products that act as massive energetic charged particles. In many materials, including micaceous minerals, the fission fragments leave linear trails of damaged material. By etching the material with a suitable reagent the fission tracks become sufficiently enlarged that they can be observed with an optical microscope and counted. Low concentrations of uranium in micaceous minerals were measured in this way by Price and Walker.[222] The uranium content of other materials can be determined by mounting the sample on the surface of a detector such as mica, silicate glass or a polycarbonate polymer prior to irradiation. Hydrofluoric acid solutions are suitable for track etching in mica or glass; sodium or potassium hydroxide solutions can be used for etching polycarbonate.

A total neutron dose of approximately 10^{15} to 10^{16} cm^{-2} has been found suitable for the measurement of uranium concentrations in the 0.01 to 1 μg/g range, using solid samples of 10 to 100 mg. Absolute detection limits for uranium reported by various workers are in the range 10^{-9} to 10^{-11} g.

Fleischer et al.[223] measured uranium concentrations from 4000 μg/g to less than 0.001 μg/g in various meteoritic minerals. Techniques have been described for determination of uranium in powdered rock[224,225] and soil[226] samples, and in solutions,[227,228] which can be evaporated on the detector surface. Where powdered solids are irradiated, the method is capable of indicating the degree of homogeneity of the uranium distribution in the sample. For example, Fisher[224] found considerable inhomogeneity of the uranium in the U. S. Geological Survey Standard rocks G-2 and GSP-1.

It should be noted that fission tracks can also be observed in uranium-bearing geological materials not irradiated by man. In most cases these arise from the spontaneous fission of ^{238}U (a very small proportion—about one in two million—of ^{238}U atoms decays in this way rather than by α decay). Measurement of naturally occurring fission tracks is of importance in fields such as geochronology and archaeology.[229]

13.3.6 Sensitivity of Neutron Activation Analysis

It is apparent from equations 13.12 and 13.20 that the activity induced in a given mass of a given element depends on the neutron flux, the fractional abundance of the isotope activated, the isotopic neutron capture cross section, the molar mass of the element, and the ratio of the irradiation time to the half-life of the radioisotope produced. The activity actually measured is lower than that produced at the end of bombardment because limitations of counting geometry and detector efficiency do not enable all disintegrations to be counted. In some cases significant decay occurs before counting is undertaken.

Neutron activation sensitivities for many elements have been given by Guinn,[45] based on the following conditions. (a) The sample is irradiated for no more than 1 hr with a flux of 10^{13} thermal neutrons $cm^{-2} s^{-1}$. (b) Radiochemical separation is assumed prior to β^- counting to obtain β^- detection limits, the minimum detectable counting rate being taken as 100 counts/min if $T_{1/2} < 1$ hr and 10 counts/min if $T_{1/2} > 1$ hr. (c) Gamma-spectrometry, using a 3 × 3-in. NaI(Tl) detector and a mean 2-cm sample-crystal distance, is assumed for γ detection limits, the minimum detectable photopeak counting rate being taken as 1000 counts/min ($T_{1/2} < 1$ min), 100 counts/min (1 min < $T_{1/2} < 1$ hr), or 10 counts/min ($T_{1/2} > 1$ hr). The sensitivities calculated on this basis, which generally agree well with experimental values, are given in Table 13.6, together with other data on the isotopes involved.

Better sensitivity than that listed can be achieved if higher neutron fluxes are available, and in the case of elements giving rise to activities with half-lives of more than about an hour, significant increases in activity will result from longer bombardment (Fig. 13.8).

13.3.7 Sources of Error

In common with other methods of analysis, neutron activation analysis is subject to systematic errors, affecting the accuracy of the result, and to random errors, which determine the precision.

Important sources of systematic error include the following. (a) Differences may exist in the neutron flux at the positions where the samples and standards are irradiated. This can be checked with flux monitor samples placed at the various locations. (b) Self-shielding errors may occur if the sample itself absorbs neutrons strongly enough to cause a significant attenuation of the thermal neutron flux throughout its volume, and a corresponding attenuation does not occur in the standard. (c) Systematic errors can occur during chemical separation procedures, for example, if care is not taken to ensure that all the radioisotope of interest is in the same chemical form for equilibration with the added carrier. (d) There may be differences in counting efficiency between samples and standards, caused by different sample-detector geometry. (e) There may be some self-absorption of the radiation emitted, particularly in the case of β^- activity. Where β^- counting of solids is used, both self absorption and backscatter from the sample support may be significant; matching of the precipitates prepared from samples and standards is important if the need to make corrections is to be avoided. In γ-emitting samples of high-density, low-energy γ-ray photons suffer Compton scattering,

TABLE 13.6 Limits of Detection for 1-hr Activation (Flux of 10^{13} Thermal Neutrons $cm^{-2}\ s^{-1}$)

				Limit of detection (μg)	
Element	Radionuclide	Half-life	γ-ray energy (MeV)	β^- counting	γ-ray spectrometry
F	^{20}F	11.56 s	1.63	—	0.1
Ne	^{23}Ne	37.6 s	0.439	—	0.5
Na	^{24}Na	14.96 hr	1.369	0.0004	0.0005
Mg	^{27}Mg	9.46 min	0.842	0.04	0.03
Al	^{28}Al	2.31 min	1.780	0.02	0.0009
Si	^{31}Si	2.62 hr	1.26	0.009	30.
P	^{32}P	14.28 days	—	0.02	—
S	^{35}S	87.9 days	—	2.	—
	^{37}S	5.07 min	3.09	10.	20.
Cl	^{38}Cl	37.29 min	1.60	0.002	0.01
Ar	^{41}Ar	1.83 hr	1.293	—	0.004
K	^{42}K	12.36 hr	1.524	0.002	0.04
Ca	^{49}Ca	8.8 min	3.10	0.09	0.2
Sc	^{46m}Sc	19.5 s	0.142	—	0.0003
	^{46}Sc	83.9 days	0.889	0.002	0.005
Ti	^{51}Ti	5.79 min	0.320	0.07	0.01
V	^{52}V	3.75 min	1.434	0.0004	0.00008
Cr	^{51}Cr	27.8 days	0.320	—	0.1
Mn	^{56}Mn	2.576 hr	0.847	0.000004	0.000007
Fe	^{59}Fe	45.6 days	1.095	9.	20.
Co	^{60m}Co	10.47 min	0.059	—	0.0002
	^{60}Co	5.263 yr	1.173	0.04	0.07
Ni	^{65}Ni	2.564 hr	1.481	0.004	0.04
Cu	^{64}Cu	12.80 hr	$(0.511)\beta^+$	0.0002	0.0006
	^{66}Cu	5.10 min	1.039	0.002	0.005
Zn	^{65}Zn	245. days	1.115	0.9	3.
	^{69m}Zn	13.8 hr	0.439	0.02	0.01
Ga	^{70}Ga	21.1 min	1.040	0.0004	0.1
	^{72}Ga	14.12 hr	0.835	0.0002	0.0004
Ge	^{75}Ge	82. min	0.265	0.0004	0.002
As	^{76}As	26.4 hr	0.559	0.0002	0.0006
Se	^{75}Se	120.4 days	0.265	—	0.1
	^{77m}Se	17.5 s	0.161	—	0.04
	^{81}Se	18.6 min	—	0.001	—
	^{81m}Se	56.8 min	0.103	—	0.009
Br	^{80}Br	17.6 min	0.618	0.00005	0.005
	^{80m}Br	4.38 hr	0.037	0.0001	0.0003
	^{82}Br	35.34 hr	0.554	0.0009	0.0008
Kr	^{81m}Kr	13. s	0.190	—	0.003
	^{85m}Kr	4.4 hr	0.150	—	0.001

TABLE 13.6 (*continued*)

Element	Radionuclide	Half-life	γ-ray energy (MeV)	Limit of detection (μg) β⁻ counting	Limit of detection (μg) γ-ray spectrometry
Rb	^{86}Rb	18.66 days	1.078	0.04	0.6
	^{86m}Rb	1.02 min	0.56	—	0.06
	^{88}Rb	17.8 min	1.863	0.009	0.09
Sr	^{87m}Sr	2.83 hr	0.388	0.0009	0.0006
Y	^{90}Y	64.0 hr	—	0.002	—
Zr	^{97}Zr	17.0 hr	0.747	0.2	0.3
Nb	^{94m}Nb	6.29 min	0.871	—	0.2
Mo	^{99}Mo	66.7 hr	0.181	0.02	0.1
	^{101}Mo	14.6 min	0.191	0.02	0.03
Ru	^{103}Ru	39.5 days	0.497	0.09	0.09
	^{105}Ru	4.44 hr	0.726	0.002	0.002
Rh	^{104m}Rh	4.41 min	0.051	—	0.00003
	^{104}Rh	43. s	0.56	—	0.002
Pd	^{109m}Pd	4.69 min	0.188	—	0.004
	^{109}Pd	13.47 hr	0.088	0.0002	0.0008
Ag	^{108}Ag	2.42 min	0.632	0.0009	0.002
	^{110}Ag	24.4 s	0.658	—	0.00006
Cd	^{111m}Cd	48.6 min	0.150	—	0.03
	^{115}Cd	53.5 hr	0.53	0.009	0.03
In	^{116ml}In	54.0 min	1.293	0.000005	0.00001
Sn	^{123m}Sn	39.5 min	0.160	0.07	0.02
	^{125m}Sn	9.5 min	0.325	0.05	0.03
Sb	^{122m}Sb	4.2 min	0.061	—	0.004
	^{122}Sb	2.80 days	0.564	0.0009	0.001
Te	^{131}Te	24.8 min	0.150	0.005	0.004
I	^{128}I	24.99 min	0.441	0.0002	0.0004
Xe	^{135}Xe	9.14 hr	0.250	—	0.01
Cs	^{134m}Cs	2.895 hr	0.128	—	0.00009
	^{134}Cs	2.046 yr	0.605	0.05	0.04
Ba	^{139}Ba	82.9 min	0.166	0.005	0.0002
La	^{140}La	40.22 hr	1.596	0.0004	0.0008
Ce	^{141}Ce	32.5 days	0.145	0.2	0.03
	^{143}Ce	33. hr	0.293	0.02	0.04
Pr	^{142}Pr	19.2 hr	1.57	0.0001	0.006
Nd	^{147}Nd	11.06 days	0.533	0.02	0.04
Sm	^{153}Sm	46.8 hr	0.103	0.00009	0.00006
Eu	^{152ml}Eu	9.3 hr	0.963	0.0000009	0.00001
Gd	^{159}Gd	18.0 hr	0.363	0.001	0.007
Tb	^{160}Tb	72.1 days	0.966	0.005	0.1
Dy	^{165}Dy	139.2 min	0.361	0.0000004	0.000004

TABLE 13.6 (*continued*)

				Limit of detection (μg)	
Element	Radionuclide	Half-life	γ-ray energy (MeV)	β^- counting	γ-ray spectrometry
Ho	^{165ml}Dy	1.26 min	0.108	—	0.000002
	^{166}Ho	26.9 hr	0.081	0.00004	0.00004
Er	^{171}Er	7.52 hr	0.308	0.0005	0.0009
Tm	^{170}Tm	134. days	0.084	0.002	0.03
Yb	^{169}Yb	31.8 days	0.198	—	0.004
	^{175}Yb	101. hr	0.396	0.0004	0.003
	^{177}Yb	1.9 hr	0.151	0.0002	0.002
Lu	^{176m}Lu	3.96 hr	0.088	0.000009	0.00004
	^{177}Lu	6.74 days	0.208	0.0001	0.0005
Hf	^{175}Hf	70. days	0.343	—	0.04
	^{178m}Hf	4.3 s	0.326	—	0.00005
	^{179m}Hf	18.6 s	0.217	—	0.004
	^{180m}Hf	5.5 hr	0.444	—	0.00005
	^{181}Hf	42.5 days	0.482	0.02	0.03
Ta	^{182m}Ta	16.5 min	0.172	—	0.02
	^{182}Ta	115.1 days	1.122	0.009	0.03
W	^{187}W	23.9 hr	0.686	0.0002	0.0004
Re	^{186}Re	88.9 hr	0.137	0.0002	0.0003
	^{188}Re	16.7 hr	0.155	0.00004	0.0001
Os	^{191}Os	15.0 days	0.129	0.02	0.09
	^{193}Os	31.5 hr	0.28	0.004	0.02
Ir	^{192ml}Ir	1.42 min	0.058	—	0.002
	^{192}Ir	74.2 days	0.317	0.0005	0.0004
	^{194}Ir	17.4 hr	0.328	0.00002	0.00005
Pt	^{199}Pt	31. min	0.197	0.004	0.01
Au	^{199}Au	3.15 days	0.158	0.02	0.02
	^{198}Au	2.697 days	0.412	0.00005	0.00004
Hg	^{197}Hg	65. hr	0.077	—	0.006
	^{197m}Hg	24. hr	0.134	—	0.002
	^{199m}Hg	43. min	0.158	—	0.3
	^{203}Hg	46.9 days	0.279	0.08	0.05
	^{205}Hg	5.5 min	0.205	0.06	0.04
Tl	^{204}Tl	3.81 yr	—	0.5	—
	^{206}Tl	4.19 min	—	0.04	—
Pb	^{207m}Pb	0.80 s	0.570	—	0.04
	^{209}Pb	3.30 hr	—	2.	—
Bi	^{210}Bi	5.013 days	—	0.05	—
Th	^{233}Pa	27.0 days	0.31	0.007	0.01
U	^{239}Np	2.346 days	0.106	0.0007	0.0004

From Ref. 45.

reach the detector with less than their characteristic energy, and do not contribute to the total-absorption photopeak. (f) The measured activity may be radiochemically impure, for example, through insufficient resolution of a γ-ray spectrum, or through imperfect separation of a desired β^- emitter. Within limits, this can often be checked by half-life studies on the activity.

Where chemical manipulations are carried out *after* activation, the presence of impurities in the reagents used has a negligible effect on the result. Furthermore, small and variable losses of the radioisotope during separation steps after equilibration with the carrier are not important, as the results are normalized by the chemical yield measurement. Nevertheless, it is desirable to maximize the yield to obtain a sample as active as possible, reducing the statistical errors in counting.

Random errors can also occur at all stages of the analysis—in sampling and preparation for irradiation, in the irradiation itself, in the subsequent chemical treatment (if any), and in the counting. The error arising from the random nature of the radioactive decay process can be estimated most readily.

If a radioactive sample is counted for a time t, and N counts are recorded, the standard deviation of N is $\sigma_N = \sqrt{N}$, and the standard deviation of the count rate is $\sigma_{N/t} = \sqrt{N}/t$, assuming a negligible error in the measurement of t. The statistics of the radioactive decay process are such that there is a 68.3% probability that a given observation of the count rate lies within one standard deviation of the true value, and a 95.5% probability that it lies within two standard deviations. In general, measurements are made of the extraneous background radiation (N_b counts in time t_b) and of sample plus background, (N_{s+b} counts in time t), the net sample count rate, a', being given by

$$a' = \frac{N_{s+b}}{t} - \frac{N_b}{t_b}$$

with standard deviation

$$\sigma = [\sigma_{s+b}^2 + \sigma_b^2]^{1/2}$$

$$= \left[\frac{N_{s+b}}{t^2} + \frac{N_b}{t_b^2}\right]^{1/2}$$

Short discussions of statistical considerations in γ-ray detection systems are given by Nielsen[4] and Guinn.[45]

Where a sample concentration is determined by comparing the sample activity with that of a standard, the fractional (or percentage) standard deviation of each measurement must be considered. If the percentage standard deviations of the net counting rates of sample and standard are $\%\sigma$ and $\%\sigma_s$, respectively, then the percentage standard deviation in the quotient of activities (equation 13.21) is

$$\%\sigma_q = [(\%\sigma)^2 + (\%\sigma_s)^2]^{1/2}$$

Repeated analysis of a given sample (n times) by the *same* procedure allows one to calculate the uncertainty $\sigma_{\bar{c}}$ of the mean concentration $\bar{c}$. This uncertainty reflects the random errors from all stages of the procedure, and is given by

$$\sigma_{\bar{c}} = \left[\sum_1^n \frac{(\bar{c} - c_i)^2}{n-1}\right]^{1/2}$$

By comparing the observed fractional standard deviation of the analytical results with that calculated for the counting alone, it is possible to assess whether or not the counting statistics are the major factor influencing the overall precision.

In the most careful neutron activation work, it is possible to achieve both precision and absolute accuracy of better than 1%, provided adequate activity can be induced. Precision of 1% to 5% is more typical of trace element determinations in geological materials, becoming poorer as the amounts to be measured approach those indicated in Table 13.6, or if less favorable activation or counting conditions are used.

An extensive discussion of systematic and random errors in neutron activation analysis, and of the statistical interpretation of results, is given by De Soete et al. (Ref. 2, Chapters 10, 11).

13.4 OTHER FORMS OF ACTIVATION ANALYSIS

Although most nuclear activation analysis is carried out with thermal, epithermal, or fast neutrons, the use of photons and charged particles has also been studied extensively. In some respects, photon activation and charged-particle activation are complementary to neutron activation, particularly in permitting determination of some elements for which neutron activation analysis is impossible or insufficiently sensitive.

γ-Ray Photon Activation

The use of low-energy γ-ray photons for activation analysis at trace levels has been restricted to the reaction $^9\text{Be}(\gamma, n)^8\text{Be} \rightarrow 2\,^4\text{He}$ which has a relatively low threshold energy (1.67 MeV). Beryllium at concentrations down to 60 μg/g has been determined by activation with a flux of 2.09 MeV γ-ray photons emitted by a ^{124}Sb source of 0.3 Ci.[230]

Many elements, however, undergo (γ, n) reactions and yield radioactive products when irradiated with photons of 15–35 MeV. Each (γ, n) reaction has a characteristic threshold energy and a major resonance at an energy E_{max} where the cross section is greatest. E_{max} lies in the range 20–25 MeV for light nuclei, decreasing to about 15 MeV for heavy nuclei. Photons of suitable energy are obtained as bremsstrahlung from the irradiation of targets such as platinum, gold, or tungsten with high-energy electrons from a source such as a linear accelerator.

Submicrogram amounts of many elements can be determined following photon irradiation for 1 hr or one half-life of the induced activity (whichever is the shorter). Sensitivities quoted by Baker[231] include the following. (The list shows the isotope activated, and a following asterisk indicates that the activity produced has $T_{1/2} < 10$ min.)

10^{-9} g: ^{63}Cu*

10^{-8}–10^{-7} g: ^{12}C, ^{14}N, ^{16}O*, ^{19}F, ^{31}P*, ^{35}Cl, ^{46}Ti, ^{50}Cr, ^{59}Co, ^{64}Zn, ^{79}Br*, ^{86}Sr, ^{90}Zr*, ^{93}Nb, ^{107}Ag, ^{123}Sb*, ^{181}Ta, ^{197}Au

10^{-7}–10^{-6} g: ^{24}Mg*, ^{27}Al*, ^{28}Si*, ^{32}S*, ^{40}Ca*, ^{54}Fe*, ^{58}Ni, ^{92}Mo, ^{127}I

Lutz[232] has tabulated sensitivities for many elements, for various irradiation periods and photon energies and has reviewed applications of photon activation analysis.[233] Geochemical applications have included the determination of low concentrations of fluorine in terrestrial and lunar material,[234] and of the halogens in seawater.[235] Although these particular applications involved radiochemical separations, the combination of photon activation and direct Ge(Li) γ spectrometry is also being exploited.

Charged-Particle Activation

Activation analysis using charged particles such as protons, deuterons, α particles, and ^{3}He nuclei has also been investigated. Particles are provided by isotopic sources (α-particles) or by charged-particle accelerators. Measurements are made of the prompt γ radiation emitted as a result of reactions such as (p, γ), of the activated nuclei (if radioactive), or of elastically or inelastically scattered particles. In some cases the reaction cross sections, particularly for elements of low atomic number, are large enough to make trace determinations feasible. Pierce et al.[236] have given sensitivities for activation by protons from a 0.5 MeV Cockroft-Walton set, and Engelmann[237] has shown that under favorable conditions, sensitivities of the order of 1 ng/g can be achieved in charged-particle activation of Be, B, C, N, O, and F in a variety of types of matrix. Up to the present, trace analysis by charged-particle activation has been applied more to materials such as high-purity metals and semiconductors than to geological samples.

Because of the limited penetration of the sample by charged particles (protons up to 3 MeV penetrate no deeper than about 50 μm into most materials), charged-particle activation is best regarded as a surface analysis technique. As the nuclear reaction cross sections are dependent on the particle energy, the extent to which different activation reactions occur can vary with depth in the sample. This situation requires obvious assumptions to be made if a bulk analysis is sought of a trace impurity. On the other hand, the possibility exists of using this technique to investigate changes of composition with depth.

13.5 RADIOCHEMICAL ISOTOPE DILUTION

Analysis by isotope dilution may be carried out with radioactive or inactive isotopes; the latter technique is described in Chapter 15. The radiochemical version involves taking a weighed amount of sample, containing an unknown mass w g of the element to be determined (X), and adding a known mass w_X of X in which some of the atoms of X are radioactive. The mixture is treated to ensure complete equilibration of the active and inactive forms, and a suitable compound of X is then extracted and purified. If the specific activity of element X in the radioactive addition is S_i counts s^{-1} g^{-1}, and that in the final product is S_f counts s^{-1} g^{-1}, then

$$w_X S_i = (w_X + w) S_f \qquad (13.27)$$

assuming negligible decay between the two measurements of specific activity. Hence

$$w = w_X \left(\frac{S_i}{S_f} - 1 \right) \qquad (13.28)$$

The ratio S_i/S_f is determined from two measurements of activity and two of mass.

Isotope dilution procedures for many elements have been listed by Tolgyessy et al.[238] An early application of this technique was that of von Hevesy and Hobbie,[239] who determined lead in rocks at levels of a few micrograms per gram.

In general, difficulties may arise in trace analysis if a large sample is not available. In this case, w will be very small. Unless w_X is also made very small there will be little difference between S_i and S_f and the possible error in the determination of $(S_i/S_f - 1)$ may make the analysis meaningless. If w_X is made very small (comparable to w) the errors and inconvenience associated with handling microscopic amounts of substances are introduced.

Substoichiometric extraction methods are valuable in overcoming this problem. Equal small (but unmeasured) amounts of the element are separated both from the original radioactive tracer solution and from the sample with tracer added, by using a substoichiometric amount of the extracting reagent (e.g., a complexing reagent or precipitant). Equal masses of the element having been separated, the ratio of the measured activities is therefore identical to the ratio of specific activities, S_i/S_f. To avoid interference, extraction conditions must be chosen so that the equilibrium constant of the extraction process is much higher for the element being determined than for any other constituent of the sample.

The principles of substoichiometric isotope dilution based on solvent extraction were outlined by Ruzicka and Stary,[240] who have given methods and applications for many elements in a book[174] and a review.[177] Recent geochemical applications have included the determination in rocks of palladium[241] in amounts down to 10 ng, indium[242] (0.01–0.14 µg/g with 4–7% precision using 0.5 g samples) and lead[243] (7–60 µg/g with 3–10% precision using 0.1 g samples). An automated substoichiometric procedure suitable for determining mercury in rocks and low grade ores has also been devised.[244]

References

1. C. M. Lederer, J. M. Hollander, and I. Perlman, *Table of Isotopes*, 6th ed., Wiley, New York, 1967.
2. D. De Soete, R. Gijbels, and J. Hoste, *Neutron Activation Analysis*, Wiley-Interscience, New York, 1972.
3. A. J. Tavendale and G. T. Ewan, *Nucl. Instr. Meth.*, **25**, 185 (1963).
4. J. M. Nielsen, in *Physical Methods of Chemistry* (eds. A. Weissberger and B. W. Rossiter), Vol. I, Part III D, Wiley-Interscience, New York, 1972.
5. H. M. Clark, in *Physical Methods of Chemistry* (eds. A. Weissberger and B. W. Rossiter), Vol. I, Part III D, Wiley-Interscience, New York, 1972.
6. J. Birks, *The Theory and Practice of Scintillation Counting*, Pergamon, London, 1964.
7. I. C. Brownridge, *Lithium-Drifted Germanium Detectors*, Plenum, London, 1972.
8. J. B. A. England, *Techniques in Nuclear Structure Physics*, Part 1, Macmillan, London, 1974.
9. J. B. A. England, *J. Phys. E: Sci. Instrum.*, **9**, 233 (1976).
10. R. L. Heath, *Scintillation Spectrometry Gamma-Ray Spectrum Catalogue*, 2nd ed., Office of Technical Services, U. S. Department of Commerce, Washington, D. C., 1964.
11. K. S. Heier and J. J. W. Rogers, *Geochim. Cosmochim. Acta*, **27**, 137 (1963).

12. R. D. Cherry and J. A. S. Adams, *Geochim. Cosmochim. Acta*, **27**, 1089 (1963).
13. J. W. Morgan and K. S. Heier, *Earth Planet. Sci. Lett.*, **1**, 158 (1966).
14. L. Civetta, P. Gasparini, and J. A. S. Adams, *Anal. Chem.*, **41**, 1319 (1969).
15. P. M. Hurley, *Bull. Geol. Soc. Amer.*, **67**, 395, 405 (1956).
16. K. S. Heier, *Geochim. Cosmochim. Acta*, **27**, 849 (1963).
17. L. Rybach and J. A. S. Adams, *Geochim. Cosmochim. Acta*, **33**, 1101 (1969).
18. J. A. S. Adams, G. E. Fryer, and J. J. W. Rogers, *Anal. Chem.*, **41** (6), 22A (1969).
19. L. Lövborg, H. Wollenberg, P. Sörensen, and J. Hansen, *Econ. Geol.*, **66**, 368 (1971).
20. L. Rybach, in *Modern Methods of Geochemical Analysis* (eds. R. E. Wainerdi and E. A. Uken), Plenum, New York, 1971.
21. G. Mathévon, *C. R. Acad. Sci., Paris, Ser. A, B*, **B268**, 353 (1969).
22. G. E. Coote, N. E. Whitehead, and N. E. Cohen, *N. Z. J. Sci.*, **13**, 610 (1970).
23. R. D. Cherry, *Nature*, **195**, 1184 (1962).
24. R. D. Cherry, *Geochim. Cosmochim. Acta*, **27**, 183 (1963).
25. E. D. Goldberg and M. Koide, *Geochim. Cosmochim. Acta*, **26**, 417 (1962).
26. K. M. Wong, *Anal. Chim. Acta*, **56**, 355 (1971).
27. N. A. Talvitie, *Anal. Chem.*, **44**, 280 (1972).
28. K. W. Puphal and D. R. Olsen, *Anal. Chem.*, **44**, 284, 1301 (1972).
29. C. W. Sill, K. W. Puphal, and F. D. Hindman, *Anal. Chem.*, **46**, 1725 (1974).
30. World Health Organization, *Methods of Radiochemical Analysis*, Geneva, 1966.
31. K. G. Darrall, P. J. Richardson, and J. F. C. Tyler, *Analyst*, **98**, 610 (1973).
32. I. Kobal and J. Kristan, *Radiochem. Radioanal. Lett.*, **10**, 291 (1972).
33. A. S. Goldin, *Anal. Chem.*, **33**, 406 (1961).
34. M. R. Scott, *Earth Planet. Sci. Lett.*, **4**, 245 (1968).
35. D. R. Percival and D. B. Martin, *Anal. Chem.*, **46**, 1742 (1974).
36. J. A. S. Adams and W. M. Lowder (eds.), *The Natural Radiation Environment*, University of Chicago Press, Chicago, 1964.
37. G. T. Seaborg and J. Livingood, *J. Am. Chem. Soc.*, **60**, 1784 (1938).
38. W. Rubinson, *J. Chem. Phys.*, **17**, 542 (1949).
39. D. J. Hughes, *Neutron Cross Sections*, Pergamon, London (1957).
40. D. Mapper, in *Methods in Geochemistry* (eds. A. A. Smales and L. R. Wager), Interscience, New York, 1960.
41. B. T. Kenna and P. E. Harrison, Rep. U. S. Atomic Energy Commission, SLA-73-637 (1973).
42. E. Steinnes and D. Brune, *Talanta*, **16**, 1326 (1969).
43. A. O. Brunfelt and E. Steinnes, *Anal. Chim. Acta*, **48**, 13 (1969).
44. H. G. Meyer, *J. Radioanal. Chem.*, **7**, 67 (1971).
45. V. P. Guinn, in *Physical Methods of Chemistry* (eds. A. Weissberger and B. W. Rossiter), Vol. I, Part III D, Wiley-Interscience, New York, 1972.

46. S. S. Nargolwalla and E. P. Przybylowicz, *Activation Analysis with Neutron Generators*, Wiley, Chichester, 1974.
47. L. E. Fite, E. A. Schweikert, R. E. Wainerdi, and E. A. Uken, in *Modern Methods of Geochemical Analysis* (eds. R. E. Wainerdi and E. A. Uken), Plenum, New York, 1971.
48. J. S. Hislop and R. E. Wainerdi, *Anal. Chem.*, **39** (2), 29A (1967).
49. M. Y. Cuypers and J. Cuypers, *J. Radioanal. Chem.*, **1**, 243 (1968).
50. E. A. Uken, *Miner. Sci. Eng.*, **2**, 24 (1970).
51. D. F. C. Morris and R. A. Killick, *Anal. Chim. Acta*, **20**, 587 (1959).
52. D. F. C. Morris and R. A. Killick, *Talanta*, **4**, 51 (1960).
53. U. Schindewolf and M. Wahlgren, *Geochim. Cosmochim. Acta*, **18**, 36 (1960).
54. S. Gohda, *Bull. Chem. Soc. Jap.*, **45**, 1704 (1972).
55. E. Steinnes, *Analyst*, **97**, 241 (1972).
56. E. D. Goldberg and H. S. Brown, *Anal. Chem.*, **22**, 308 (1950).
57. A. A. Smales, *Geochim. Cosmochim. Acta*, **8**, 300 (1955).
58. E. A. Vincent and J. H. Crocket, *Geochim. Cosmochim. Acta*, **18**, 130, 143 (1960).
59. A. R. Degrazia and L. A. Haskin, *Geochim. Cosmochim. Acta*, **28**, 559 (1964).
60. P. A. Baedecker and W. D. Ehmann, *Geochim. Cosmochim. Acta*, **29**, 329 (1965).
61. D. A. Beardsley, G. B. Briscoe, J. Ruzicka, and M. Williams, *Talanta*, **12**, 829 (1965).
62. A. Chow and F. E. Beamish, *Talanta*, **14**, 219 (1967).
63. E. N. Gil'bert, G. V. Glukhova, G. G. Glukhov, V. A. Mikhailov, and V. G. Torgov, *J. Radioanal. Chem.*, **8**, 39 (1971).
64. H. Hamaguchi, G. W. Reed, and A. Turkevich, *Geochim. Cosmochim. Acta*, **12**, 337 (1957).
65. G. W. Reed, K. Kigoshi, and A. Turkevich, *Geochim. Cosmochim. Acta*, **20**, 122 (1960).
66. W. D. Ehmann and J. R. Huizenga, *Geochim. Cosmochim. Acta*, **17**, 125 (1959).
67. J. Ruzicka, A. Zeman, and I. Obrusnik, *Talanta*, **12**, 401 (1965).
68. G. Marowsky, *Z. Anal. Chem.*, **253**, 267 (1971).
69. P. M. Santoliquido and W. D. Ehmann, *Geochim. Cosmochim. Acta*, **36**, 897 (1972).
70. R. H. Filby, *Anal. Chim. Acta*, **31**, 434 (1964).
71. R. H. Filby, *Geochim. Cosmochim. Acta*, **28**, 49 (1965).
72. A. Wyttenbach, H. R. von Gunten, and W. Scherle, *Geochim. Cosmochim. Acta*, **29**, 467 (1965).
73. E. A. Vincent and L. I. Bilefield, *Geochim. Cosmochim. Acta*, **19**, 63 (1960).
74. L. I. Bilefield and E. A. Vincent, *Analyst*, **86**, 386 (1961).
75. R. A. Schmitt, R. H. Smith, and D. A. Olehy, *Geochim. Cosmochim. Acta*, **27**, 1077 (1963).

76. P. Rey, H. Wakita, and R. A. Schmitt, *Anal. Chim. Acta*, **51**, 163 (1970).
77. M. Gillberg, *Geochim. Cosmochim. Acta*, **28**, 495 (1964).
78. A. Wyttenbach, H. R. von Gunten, and W. Scherle, *Geochim. Cosmochim. Acta*, **29**, 475 (1965).
79. A. A. Smales, D. Mapper, and A. J. Wood, *Analyst*, **82**, 75 (1957).
80. A. A. Smales, D. Mapper, and A. J. Wood, *Geochim. Cosmochim. Acta*, **13**, 123 (1958).
81. K. K. Turekian and M. H. Carr, *Geochim. Cosmochim. Acta*, **24**, 1 (1961).
82. O. Johansen and E. Steinnes, *Talanta*, **17**, 407 (1970).
83. G. L. Bate, H. A. Potratz, and J. R. Huizenga, *Geochim. Cosmochim. Acta*, **18**, 101 (1960).
84. M. H. Carr and K. K. Turekian, *Geochim. Cosmochim. Acta*, **26**, 411 (1962).
85. A. A. Smales and L. Salmon, *Analyst*, **80**, 37 (1955).
86. M. J. Cabell and A. A. Smales, *Analyst*, **82**, 390 (1957).
87. A. A. Smales, T. C. Hughes, D. Mapper, C. A. J. McInnes, and R. K. Webster, *Geochim. Cosmochim. Acta*, **28**, 209 (1964).
88. R. A. Nadkarni and B. C. Haldar, *Anal. Chem.*, **44**, 1504 (1972).
89. A. O. Brunfelt, O. Johansen, and E. Steinnes, *Anal. Chim. Acta*, **37**, 172 (1967).
90. H. S. Brown and E. D. Goldberg, *Science*, **109**, 347 (1949).
91. D. F. C. Morris and M. E. Chambers, *Talanta*, **5**, 147 (1960).
92. W. D. Ehmann and J. L. Setser, *Science*, **139**, 594 (1963).
93. E. Merz, *Geochim. Cosmochim. Acta*, **26**, 347 (1962).
94. E. Merz and E. Schrage, *Geochim. Cosmochim. Acta*, **28**, 1873 (1964).
95. R. A. Schmitt, E. Bingham, and A. A. Chodos, *Geochim. Cosmochim. Acta*, **28**, 1961 (1964).
96. J. L. Setser and W. D. Ehmann, *Geochim. Cosmochim. Acta*, **28**, 769 (1964).
97. J. R. Butler and A. J. Thompson, *Geochim. Cosmochim. Acta*, **29**, 167 (1965).
98. D. F. C. Morris and R. A. Killick, *Talanta*, **11**, 781 (1964).
99. W. D. Ehmann and J. F. Lovering, *Geochim. Cosmochim. Acta*, **31**, 357 (1967).
100. G. G. Goles and E. Anders, *Geochim. Cosmochim. Acta*, **26**, 723 (1962).
101. A. A. Smales, J. van R. Smit, and H. Irving, *Analyst*, **82**, 539 (1957).
102. L. R. Wager, J. van R. Smit, and H. Irving, *Geochim. Cosmochim. Acta*, **13**, 81 (1958).
103. T. B. Pierce and P. F. Peck, *Analyst*, **86**, 580 (1961).
104. D. Mapper and J. R. Freyer, *Analyst*, **87**, 297 (1962).
105. O. Johansen and E. Steinnes, *Talanta*, **13**, 1177 (1966).
106. P. R. Rushbrook and W. D. Ehmann, *Geochim. Cosmochim. Acta*, **26**, 649 (1962).
107. W. Herr and R. Wolfle, *Z. Anal. Chem.*, **209**, 213 (1965).
108. J. H. Crocket, *Geochim. Cosmochim. Acta*, **36**, 517 (1972).

109. R. A. Schmitt, A. W. Mosen, C. S. Suffredini, J. E. Lasch, R. A. Sharp, and D. A. Olehy, *Nature*, **186**, 863 (1960).
110. A. W. Mosen, R. A. Schmitt, and J. Vasilevskis, *Anal. Chim. Acta*, **25**, 10 (1961).
111. R. A. Schmitt, R. H. Smith, J. E. Lasch, A. W. Mosen, D. A. Olehy, and J. Vasilevskis, *Geochim. Cosmochim. Acta*, **27**, 577 (1963).
112. J. W. Chase, J. W. Winchester, and C. D. Coryell, *J. Geophys. Res.*, **68**, 567 (1963).
113. J. W. Chase, C. C. Schnetzler, G. K. Czamanske, and J. W. Winchester, *J. Geophys. Res.*, **68**, 577 (1963).
114. L. A. Haskin and M. Gehl, *Science*, **139**, 1056 (1963).
115. H. B. Desai, R. Krishnamoorthy Iyer, and M. Sankar Das, *Talanta*, **11**, 1249 (1964).
116. K. Rengan and W. W. Meinke, *Anal. Chem.*, **36**, 157 (1964).
117. D. G. Towell, R. Volfousky, and J. W. Winchester, *Geochim. Cosmochim. Acta*, **29**, 569 (1965).
118. U. Krähenbühl, H. P. Rolli, and H. R. von Gunten, *Helv. Chim. Acta*, **55**, 697 (1972).
119. H. Hamaguchi, R. Kuroda, T. Shimizu, I. Tsukahara, and R. Yamamoto, *Geochim. Cosmochim. Acta*, **26**, 503 (1962).
120. C. K. Kim and W. W. Meinke, *Anal. Chem.*, **35**, 2135 (1963).
121. J. C. Ricq, J. P. Vidal, M. Capitant, and G. Troly, *Chim. Anal.*, **47**, 77 (1965).
122. W. D. Ehmann, *Geochim. Cosmochim. Acta*, **19**, 149 (1960).
123. G. L. Bate and J. R. Huizenga, *Geochim. Cosmochim. Acta*, **27**, 345 (1963).
124. J. W. Morgan and J. F. Lovering, *Science*, **144**, 835 (1964).
125. J. W. Morgan, *Anal. Chim. Acta*, **32**, 8 (1965).
126. D. F. C. Morris, N. Hill, and B. A. Smith, *Mikrochim. Acta*, 962 (1963).
127. J. H. Crocket and G. B. Skippen, *Geochim. Cosmochim. Acta*, **30**, 129 (1966).
128. D. F. C. Morris and F. W. Fifield, *Talanta*, **8**, 612 (1961).
129. K. Ishida, R. Kuroda, and K. Kawabuchi, *Anal. Chim. Acta*, **36**, 18 (1966).
130. K. Terada, Y. Yoshimura, S. Osaki, and T. Kiba, *Talanta*, **14**, 53 (1967).
131. R. A. Nadkarni and B. C. Haldar, *Radiochem. Radioanal. Lett.*, **12**, 223 (1972).
132. J. T. Tanner and W. D. Ehmann, *Geochim. Cosmochim. Acta*, **31**, 2007 (1967).
133. D. M. Kemp and A. A. Smales, *Anal. Chim. Acta*, **23**, 410 (1960).
134. D. M. Kemp and A. A. Smales, *Geochim. Cosmochim. Acta*, **18**, 149 (1960).
135. H. Hamaguchi, T. Watanabe, N. Onuma, K. Tomura, and R. Kuroda, *Anal. Chim. Acta*, **33**, 13 (1965).
136. A. V. Lavrukhina, I. S. Kalicheva, and G. M. Kolesov, *Geokhimiya*, 651 (1967).
137. U. Schindewolf, *Geochim. Cosmochim. Acta*, **19**, 134 (1960).
138. V. Lavrakas, T. J. Golembeski, G. Pappas, J. E. Gregory, and H. L. Wedlick, *Anal. Chem.*, **46**, 952 (1974).

139. H. Hamaguchi, K. Kawabuchi, N. Onuma, and R. Kuroda, *Anal. Chim. Acta*, **30**, 335 (1964).
140. H. Hamaguchi, R. Kuroda, N. Onuma, K. Kawabuchi, T. Mitsubayashi, and K. Hosohara, *Geochim. Cosmochim. Acta*, **28**, 1039 (1964).
141. B. A. Loveridge, R. K. Webster, J. W. Morgan, A. M. Thomas, and A. A. Smales, *Anal. Chim. Acta*, **23**, 154 (1960).
142. P. J. Magno and F. E. Knowles, *Anal. Chem.*, **37**, 1112 (1965).
143. D. H. F. Atkins and A. A. Smales, *Anal. Chim. Acta*, **22**, 462 (1960).
144. D. F. C. Morris and A. Olya, *Talanta*, **4**, 194 (1960).
145. J. R. Butler and A. J. Thompson, *Geochim. Cosmochim. Acta*, **26**, 516 (1962).
146. W. D. Ehmann, *Geochim. Cosmochim. Acta*, **29**, 43 (1965).
147. E. N. Jenkins, *Analyst*, **80**, 301 (1955).
148. G. L. Bate, J. R. Huizenga, and H. A. Potratz, *Geochim. Cosmochim. Acta*, **16**, 88 (1959).
149. J. W. Morgan and J. F. Lovering, *Anal. Chim. Acta*, **28**, 405 (1963).
150. J. F. Lovering and J. W. Morgan, *J. Geophys. Res.*, **69**, 1979 (1964).
151. J. W. Morgan and J. F. Lovering, *J. Geophys. Res.*, **69**, 1989 (1964).
152. K. Bachmann, *Z. Anal. Chem.*, **219**, 340 (1966).
153. H. Stärk and C. Turkowsky, *Radiochim. Acta*, **5**, 16 (1966).
154. M. Mantel, P. Sung-Tung, and S. Amiel, *Anal. Chem.*, **42**, 267 (1970).
155. P. J. Aruscavage and H. T. Millard, *J. Radioanal. Chem.*, **11**, 67 (1972).
156. C. K. Kim and W. W. Meinke, *Talanta*, **10**, 83 (1963).
157. H. A. Mahlmann and G. W. Leddicotte, *Anal. Chem.*, **27**, 823 (1955).
158. K. H. Ebert, H. König, and H. Wanke, *Z. Naturforsch.* **12A**, 763 (1957).
159. L. A. Haskin, H. W. Fearing, and F. S. Rowland, *Anal. Chem.*, **33**, 1298 (1961).
160. O. Bobleter and I. Musyl, *Radiochim. Acta*, **3**, 57 (1964).
161. A. Alian and P. Parthasarathy, *Anal. Chim. Acta*, **35**, 69 (1966).
162. D. M. Kemp and A. A. Smales, *Anal. Chim. Acta*, **23**, 397 (1960).
163. A. Amiruddin and W. D. Ehmann, *Geochim. Cosmochim. Acta*, **26**, 1011 (1962).
164. T. B. Pierce and P. F. Peck, *Analyst*, **87**, 369 (1962).
165. R. H. Filby, *Anal. Chem.*, **36**, 1597 (1964).
166. R. H. Filby, *Anal. Chim. Acta*, **31**, 557 (1964).
167. T. K. Ball and R. H. Filby, *Geochim. Cosmochim. Acta*, **29**, 737 (1965).
168. T. K. Choy, H. R. Lukens, and G. H. Andersen, *Nucl. Appl.*, **1**, 179 (1965).
169. Y. Kamemoto and S. Yamagishi, *Bull. Chem. Soc. Jap.*, **36**, 1411 (1963).
170. Y. Kamemoto and S. Yamagishi, *Talanta*, **11**, 27 (1964).
171. J. Ruzicka and J. Stary, *Talanta*, **10**, 287 (1963).
172. N. Suzuki and K. Kudo, *Anal. Chim. Acta*, **32**, 456 (1965).
173. T. B. Pierce and P. F. Peck, *Analyst*, **88**, 603 (1963).

174. J. Ruzicka and J. Stary, *Substoichiometry in Radiochemical Analysis*, Pergamon, New York, 1968.
175. R. A. Nadkarni, *J. Radioanal. Chem.*, **20**, 139 (1974).
176. R. A. Nadkarni, *Radiochem. Radioanal. Lett.*, **17**, 207 (1974).
177. J. Stary and J. Ruzicka, *Talanta*, **18**, 1 (1971).
178. D. F. Schutz and K. K. Turekian, *Geochim. Cosmochim. Acta*, **29**, 259 (1965).
179. A. Alian and R. Shabana, *Microchem. J.*, **12**, 427 (1967).
180. F. Girardi and E. Sabbioni, *J. Radioanal. Chem.*, **1**, 169 (1968).
181. G. H. Morrison, J. T. Gerard, A. Travesi, R. L. Currie, S. F. Peterson, and N. M. Potter, *Anal. Chem.*, **41**, 1633 (1969).
182. T. E. Gills, W. F. Marlow, and B. A. Thompson, *Anal. Chem.*, **42**, 1831 (1970).
183. R. H. Filby, W. A. Haller, and K. R. Shah, *J. Radioanal. Chem.*, **5**, 277 (1970).
184. R. O. Allen, L. A. Haskin, M. R. Anderson, and O. Müller, *J. Radioanal. Chem.*, **6**, 115 (1970).
185. K. Tomura, H. Higuchi, N. Miyaji, N. Onuma, and H. Hamaguchi, *Anal. Chim. Acta*, **41**, 217 (1968).
186. J. C. Laul, D. R. Case, M. Wechter, F. Schmidt-Bleek, and M. E. Lipschutz, *J. Radioanal. Chem.*, **4**, 241 (1970).
187. E. Steinnes, *J. Radioanal. Chem.*, **10**, 65 (1972).
188. A. Elek, G. Perneczki, E. Szabo, and N. N. Dogadkin, *Radiochem. Radioanal. Lett.*, **15**, 123 (1973).
189. A. O. Brunfelt, I. Roelandts, and E. Steinnes, *Analyst*, **99**, 277 (1974).
190. R. A. Nadkarni and G. H. Morrison, *Anal. Chem.*, **46**, 232 (1974).
191. R. R. Keays, R. Ganapathy, J. C. Laul, U. Krähenbühl, and J. W. Morgan, *Anal. Chim. Acta*, **72**, 1 (1974).
192. A. O. Brunfelt and E. Steinnes, *Geochim. Cosmochim. Acta*, **30**, 921 (1966).
193. R. M. Stueber and G. G. Goles, *Geochim. Cosmochim. Acta*, **31**, 75 (1967).
194. J. R. DeVoe (ed.), *Modern Trends in Activation Analysis*, National Bureau of Standards, Special Publication 312, Vol. 2, Washington D. C. (1969).
195. L. E. Fite and R. E. Wainerdi, *Miner. Sci. Eng.*, **2**, 3 (1970).
196. G. L. Schroeder, R. D. Evans, and R. C. Ragaini, *Anal. Chem.*, **38**, 432 (1966).
197. J. Turkstra, P. J. Pretorius, and W. J. de Wet, *Anal. Chem.*, **42**, 835 (1970).
198. J. F. Lamb, S. G. Prussin, J. A. Harris, and J. M. Hollander, *Anal. Chem.*, **38**, 813 (1966).
199. J. C. Cobb, *Anal. Chem.*, **39**, 127 (1967).
200. G. E. Gordon, K. Randle, G. G. Goles, J. B. Corliss, M. H. Beeson, and S. Oxley, *Geochim. Cosmochim. Acta*, **32**, 369 (1968).
201. I. M. Dale, P. Henderson, and A. Walton, *Radiochem. Radioanal. Lett.*, **5**, 91 (1970).
202. J. Kuncir, J. Benada, Z. Randa, and M. Vobecky, *J. Radioanal. Chem.*, **5**, 369 (1970).

203. S. M. Lombard, K. W. Marlow, and J. T. Tanner, *Anal. Chim. Acta*, **55**, 13 (1971).
204. H. H. Schock, *Z. Anal. Chem.*, **263**, 100 (1973).
205. W. H. Zoller and G. E. Gordon, *Anal. Chem.*, **42**, 257 (1970).
206. R. Dams, J. A. Robbins, K. A. Rahn, and J. W. Winchester, *Anal. Chem.*, **42**, 861 (1970).
207. J. Rosenberg and H. B. Wiik, *Radiochem. Radioanal. Lett.*, **6**, 45 (1971).
208. R. J. Rosenberg, *Acta Chem. Fenn.*, **45**, 399 (1972).
209. D. Y. Jérome, J.-C. Philippot, and E. Brichet, *Earth Planet. Sci. Lett.*, **13**, 436 (1972).
210. D. Y. Jérome and J.-C. Philippot, *Geochim. Cosmochim. Acta*, **37**, 909 (1973).
211. P. A. Helmke, D. P. Blanchard, J. W. Jacobs, and L. A. Haskin, *Geochim. Cosmochim. Acta*, **37**, 869 (1973).
212. J. W. Morgan, U. Krähenbühl, R. Ganapathy, and E. Anders, *Geochim. Cosmochim. Acta*, **37**, 953 (1973).
213. J. C. Laul and R. A. Schmitt, *Geochim. Cosmochim. Acta*, **37**, 927 (1973).
214. R. F. Coleman and T. B. Pierce, *Analyst*, **92**, 1 (1967).
215. J. Op de Beeck and J. Hoste, *Analyst*, **99**, 973 (1974).
216. W. S. Lyon, E. Ricci, and H. H. Ross, *Anal. Chem.*, **44**, 438R (1972); **46**, 431R (1974).
217. W. S. Lyon and H. H. Ross, *Anal. Chem.*, **48**, 96R (1976).
218. R. Gijbels, *Miner. Sci. Eng.*, **5**, 304 (1973).
219. D. L. Showalter and R. A. Schmitt, *Adv. Act. Anal.*, **2**, 185 (1972).
220. R. H. Filby and K. R. Shah, *Toxic. Environ. Chem. Rev.*, **2**, 1 (1974).
221. *International Colloquium on Activation Analysis of Very Low Amounts of Elements* (Saclay, France, October 1972), *J. Radioanal. Chem.*, **18**, 9–185 (1973).
222. P. B. Price and R. M. Walker, *Appl. Phys. Lett.*, **2**, 23 (1963).
223. R. L. Fleischer, C. W. Naesser, P. B. Price, R. M. Walker, and U. B. Marvin, *Science*, **148**, 629 (1965).
224. D. E. Fisher, *Anal. Chem.*, **42**, 414 (1970).
225. T. Hashimoto, S. Hayashi, and S. Iwata, *Jap. Anal.*, **19**, 1538 (1970).
226. A. Wiechen, *Radiochem. Radioanal. Lett.*, **17**, 389 (1974).
227. A. V. Murali, P. P. Parekh, and M. Sankar Das, *Anal. Chim. Acta*, **50**, 71 (1970).
228. H. Matsuda, Y. Tsutsui, S. Nakano, and S. Umemoto, *Talanta*, **19**, 851 (1972).
229. R. L. Fleischer, P. B. Price, and R. M. Walker, *Science*, **149**, 383 (1965).
230. G. Goldstein, *Anal. Chem.*, **35**, 1620 (1963).
231. C. A. Baker, *Analyst*, **92**, 601 (1967).
232. G. J. Lutz, *Anal. Chem.*, **41**, 424 (1969).
233. G. J. Lutz, *Anal. Chem.*, **43**, 93 (1971).
234. J. Hislop, A. G. Pratchett, and D. R. Williams, *Analyst*, **96**, 117 (1971).
235. P. Wilkniss, *Radiochim. Acta*, **11**, 138 (1969).
236. T. B. Pierce, P. F. Peck, and D. R. A. Cuff, *Analyst* **92**, 143 (1967).

237. C. Engelmann, *J. Radioanal. Chem.*, **7**, 89 (1971).
238. J. Tolgyessy, T. Braun, and T. Kyrs, *Isotope Dilution Analysis*, Pergamon, New York, 1972.
239. G. von Hevesy and R. Hobbie, *Z. Anal. Chem.*, **88**, 1 (1932).
240. J. Ruzicka and J. Stary, *Talanta*, **8**, 228 (1961).
241. G. B. Briscoe and S. Humphries, *Talanta*, **18**, 39 (1971).
242. L. P. Greenland and E. Y. Campbell, *Anal. Chim. Acta*, **67**, 29 (1973).
243. P. Aruscavage, *Anal. Chim. Acta*, **82**, 343 (1976).
244. G. B. Briscoe, B. G. Cooksey, J. Ruzicka, and M. Williams, *Talanta*, **14**, 1457 (1967).

CHAPTER

14

ELECTROANALYTICAL METHODS

Electrochemical methods have not, in general, been used as widely in the analysis of geological materials as in some other fields, such as metallurgical analysis. However, the increasing use of potentiometric measurements with ion-selective electrodes, and of potential-scanning electrolytic techniques such as polarography and anodic stripping voltammetry, is sufficient to justify a brief outline of these methods and some notes on their application. Many other electroanalytical techniques are of little value for trace-element analysis or have seldom been applied in trace analysis of geological materials. Accounts of theory and instrumentation for various electroanalytical methods can be found in specialized texts.[1-14] Basic principles of galvanic cells and of electrolysis are discussed by Lingane,[1] and a brief account of the IUPAC convention on cell electromotive forces (emf) is given by Ives and Janz.[6]

14.1 POTENTIOMETRY WITH ION-SELECTIVE ELECTRODES

14.1.1 Potentiometry and Potentiometric Titrations

Potentiometry involves measuring the equilibrium (zero current) electrode potential of an indicating electrode system relative to that of a reference electrode system. The reaction tending to occur at the indicating electrode can be represented by

$$\text{reduced form} \rightleftharpoons \text{oxidized form} + ne^-$$

The electrode potential E of the indicating electrode (relative to the standard hydrogen electrode) depends on the activities of the reduced and oxidized forms, a_{red} and a_{ox}, according to the Nernst equation:

$$E = E^o + \frac{RT}{nF} \ln \frac{a_{\text{ox}}}{a_{\text{red}}} \qquad (14.1)$$

where R is the gas constant, F is the Faraday constant, and E^o is the standard electrode potential of the indicating electrode.

Because the analytical chemist is more often interested in the concentration ratio c_{ox}/c_{red} than in the activities it is convenient to replace each activity by the product of concentration and activity coefficient, y;

$$E = E^o + \frac{RT}{nF} \ln \frac{y_{ox}}{y_{red}} + \frac{RT}{nF} \ln \frac{c_{ox}}{c_{red}} \quad (14.2)$$

Information about concentrations can therefore be derived directly from potentiometric measurements made under conditions where the activity coefficients are known or can be calculated. Alternatively, for a series of measurements made in identical media, the ionic strength and the activity coefficient ratio y_{ox}/y_{red} are constant. In this case equation 14.2 can be written

$$E = E^{o\prime} + \frac{RT}{nF} \ln \frac{c_{ox}}{c_{red}} \quad (14.3)$$

in which $E^{o\prime}$ is the formal potential of the electrode system. $E^{o\prime}$ is the electrode potential observed *in a defined medium* for a solution containing equal formal concentrations of the oxidized and reduced species.

Titrations involving a red-ox reaction or involving processes such as precipitation or complex formation that remove an electroactive species can be followed potentiometrically. Such titrations are characterized by a rapid change of E with volume of titrant near the equivalence point. In this vicinity small additions of titrant produce relatively large changes of the ratios a_{ox}/a_{red} and c_{ox}/c_{red}. The problem of obtaining concentrations from emf measurements is largely eliminated when the measurements are used for detecting equivalence points.

14.1.2 Ion-Selective Electrodes

The most important electrodes used in potentiometric analysis are those described as membrane electrodes, the most familiar being the glass electrode sensitive to hydrogen ions. Since the 1950s, glass membranes of different composition sensitive to other cations (e.g., Na^+, Ag^+) have been developed, and more recently a wide range of different kinds of membrane electrode (ion-selective electrodes) have been made that are more or less specific in their response.

Solid-state electrodes, as shown in Fig. 14.1, contain a single crystal of a sparingly soluble salt as the active membrane material. Electrodes of this type include those made with lanthanum fluoride (selective for F^-), silver halides (Cl^-, Br^-, I^-), silver sulfide (S^{2-}, Ag^+), and mixtures of silver sulfide with other sparingly soluble sulfides (Cu^{2+}, Cd^{2+}, Pb^{2+}).

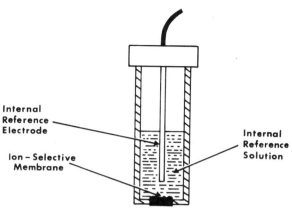

Fig. 14.1. Solid-state membrane electrode. In the case of the fluoride electrode, the membrane consists of a single LaF_3 crystal doped with Eu^{2+} to improve its conductivity. The internal reference solution contains NaF and NaCl, and the internal reference electrode is Ag/AgCl.

Heterogeneous membrane electrodes contain sparingly soluble salts in powdered form, dispersed in an inert binder such as silicone rubber. They are generally similar in their properties to the single-crystal electrodes and are available for a similar range of ions.

Liquid membrane electrodes contain a solution of a liquid ion exchanger or some other substance with a strong affinity for a particular ion or group of ions. Both the active material and its organic solvent have low solubility in water. The ion-selective layer is retained between permeable membranes or within sintered glass disks. Examples include a calcium-selective electrode containing the calcium salt of didecylphosphoric acid as active material and a potassium-selective electrode containing the cyclic polypeptide valinomycin, which binds potassium, rubidium, and cesium strongly.

If a membrane is selective for ion i, then a potential difference E_M is developed across the membrane when the activity a_i of ion i on one side differs from its activity a_i' on the other side:

$$E_M = \frac{RT}{z_i F} \ln \frac{a_i}{a_i'} \tag{14.4}$$

where z_i is the charge number of ion i. An electrode such as that shown in Fig. 14.1, containing an internal reference of constant potential and a constant activity a_i' in the internal solution, exhibits a potential E that depends on the activity of i in the solution in which it is immersed.

$$E = \text{const.} + \frac{2.303 RT}{z_i F} \log a_i \tag{14.5}$$

The ion-selective electrode is combined with an external reference electrode to give a cell whose emf can be measured. (Some commercial products are marketed as a "combination electrode" with the reference electrode incorporated with the ion-selective electrode in a single unit.)

Unfortunately, no membrane electrode is entirely specific in its response. The potential of an electrode specific for a univalent ion i, for example, is influenced by a foreign univalent ion j, according to

$$E = \text{const.} + \frac{2.303RT}{F} \log(a_i + K_{ij}a_j) \tag{14.6}$$

where K_{ij} is the selectivity ratio of ion i over ion j. A more general empirical expression[15] where there are several interfering ions j of valency z_j is

$$E = \text{const.} + \frac{2.303RT}{z_i F} \log\left(a_i + \sum_j K_{ij} a_j^{z_i/z_j}\right) \tag{14.7}$$

If ion j is not to interfere in the measurement of ion i when both ions are present at comparable concentrations, then K_{ij} should be much less than unity.

In principle, the potentiometric measurements yield information about the activity of ion i. Concentrations are obtained by ensuring that samples and standards are measured in a medium of constant ionic strength and by drawing a calibration graph of E vs concentration for a series of standard solutions. It is often possible, especially in the analysis of nonsaline waters, to maintain constant ionic strength by adding to all samples and standards an excess of a neutral electrolyte. In the measurement of fluoride, for example, an electrolyte medium ("total ionic strength adjustment buffer," or TISAB) consisting of 1 M sodium chloride, 0.25 M acetic acid, 0.75 M sodium acetate, and 0.001 M sodium citrate, has been used.[16] Equal volumes of sample and buffer are mixed, giving a solution of suitable pH and well-defined ionic strength. The presence of citrate ensures preferential formation of citrato-complexes of metals such as aluminum and iron, releasing any fluoride originally bound in fluorocomplexes. Other complexing agents, such as trans-1,2-diaminocyclohexanetetra-acetic acid (cyclohexylenedinitrilotetra-acetic acid, CDTA), have been used for the same purpose.

As indicated in the above example, an ion-selective electrode responds only to the free ion i itself, and not to complexed forms. Information obtained with an ion-selective electrode may therefore be different from, and complementary to, that obtained by other methods of analysis.

As an alternative to a calibration curve the method of standard additions may be used, being particularly suitable for solutions of differing ionic strength. The initial sample of volume V_x and concentration c_x gives an emf

E_1. After addition of volume V_s of a standard solution of concentration c_s, the emf becomes E_2. Provided the ionic strength is not substantially altered by the addition, the unknown concentration can be calculated from the change of emf $\Delta E(= E_2 - E_1)$:

$$c_x = \frac{c_s V_s}{V_x + V_s} \left[\exp \frac{z_i F \Delta E}{RT} - \frac{V_x}{V_x + V_s} \right]^{-1} \tag{14.8}$$

If a series of measurements of ΔE is made with varying standard additions V_s, then

$$\exp \frac{z_i F \Delta E}{RT} - 1 = \left(\frac{c_s}{c_x} - 1 \right) \frac{V_s}{V_x + V_s} \tag{14.9}$$

and c_x can be obtained from the slope of a plot of $\exp(z_i F \Delta E / RT) - 1$ vs $V_s/(V_x + V_s)$. If $V_s \ll V_x$, the use of simpler forms of equations 14.8 and 14.9 becomes possible (see, for example, Ref. 17).

An alternative procedure is based on the fact that, in many situations, the function $\mathscr{F}$ defined by

$$\mathscr{F} = (V_x + V_s) \exp \frac{z_i EF}{RT}$$

is proportional to $(V_x c_x + V_s c_s)$. A standard addition titration can be carried out, and values of $\mathscr{F}$ plotted vs V_s. The graph is extrapolated to $\mathscr{F} = 0$, at which point V_s has the value $-(V_x c_x/c_s)$, from which c_x can be obtained directly.[18,19]

Because of the logarithmic nature of the emf-activity relation, a 10-fold change in activity gives only a 59.2 mV change in emf for a univalent ion and 29.6 mV change for a divalent ion at 298 K. Consequently, a 1 mV error in emf leads to concentration errors of approximately 3.9% for a univalent ion and 7.8% for a divalent ion. It is therefore desirable to make the emf measurements to within ± 0.1 mV, and to minimize other sources of error such as variable liquid-junction potentials.[20] In the measurement of low concentrations, attention must also be paid to the response time of the electrode, as the attainment of equilibrium may take up to several minutes.

More detailed discussion of the theory, construction, behavior, and use of ion-selective electrodes can be found in several monographs[7,8,21] and reviews.[22-27,84,85]

14.1.3 Applications of Ion-Selective Electrodes

In the field of trace-element analysis of waters and rocks, the development of the fluoride ion-selective electrode,[28] based on a Eu^{2+}-doped lanthanum fluoride crystal, has been particularly important. The potential of this electrode is given by the appropriate form of equation 14.5:

$$E = \text{const.} - \frac{2.303RT}{F} \log a_{F^-} \qquad (14.10)$$

The only significant interference, arising from the presence of hydroxide ion, can be controlled by adjustment of pH to ensure that the hydroxide ion activity is at least tenfold smaller than the fluoride activity. Measurements are normally carried out in a medium with pH 5–6; at lower pH, fluoride becomes bound as HF and HF_2^-. Fluoride levels down to about 10^{-6} to 10^{-7} mol/l can be measured if the ionic strength and buffer reagents can be obtained sufficiently free of fluoride.

The determination of fluoride in drinking waters, and in seawater and other media containing chloride, has been studied extensively.[16,29–33] Crosby et al.[29] made a detailed comparison of the results of electrode measurements with those obtained by several standard spectrophotometric methods, and concluded that the fluoride electrode was superior in speed, accuracy, and convenience.

Fluoride has been determined in solutions prepared from phosphate rock samples (which frequently contain 2% to 4% fluoride),[34–36] and the use of appropriate ion-selective electrodes for the successive determination of iodide, bromide, chloride, and fluoride in phosphate rock has also been reported.[37] Fluoride can be determined in silicate rocks without prior separation of aluminum if the measurements are made in a sodium citrate-potassium nitrate medium.[19,38]

The fluoride electrode has been used for the determination of aluminum, potentiometric titration being carried out with a standard fluoride solution. For example, aluminum at levels of 10 to 100 $\mu g/ml$ in paper-mill waters has been estimated in this way.[39]

The determination of traces of sulfide ion in water can be carried out with a silver sulfide electrode.[40,41] Concentrations of sulfide well below 1 $\mu g/ml$ can be determined directly, and preconcentration of the sulfide (e.g., by coprecipitation as zinc sulfide with zinc hydroxide[40]) can be used to measure sulfide concentrations as low as 3 ng/ml.[41]

Calcium ion-selective electrodes have been used to measure calcium in seawater[42] and in soil extracts,[43,44] and a silver sulfide-copper sulfide electrode has enabled copper to be determined at levels of a few nanograms per milliliter in seawater[45] and tap waters.[17] Many other applications, including those falling somewhat beyond the scope of the present work, have been summarized in recent reviews.[26,27,85]

14.2 CONTROLLED POTENTIAL METHODS

14.2.1 Classical Polarography

In polarography the electrolysis of an analyte takes place at a dropping mercury electrode (DME), which consists of uniform droplets of mercury emerging at a constant rate from the tip of a capillary tube. The potential at the DME, relative to that at a reference electrode, is caused to undergo a slow linear variation with time, and the electrolysis current is recorded, as shown in Fig. 14.2. The electrolysis current from a reduction, such as that of divalent cadmium to a cadmium-mercury analgam, eventually reaches a value i_d, limited by the rate at which the reducible species diffuses to the electrode. This limitation applies until a potential is reached at which reduction of another species occurs, such as that of hydrogen ions to hydrogen. The potential, $E_{1/2}$, at which the polarographic wave reaches half its limiting height, is characteristic of a given electrode reaction under given conditions of solvent medium and temperature. Tables of half-wave potentials can be found in specialized texts on polarography (e.g., Refs. 10 and 12).

Oscillations of the diffusion current result from the periodic change in effective area of the DME. Use of a fast recording system shows that at the end of each drop lifetime, the current reaches a peak value $(i_d)_{max}$ given by

$$(i_d)_{max} = (\tfrac{7}{6})\bar{i}_d$$

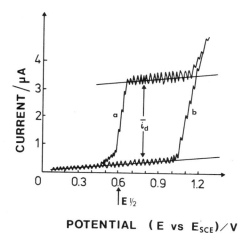

Fig. 14.2. Polarograms of 1 M HCl containing Cd(II), 20 μg/ml (a) and 1 M HCl alone (b).

In quantitative analysis use is made of the proportionality between $\bar{i}_d$ and c, the concentration of the analyte in the bulk of the solution, as given by the Ilkovic equation:

$$\bar{i}_d = knm^{2/3}t^{1/6}D^{1/2}c \qquad (14.5)$$

where n is the number of electrons involved in the electrode reaction, m is the average mercury flow rate, t is the lifetime of each mercury drop, and D is the diffusion coefficient of the analyte. If $\bar{i}_d$ is expressed in μA, c in mmol/l, m in mg s^{-1}, t in s, and D in cm^2 s^{-1}, the numerical value of the constant k is 607 at 25°C. In practice, an analytical calibration curve is obtained by measuring $\bar{i}_d$ for several standard solutions, eliminating the need to know m, t, and D accurately. Alternatively, the method of standard additions can be used. The sample of volume V_x and concentration c_x gives a diffusion current $(\bar{i}_d)_1$. After addition of volume V_s of standard solution with concentration c_s, the diffusion current is $(\bar{i}_d)_2$. The unknown concentration c_x is then obtained from

$$c_x = \frac{c_s V_s (\bar{i}_d)_1}{(V_s + V_x)(\bar{i}_d)_2 - V_x (\bar{i}_d)_1} \qquad (14.6)$$

14.2.2 Solid Electrode Voltammetry

The electrolysis of solutions can be studied under conditions similar to those of polarography, in which the DME is replaced by a solid electrode of a material such as platinum, gold, pyrolytic or wax-impregnated graphite, or boron carbide, rotated at constant speed. In this case transport of the electroactive species to the electrode occurs by both convection and diffusion. The current-voltage curves have a general form similar to the polarograms of Fig. 14.2, except that the periodic current oscillations are replaced by small erratic variations arising from turbulence at the electrode surface. As in polarography, the quantitative application is dependent on the proportionality between the limiting current, i_{lim}, and c. Solid electrode voltammetry can give rather greater sensitivity than polarography and is useful in regions of potential not accessible with the DME.

14.2.3 Modifications of Polarography

Some properties of the DME make it particularly suitable for electroanalytical work. The high overpotential for liberation of hydrogen at a mercury electrode allows the study of many metal-ion reductions for which the electrode potentials are quite negative. In addition, the steady emergence of mercury droplets from the capillary tip provides a continual renewal of the electrode surface. However, the classical polarographic method of obtaining current-voltage curves has limitations, particularly in trace analysis. One disadvantage lies in the steady fluctuation of electrode area, the effects of which are only partly eliminated by damping the output signal. A most important limitation arises from the existence of a capacitive (or non-Faradaic) contribution to the total current. A current flow is required to develop the electrical double-layer at the electrode-solution interface, a process that is somewhat analogous to the charging of a capacitor. This charging current is highest during the early part of the lifetime of each drop, when the electrode area is increasing most rapidly. The magnitude of the capacitive contribution increases with increasing applied potential, as the "capacitor" must be charged to higher voltages. This is the major cause of the sloping baseline or residual current (see Fig. 14.2), and provides a background (often of the order of 0.1 μA) above which the diffusion current must be measured. For this reason classical polarography operates most satisfactorily in the 10^{-3} to 10^{-4} mol/l range, although determinations below 10^{-5} mol/l can be made in favorable cases.

To make lower concentration ranges accessible, classical polarography has been modified in various ways that minimize the effects of the capacitive current and drop-area fluctuations. Some of these modifications are summarized in Table 14.1.

The development of differential pulse polarography is of particular interest for trace metal analysis. This technique involves modifying the linear potential scan by adding periodic potential pulses of constant height, ΔE_p, as illustrated in Fig. 14.3. A pulse height of 5 to 100 mV is generally chosen. The instrument samples the current twice during each operating interval, once just before application of the pulse, and again near the end of the pulse when the capacitive contribution has decayed to a minimum. The difference Δi between the two currents is determined, and a voltage proportional to Δi is plotted. Repetition of this pulsing and sampling process over the voltage range of a polarographic wave leads to a derivative curve of the type shown in Fig. 14.4. In favorable cases analyte concentrations of 10^{-8} mol/l can be measured, and the derivative form of the output allows peaks 40 mV apart to be resolved.

14.2.4 Stripping Voltammetry

In anodic stripping voltammetry (ASV), a controlled-potential electrolysis is used to deposit a metal of interest from a stirred solution on a suitable electrode, such as a hanging mercury drop. The electrolysis acts as a concentration process, as the metal accumulates in the form of an amalgam. After an electrolysis time of 1 to 30 min, the solution is allowed to become quiescent

TABLE 14.1 Some Electrolysis Techniques Based on the Dropping Mercury Electrode

Technique	Voltage program	Scan rate (mV s^{-1})	Relation of current to bulk conc.	Most favorable performance		
				Lowest conc. (mol l^{-1})	Resolution[a] (mV)	Separability[b]
1. Classical (d.c.) polarography	Slow linear ramp over life of many drops	3–5	Diffusion current $i_d \propto c$	1×10^{-5}	100	10:1
2. Single-sweep (oscilloscopic) polarography	Fast linear ramp over life of single drop	250–500	Peak current $i_p \propto c$	3×10^{-7} (single channel instrument) 5×10^{-8} (double channel instrument)	40 40	400:1 1000:1

3. Pulse polarography (normal mode)	D.c. pulses of increasing amplitude (one pulse per drop near end of drop life)	0	Current in last fraction of pulse $i_p \propto c$	1×10^{-7}	40	10,000:1
4. Pulse polarography (differential mode)	D.c. pulses of constant amplitude on slow linear ramp (one pulse per drop near end of drop life)	5–15	Current sampled just before pulse, and at end of pulse; $\Delta i / \Delta E_p$ measured	1×10^{-8}	30–40	50,000:1

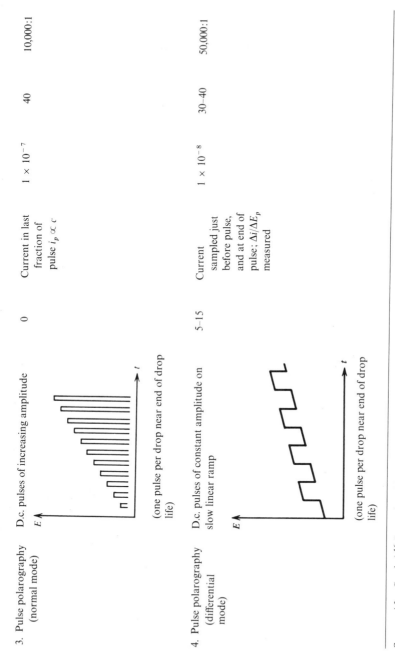

Source: After Strobel.[46] Reproduced by permission, from *Chemical Instrumentation*, Addison-Wesley, Reading, Mass., 1973.
[a] Resolution: Minimum difference in $E_{1/2}$ for distinguishing two species at equal concentrations with less than 1 % interference.
[b] Separability: Maximum concentration ratio c_1/c_2 tolerable for two species (with optimum $\Delta E_{1/2}$) before waves are no longer well enough resolved to allow calculation of concentrations.

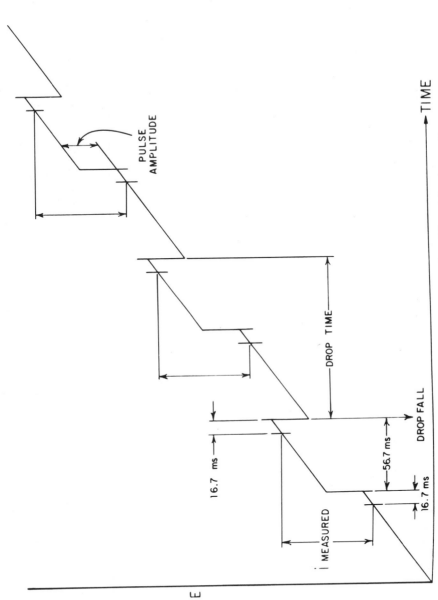

Fig. 14.3. Voltage-time program for differential pulse polarography and differential pulse anodic stripping voltammetry. For metal analysis, the voltage is scanned toward more negative potentials in differential pulse polarography and toward more positive potentials in anodic stripping voltammetry. *Source*: Flato.[47] Reprinted by permission from *Analytical Chemistry*. Copyright by the American Chemical

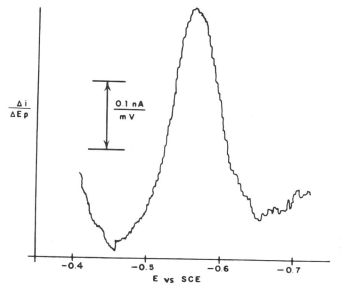

Fig. 14.4. Differential pulse polarography of Cd^{2+} (5×10^{-8} mol/l) at a stationary mercury electrode. Electrolyte: 0.01 M KNO_3. Pulse height, ΔE_p: 50 mV. *Source*: reproduced by permission of Princeton Applied Research Corp., Princeton, N.J.

and the working electrode is scanned slowly (e.g., 5–20 mV s^{-1}) through a potential range that allows the metal to be oxidized back into solution. Because of the preconcentration of the analyte, a greater current (and greater sensitivity) is achieved than in the normal voltammetric procedure, which examines reduction of the metal from a dilute solution.

With a hanging mercury drop electrode, several elements can be determined simultaneously if their peak potentials are sufficiently well separated, as shown in Fig. 14.5. Some selectivity can be achieved by careful choice of the initial electrolysis potential. For a given ion, examined under carefully controlled deposition and stripping conditions, the peak current is proportional to its concentration in the initial solution. Anodic stripping voltammetry allows many elements to be determined at concentrations in the range 10^{-5} to 10^{-8} mol/l.

The technique can be used in an analogous way to analyze a substance that can first be deposited in an anodic process and then stripped by a cathodic potential scan. With a silver electrode at about $+0.35$ V (vs SCE) for example, chloride ions undergo the reaction

$$Cl^-(aq) + Ag(s) \longrightarrow AgCl(s) + e^-$$

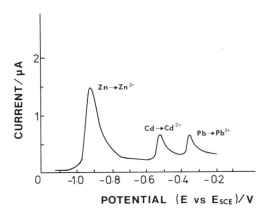

Fig. 14.5. Anodic stripping voltammetry of a solution containing Zn^{2+}, Cd^{2+}, and Pb^{2+} (each at 1 µg/ml) in 1 M KNO_3. Hanging mercury drop electrode; electrolysis for 60 s at -1.1 V; stripping scan from -1.1 to -0.2 V at 20 mV/s; potential of mercury drop measured relative to that of saturated calomel electrode (SCE).

and then pass back into solution in the reverse (cathodic) process during a voltage scan from $+0.35$ to -0.15 V.

Differential Pulse Anodic Stripping Voltammetry

By using the normal electrodeposition step of anodic stripping voltammetry, and an anodic stripping scan with the differential pulse waveform of the type shown in Fig. 14.3, the influence of capacitive current and noise contributions to the signal can be minimized, and the sensitivity of ASV can be increased by at least one order of magnitude. The higher sensitivity of this combination permits the use of shorter electrolysis times (e.g., 1 to 3 min), giving a faster analysis than normal ASV. Lower scan rates, typically about 5 mV s^{-1}, can be used, giving better reproducibility. It is possible to analyze aqueous samples for a number of elements at concentrations of 10^{-7} to 10^{-9} mol/l (i.e., about 0.1–10 ng/ml), as illustrated in Fig. 14.6.

Further discussion of techniques, instrumentation, and applications of anodic stripping voltammetry can be found in several reviews.[47-52] The respective merits of a number of voltammetric methods in trace analysis have been discussed.[53]

14.2.5 Applications of Polarographic and Voltammetric Techniques

Until quite recently, relatively little use was made of polarographic and voltammetric techniques in trace-element analysis of geological materials.

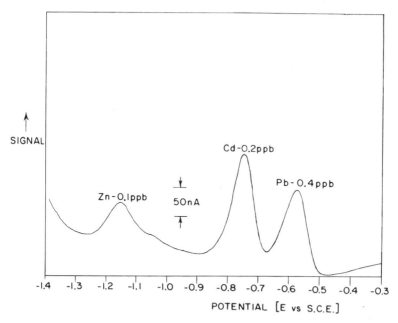

Fig. 14.6. Differential pulse anodic stripping voltammetry of a solution containing Zn^{2+}, Cd^{2+}, and Pb^{2+}, each at a concentration of 2×10^{-9} mol/l. Electrode: mercury-plated wax-impregnated graphite. *Source:* after Flato.[47] Reprinted by permission of *Analytical Chemistry*. Copyright by the American Chemical Society.

The limited selectivity and sensitivity of classical polarography made preliminary extraction steps necessary in most cases. Elements such as lead, cadmium, and copper in rocks were determined by Miholic,[54] and a method suitable for Cd, Co, Cu, Ni, and Zn in rocks at levels of 50 to 400 µg/g was developed by Smythe and Gatehouse.[55] The polarographic determination of microgram amounts of uranium was studied in some detail.[56,57] Solvent extraction was used by several workers[58–60] in the polarographic determination of cadmium in rocks at levels of 0.05 to 1 µg/g. Advantage was taken of the better sensitivity of square-wave polarography.[59] The direct determination of lead in river water at levels down to 2 ng/ml has also been carried out by square-wave polarography.[61] An extensive investigation of the concentrations of Bi, Cd, In, Pb, Tl, and Zn in many standard rocks[62] made use of pulse polarography, following selective vaporization of these elements in a hydrogen stream at 1150°C. Many of the values reported in this study are considerably lower than mean values found by other methods of analysis,[63,64] and may reflect losses or incomplete separation during processing.

Recent developments in commercial voltammetric instrumentation are bringing about a large increase in the use being made of various forms of polarography and stripping voltammetry, particularly for the analysis of natural waters. Solutions resulting from rock dissolution have also been analyzed. Detection limits for linear-scan ASV lie in the range 0.01 to 0.04 ng/ml for Bi, Cd, Cu, Pb, Tl, and Zn, and are even lower in some cases for differential pulse ASV.[52] Elements determined by ASV include antimony,[65] bismuth,[65,66] cadmium,[67-71] copper,[68,69] indium,[72] lead,[68,69,71] silver,[73] tin[74-76] and zinc.[67,68,77,78] The advantages of differential pulse ASV in the determination of Cd, Cu, Pb, and Zn in water samples have recently been discussed.[79] In general, no treatment is necessary other than the acidification recommended for sample storage. Contamination from buffers or other added reagents is thereby avoided. Bond and co-workers[80,81] have also demonstrated the use of a.c. polarography in determination of lead, tin, and uranium in minerals and geochemical samples.

Further discussion of these techniques and summaries of other applications, can be found in reviews such as those of Copeland and Skogerboe,[52] Hulanicki,[82] and Kissinger.[83]

References

1. J. J. Lingane, *Electroanalytical Chemistry*, 2nd ed., Interscience, New York, 1958.
2. A. J. Bard (ed.), *Electroanalytical Chemistry*, Vol. 1, Dekker, New York, 1966; Vol. 2, Dekker, New York, 1967; Vol. 3, Dekker, New York, 1969; Vol. 4, Dekker, New York, 1970.
3. P. Delahay, *New Instrumental Methods in Electrochemistry*, Interscience, New York, 1954.
4. D. R. Browning (ed.), *Electrometric Methods*, McGraw-Hill, Maidenhead, England, 1969.
5. A. Weissberger and B. Rossiter (eds.), *Physical Methods of Chemistry*, Vol. I, Part IIA, *Electrochemical Methods*, Wiley-Interscience, New York, 1972.
6. D. J. G. Ives and G. J. Janz (eds.), *References Electrodes, Theory and Practice*, Academic Press, New York, 1961.
7. J. Koryta, *Ion-Selective Electrodes*, Cambridge University Press, Cambridge, England, 1975.
8. N. Lakshminarayanaiah, *Membrane Electrodes*, Academic Press, New York, 1976.
9. I. M. Kolthoff and J. J. Lingane, *Polarography*, Interscience, New York, 1952.
10. L. Meites, *Polarographic Techniques*, 2nd ed., Interscience, New York, 1965.
11. J. Heyrovsky and J. Kuta, *Principles of Polarography*, Academic Press, London, 1965.

REFERENCES

12. J. Heyrovsky and P. Zuman, *Practical Polarography*, Academic Press, London, 1968.
13. J. T. Stock, *Amperometric Titrations*, Interscience, New York, 1965.
14. G. W. C. Milner and G. Phillips, *Coulometry in Analytical Chemistry*, Pergamon, New York, 1967.
15. J. W. Ross, *Science*, **156**, 1378 (1967).
16. M. S. Frant and J. W. Ross, *Anal. Chem.*, **40**, 1169 (1968).
17. M. J. Smith and S. E. Manahan, *Anal. Chem.*, **45**, 836 (1973).
18. G. Gran, *Analyst*, **77**, 661 (1952).
19. D. Jagner and V. Pavlova, *Anal. Chim. Acta*, **60**, 153 (1972).
20. A. K. Covington, in *Ion-Selective Electrodes* (ed. R. A. Durst), N. B. S. Special Publ. No. 314, U.S. Government Printing Office, Washington, D. C., 1969.
21. R. A. Durst (ed.), *Ion-Selective Electrodes*, N. B. S. Special Publ. No. 314, U.S. Government Printing Office, Washington, D. C., 1969.
22. E. Pungor, *Anal. Chem.*, **39** (13), 28A (1967).
23. E. Pungor and K. Toth, *Analyst*, **95**, 625 (1970).
24. A. K. Covington, *Chem. Brit.*, **5**, 388 (1969).
25. G. A. Rechnitz, *Anal. Chem.*, **41** (12), 109A (1969).
26. J. Tengyl, in *M.T.P. International Review of Science*, Physical Chemistry Series One, Vol. 12: Analytical Chemistry, Part I (ed. T. S. West), Butterworth, London, 1973.
27. R. P. Buck, *Anal. Chem.*, **44**, 270R (1972); **46**, 28R (1974); **48**, 23R (1976).
28. M. S. Frant and J. W. Ross, *Science*, **154**, 1553 (1966).
29. N. T. Crosby, A. L. Dennis, and J. G. Stevens, *Analyst*, **93**, 643 (1968).
30. T. B. Warner, *Science*, **165**, 178 (1969).
31. R. E. Mesner, *Anal. Chem.*, **40**, 443 (1968).
32. T. B. Warner, *Anal. Chem.*, **41**, 527 (1969).
33. T. Anfalt and D. Jagner, *Anal. Chim. Acta*, **53**, 13 (1971).
34. C. R. Edmond, *Anal. Chem.*, **41**, 1327 (1969).
35. L. Evans, R. D. Hoyle, and J. B. Macaskill, *N. Z. J. Sci.*, **13**, 143 (1970).
36. B. L. Ingram and I. May, *U.S. Geol. Surv. Prof. Pap.* No. 750-B, B180-B184 (1971).
37. E. J. Duff and J. L. Stuart, *Analyst*, **100**, 739 (1975).
38. B. L. Ingram, *Anal. Chem.*, **42**, 1825 (1970).
39. A. Homola and R. O. James, *Anal. Chem.*, **48**, 776 (1976).
40. E. W. Baumann, *Anal. Chem.*, **46**, 1345 (1974).
41. D. Weiss, *Chem. Listy.*, **68**, 528 (1974).
42. M. E. Thompson and J. W. Ross, *Science*, **154**, 1643 (1966).
43. E. A. Woolson, J. H. Axley, and P. C. Kearney, *Soil Sci.*, **109**, 279 (1970).
44. K. L. Cheng, J.-C. Hung, and D. H. Prager, *Microchem. J.*, **18**, 256 (1973).

45. R. Jasinski, I. Trachtenberg, and D. Andrychuk, *Anal. Chem.*, **46**, 364 (1974).
46. H. A. Strobel, *Chemical Instrumentation*, 2nd ed., Addison-Wesley, Reading, Mass., 1973.
47. J. B. Flato, *Anal. Chem.*, **44** (11), 75A (1972).
48. E. Barendrecht, in *Electroanalytical Chemistry*, Vol. 2 (ed. A. J. Bard), Dekker, New York, 1967.
49. H. Siegerman and G. O'Dom, *Am. Lab.*, **4** (6), 59 (1972).
50. W. D. Ellis, *J. Chem. Educ.*, **50**, A131 (1973).
51. B. Fleet and R. D. Jee, *Electrochem.*, **3**, 210 (1973).
52. T. R. Copeland and R. K. Skogerboe, *Anal. Chem.*, **46**, 1257A (1974).
53. A. M. Bond, *Anal. Chim. Acta*, **74**, 163 (1975).
54. S. Miholic, *Mikrochem. Mikrochim. Acta*, **36/37**, 393 (1951).
55. L. E. Smythe and B. M. Gatehouse, *Anal. Chem.*, **27**, 901 (1955).
56. H. I. Shalgosky, *Analyst*, **81**, 512 (1956).
57. F. Hecht, J. Korkisch, R. Patzak, and A. Thirard, *Mikrochim. Acta*, 1283 (1956).
58. I. Carmichael and A. McDonald, *Geochim. Cosmochim. Acta*, **22**, 87 (1961).
59. R. E. Stanton, A. McDonald, and I. Carmichael, *Analyst*, **87**, 134 (1962).
60. J. R. Butler and A. J. Thompson, *Geochim. Cosmochim. Acta*, **31**, 97 (1967).
61. E. B. Buchanan, T. D. Schroeder, and B. Novosel, *Anal. Chem.*, **42**, 370 (1970).
62. W. Wahler, *Neues Jahrb. Mineral., Abh.*, **108**, 36 (1968).
63. M. Fleischer, *Geochim. Cosmochim. Acta*, **33**, 65 (1969).
64. F. J. Flanagan, *Geochim. Cosmochim. Acta*, **33**, 81 (1969).
65. T. R. Gilbert and D. N. Hume, *Anal. Chim. Acta*, **65**, 451 (1973).
66. T. M. Florence, *J. Electroanal. Chem.*, **49**, 255 (1974).
67. M. Ariel and U. Eisner, *J. Electroanal. Chem.*, **5**, 362 (1963).
68. I. Sinko and J. Dolezal, *J. Electroanal. Chem.*, **25**, 299 (1970).
69. T. M. Florence, *J. Electroanal. Chem.*, **27**, 273 (1970).
70. R. G. Clem, G. Litton, and L. D. Ornelas, *Anal. Chem.*, **45**, 1306 (1973).
71. G. E. Batley and T. M. Florence, *J. Electroanal. Chem.*, **55**, 23 (1974).
72. T. M. Florence, G. E. Batley, and Y. J. Farrar, *J. Electroanal. Chem.*, **56**, 301 (1974).
73. U. Eisner and H. B. Mark, *J. Electroanal. Chem.*, **24**, 345 (1970).
74. A. M. Bond, T. A. O'Donnell, A. B. Waugh, and R. J. W. McLaughlin, *Anal. Chem.*, **42**, 1168 (1970).
75. V. P. Portretnyi, V. F. Malyuta, and V. T. Chuiko, *Z. Anal. Khim.*, **28**, 1337 (1973).
76. T. M. Florence and Y. J. Farrar, *J. Electroanal. Chem.*, **51**, 191 (1974).
77. G. Koster, U. Eisner, and M. Ariel, *Z. Anal. Chem.*, **224**, 269 (1967).
78. S. H. Lieberman and A. Zirino, *Anal. Chem.*, **46**, 20 (1974).

REFERENCES

79. H. Blutstein and A. M. Bond, *Anal. Chem.*, **48**, 759 (1976).
80. A. M. Bond, *Anal. Chem.*, **45**, 2026 (1973).
81. A. M. Bond, V. S. Biskupsky, and D. A. Wark, *Anal. Chem.*, **46**, 1551 (1974).
82. A. Hulanicki, in *M.T.P. International Review of Science*, Physical Chemistry Series One, Vol. 13: Analytical Chemistry, Part II (ed. T. S. West), Butterworth, London, 1973.
83. P. T. Kissinger, *Anal. Chem.*, **46**, 15R (1974); **48**, 17R (1976).
84. J. Koryta, *Anal. Chim. Acta*, **61**, 329 (1972).
85. J. Koryta, *Anal. Chim. Acta*, **91**, 1 (1977).

CHAPTER

15

MASS SPECTROMETRY AND SPARK-SOURCE MASS SPECTROGRAPHY

15.1 INTRODUCTION AND THEORY

Trace analysis of inorganic materials can be carried out by making measurements of isotopic abundances or ratios with a mass spectrometer or mass spectrograph. These are not spectroscopic techniques in the optical sense, but the terms *mass spectrum* and *mass spectrography* were applied by early workers whose instruments produced a photographic record resembling an optical line spectrum. The term *mass spectrometer* is applied to instruments that collect information in the form of an electrical signal that is amplified and displayed on a galvanometer or strip chart recorder. Commercial mass spectrometers become available in the 1950s, although experimental instruments were in use by 1919, following the pioneering work of J. J. Thomson, A. J. Dempster, and F. W. Aston. Since the mid-1950s, mass spectrometers and mass spectrographs have found extensive use in the analysis of geological materials.

The basic design of the mass spectrometer is illustrated schematically[1] in Fig. 15.1. The sample in gaseous form is ionized positively by an electron beam traveling from filament F to plate P_1. Plate P_2 has a positive charge and repels the ion through a slit in plate P_3. A potential difference between plates P_3 and P_4 accelerates the ions through a slit in P_4 and causes them to pass through the magnetic field of magnetic poles M_1 and M_2.

The kinetic energy imparted to an ion of mass m and charge e by the accelerating potential V is

$$\tfrac{1}{2}mv^2 = eV \tag{15.1}$$

where v is the velocity of the ion. On entering the magnetic field of strength H, the ion assumes a circular path in which the force acting on the particle, Hev, exactly opposes the centripetal force, mv^2/r, where r is the radius of the path.

$$Hev = \frac{mv^2}{r} \tag{15.2}$$

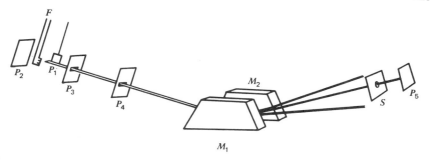

Fig. 15.1. Basic components of a mass spectrometer. *Source*: Bassett.[1] Reproduced by permission of Dowden, Hutchinson, and Ross, Inc., Stroudsburg, Pa. Copyright 1972.

From equation 15.2 we obtain

$$\frac{m}{e} = \frac{rH}{v} \tag{15.3}$$

If v is eliminated between equations 15.1 and 15.3, we have

$$\frac{m}{e} = \frac{r^2 H^2}{2V} \tag{15.4}$$

Clearly, for a given accelerating potential and field strength, singly charged ions are separated in orbits of different radius, according to their masses. When the deflected beam of ions leaves the magnetic field, only ions with a certain value of m/e pass through a slit S to a collecting plate P_5, which becomes positively charged. The charge is leaked to earth through a high resistance and the current is used to obtain an electronic readout.

In the mass spectrometer a sequence of ions of varying m/e is brought in turn to plate P_5 by varying the accelerating potential or magnetic-field strength. High sensitivity and precision can be achieved in the measurement of ion-beam intensities. The mass spectrometer has therefore found extensive use in the study of isotopic abundance ratios, which are important in dating techniques, isotope thermometry and analysis by isotope dilution.

In the mass spectrograph the magnetic field and accelerating potential are kept constant, and the dispersed ion beam impinges on a photographic plate in a manner analogous to that of the emission spectrograph (Chapter 9). Many masses can be recorded simultaneously with high resolution. Mass spectrography has been extensively used for trace analysis in conjunction with a spark source (Section 15.4).

15.2 INSTRUMENTATION

Mass spectrometric equipment consists basically of three components: ion sources, mass analyzers, and ion detectors. The various types of each component are summarized in Table 15.1. There is a tendency for instruments to be classified according to their ion sources, since this is where the greatest differences are to be found. For a more detailed description of instrumentation the reader is referred to standard works.[5,9-11]

15.2.1 Ion Sources

The purposes of the ion source are to produce monoenergetic ions representative of the sample and to form a beam suitable for injection into the analyzer. *Electron bombardment*[2] is used in most sources.

The original sample is vaporized and fed into an evacuated chamber where it meets a beam of electrons flowing at right angles to it. The electron beam energy is of the order of 70 to 80 eV. If solid, the sample has to be vaporized by heating in a crucible or Knudsen cell. For most geological applications, the operation is very tedious and inefficient but can be used successfully for large samples.

Thermal ionization[3] is one of the better methods for mass spectrometric analysis of geological materials and is particularly suited to small samples. The technique depends on the sample being placed as a slurry on a filament, dried, and then heated strongly. The efficiency of positive ion emission, n^+/n_o, is given by

$$\frac{n^+}{n_o} = \exp\frac{e(w - \phi)}{kT} \tag{15.5}$$

where ϕ is the ionization potential of the analyte element, k is the Boltzmann constant, T is the absolute temperature, and w is the work function of the material comprising the filament. When $w > \phi$ (i.e., for elements of low ionization potential), the efficiency of this source approaches 100% at quite

TABLE 15.1 Mass Spectrometric Systems

Ion sources	Mass analyzers	Ion detectors
Electron bombardment[2]	Magnetic analysis[7]	Electrometers[8]
Thermal ionization[3]	Electrostatic analysis[8]	Multipliers[8]
Ion bombardment[4]	Time-of-flight analysis[8]	Photographic detectors[8]
Spark source[5]		
Laser source[6]		

low temperatures. When $w < \phi$ the filament temperature must be raised to increase efficiency but this results in the sample being consumed too rapidly.

Ion bombardment (sputtering)[4] is a technique in which surface particles are removed from a sample by positive ion bombardment. Because only surface layers are involved, the sample must be extremely homogeneous to obtain good results and the technique is seldom used for geological samples.

Spark sources[5] are particularly suitable for trace analysis of geological samples and are discussed more fully in Section 15.4. A recent development is the appearance of *laser-source* instruments.[6]

15.2.2 Mass Analyzers

Mass analyzers[7,8] have the twofold purposes of resolving ions of different masses and of maximizing ion intensities by sharp focusing. This is analogous to the use of prisms, gratings, and lenses in optical emission spectrography.

In the case of the magnetic analyzer, described in Section 15.1, it follows from equation 15.4 that the radius of curvature of the ion path is

$$r = \left(\frac{2mV}{H^2 e}\right)^{1/2} \quad (15.6)$$

It is clear that such an analyzer separates ions according to mass (or, more strictly, according to m/e) when V and H are constant. In practice, the resolution that can be achieved with a single-focusing analyzer is limited by variations in the energies of the ions entering the magnetic field. These variations occur because the accelerating potential experienced by an ion depends on where it is formed in the source.

Higher resolution is achieved with a double focusing system. The ion beam can first be dispersed with an electrostatic analyzer, which consists of two curved metallic plates with a potential difference between them. The beam passes between the plates, experiences an electrostatic field of strength E and is deflected in a circular path of radius r given by

$$r = \frac{mv^2}{eE} = \frac{2V}{E} \quad (15.7)$$

For a given electric field the ions are dispersed according to their kinetic energies. A slit at the end of the electrostatic analyzer allows the passage of only those ions of a very narrow range of energies. The monoenergetic ion beam then passes into a magnetic analyzer, which brings all ions of a particular m/e to a common focus.

The time-of-flight mass spectrometer makes use of the fact that, in a monoenergetic ion beam, heavy ions move more slowly than light ones. If the energy is imparted in pulses, a train of ions strikes the collecting plate at different times and can be analyzed sequentially. This technique is mentioned for the sake of completeness and not for any application to trace inorganic analysis.

15.2.3 Ion Detectors

Ion detectors include electrometers and electron multipliers, the latter being used if amplification of the signal is necessary. In the measurement of isotope ratios, a dual collection system is often used, the two slits and detector units being placed in the appropriate positions for the two masses concerned. Photographic detection is discussed briefly in connection with spark-source mass spectrography (Section 15.4).

Further discussion of mass spectrometric instrumentation can be found in various standard texts[9,10] and articles.[5,11] Recent advances have been reviewed.[12]

15.3 MASS SPECTROMETRIC ISOTOPE DILUTION

15.3.1 Principles and Practice

Isotope dilution analysis may be carried out either with radioactive tracers and radiochemical methods of analysis (Section 13.5), or with stable tracers, using mass spectrometry for the analysis. Mass spectrometric isotope dilution involves determining the concentration of an element from the change in its natural isotopic composition caused by the addition of an isotopic tracer of the same element with an artificially altered isotopic composition.

Consider an element with n isotopes, isotope i having relative atomic mass M_i and a natural fractional abundance a_i. The relative atomic mass of the element in its natural state is

$$M = \sum_{i=1}^{n} a_i M_i \tag{15.8}$$

Suppose that the abundance ratio of two particular isotopes of this element, j and k, is A_{jk}:

$$A_{jk} = \frac{a_j}{a_k} \tag{15.9}$$

To x g of the natural element is added y g of a tracer in which the abundances of j and k are b_j, b_k. The abundance ratio in the tracer is

$$B_{jk} = \frac{b_j}{b_k} \tag{15.10}$$

and the relative atomic mass of the element in the tracer is

$$M' = \sum_{i=1}^{n} b_i M_i \quad (15.11)$$

In the equilibrated mixture the ratio of the number of moles (or atoms) of j and k is

$$C_{jk} = \frac{(xa_j/M + yb_j/M')}{(xa_k/M + yb_k/M')} \quad (15.12)$$

Hence

$$x = y\frac{b_k M(C_{jk} - B_{jk})}{a_k M'(A_{jk} - C_{jk})} \quad (15.13)$$

But

$$\frac{b_k M}{a_k M'} = \frac{(1/a_k)\sum_{i=1}^{n} a_i M_i}{(1/b_k)\sum_{i=1}^{n} b_i M_i} = \frac{\sum_{i=1}^{n} A_{ik} M_i}{\sum_{i=1}^{n} B_{ik} M_i} \quad (15.14)$$

Substitution in equation 15.12 gives

$$x = y\frac{\sum_{i=1}^{n} A_{ik} M_i}{\sum_{i=1}^{n} B_{ik} M_i}\frac{C_{jk} - B_{jk}}{A_{jk} - C_{jk}} \quad (15.15)$$

In the case of an element with two isotopes, equation 15.15 reduces to

$$x = y\frac{M_1 + A_{21}M_2}{M_1 + B_{21}M_2}\frac{C_{21} - B_{21}}{A_{21} - C_{21}} \quad (15.16)$$

A good example of the application of stable isotope dilution is in the determination of lithium, as quoted by Webster.[13] Normal lithium has a $^7Li/^6Li$ isotopic ratio of 12.4. If a geological sample containing x g of normal lithium is diluted with y g of a lithium tracer consisting of 99.6% ^6Li and 0.4% ^7Li, and the isotope ratio $C_{7,6}$ is then measured, x can be calculated from

$$x = 15.4y\frac{C_{7,6} - 0.004}{12.4 - C_{7,6}}$$

The mass spectra obtained at different stages of the operation of determining lithium in G-1 are illustrated in Fig. 15.2.

The following experimental steps are involved. The sample is dissolved in a suitable solvent and the stable tracer is added to this solution. The normal and tracer forms of the element are then mixed thoroughly. Conversion of all the element to the same chemical form may be necessary. The element is then purified to a stage where it is suitable for mass spectrometric analysis.

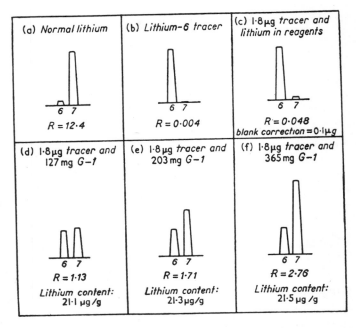

Fig. 15.2. The determination of lithium in granite (*G*-1) by mass spectrometric isotope dilution. *Source*: Webster.[13]

Quantitative separation is unnecessary. Finally, the isotope ratio is determined. The same procedure may be used when determining the reagent blank, as indicated in Fig. 15.2c.

15.3.2 Sensitivity, Precision, and Accuracy of Isotope Dilution

The working range of isotope dilution is limited by its being essentially a ratio method. For each element there are upper and lower limits to the range of accessible concentrations when a given amount of tracer is used. In the above example of lithium determination, as the concentration of lithium in the rock decreases, the composition of the mixture (sample and tracer) approaches that of the tracer and eventually becomes indistinguishable from it. At high concentrations, the mixture approaches the composition of normal lithium and again becomes indistinguishable. For a given mass of tracer (y g), the working range depends on the isotopic ratios in the naturally occurring element and in the available tracer.

If the minimum detectable mass x is regarded as that required to change the isotopic ratio of the tracer by 2%, then from equation 15.13,

$$\left(\frac{x}{y}\right)_{min} = \frac{b_k M}{a_k M'} \frac{0.02 B_{jk}}{A_{jk} - 1.02 B_{jk}} \quad (15.17)$$

In the case of the ^{85}Rb/^{87}Rb ratio, for example, where $A_{jk} = 2.59$, if the available tracer has $B_{jk} = 0.177$, the minimum ratio (x/y) is calculated to be about 1/230. In other words, y g of rubidium tracer can be used to determine as little as $y/230$ g of rubidium in a sample. An upper limit for x can also be calculated, where y g of tracer causes only a 2% change in the normal abundance. In the above example this occurs at about $x = 140 \, y$.

Other examples noted by Webster[13] are ranges for uranium (^{235}U/^{238}U) from $y/100,000$ to $5000y$, and for silver (^{107}Ag/^{109}Ag) from $y/230$ to $80y$.

When there is a large difference in natural abundances of the pair of isotopes chosen, and when the naturally rarer isotope is abundant in the tracer, the working range is very large. In cases where the natural ratio is close to unity (e.g., silver), the working range is shorter.

There is also an optimum ratio of x to y, where a given error in the determination of C_{jk} has the smallest effect on the calculated value of x. This can be shown[13] to occur with the mixture for which

$$C_{jk} = (A_{jk} B_{jk})^{1/2}$$

Absolute sensitivities for different elements are determined by the minimum x/y ratio as calculated above, and by the size of sample required for precise quantitative analysis in the mass spectrometer. Microgram quantities are often needed, although for elements ionized with high efficiency, such as the alkali metals, amounts as small as 10^{-10} g may be adequate. Thus, in the case of rubidium, isotope dilution would serve to measure amounts as low as $1/230 \times 10^{-10}$ or about 10^{-12} to 10^{-13} g. In practice, the sensitivity may also be limited by experimental factors such as the level of contamination of the analyte in the reagents used for chemical treatment of the samples. Nevertheless, sensitivities in the range 10^{-7} to 10^{-12} g have been achieved for many elements.

Stable isotope dilution is relatively free from interferences. Interference between isobars (isotopes of different elements with the same mass number) can be avoided by calculating a correction from the recorded amounts of other isotopes of the interferent, by chemical separations or by separating the elements according to their ease of volatilization in the ion source.

Isotope ratios can usually be measured with a precision better than 0.5%, and many workers have obtained relative standard deviations of 1% to 3% for replicate analyses of geological samples. The concentrations of uranium[14] and rubidium[15] in seawater, for example, were found to be 3.26 ± 0.06

and 121.4 ± 1.1 µg/l, respectively. Rubidium and cesium have been determined in meteorites,[16,17] and about 50 isotope dilution analyses of the standard rocks G-1 and W-1 have been summarized.[18] Several elements, including K, Rb, Sr, Ba, and the lanthanides, have been determined in volcanic rocks[19] and in lunar materials.[20]

Systematic errors may be introduced by any effect in which the instrument behaves differently for isotopes of different mass, such as diffusion from a sample vaporization cell, or the response of an electron multiplier. Corrections for such effects can be made by carrying out an isotope dilution on a standard.

Stable isotope dilution analysis by mass spectrometry requires relatively expensive equipment, and analysis times are long by comparison with many other instrumental methods. It can not be used for the 17 monoisotopic elements. On the other hand few analytical techniques are capable of such high precision and accuracy for many elements in geological materials at concentrations in the range 1 µg/g to 1 ng/g or less.

Mass spectrometric measurements of the ratios of isotopic abundances of lead and uranium have been used extensively in geochronological studies of terrestrial and, more recently, lunar[21] materials. The development of extremely precise mass spectrometric techniques of measuring $^{87}Sr/^{86}Sr$ ratios (to within about five parts in 10^5) has been of great importance in studies on planetary evolution.[22,23]

15.4 SPARK SOURCE MASS SPECTROGRAPHY

Although the spark source mass spectrograph has been available as an analytical tool for more than 15 years, it was not until the mid-1960s that it became established as a technique for trace analysis in geochemistry. Its acceptance in this field is largely due to the work of Brown and Wolstenholme,[24] Taylor,[25,26] and Nicholls et al.,[27] who showed that good precision and accuracy (±5%) could be obtained for most elements.

15.4.1 Analytical Details

Instrumentation for spark source mass spectrography has been discussed by Deines.[5] A typical instrument is shown in Fig. 15.3, and a typical operating procedure[28] is summarized in Table 15.2.

Quantitative analysis procedures with the spark source mass spectrograph are similar to those used in emission spectrography (Chapter 9). Detailed descriptions have been given.[5,8,11,25-28] The photographic plate is first calibrated by a graded series of exposures (usually three) analogous to the steps of the step-sector self-calibration procedure in emission spectrography

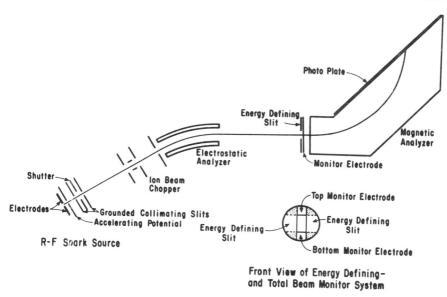

Fig. 15-3. Schematic layout of a spark source mass spectrograph. *Source*: Deines.[5]

(p. 147). Alternatively, the plate may be calibrated by measuring the density of lines from two isotopes of the same element (e.g., ^{171}Yb and ^{172}Yb). Internal standards are used, as in emission spectrography, and greatly improve the precision and accuracy of the method.

15.4.2 Features of Spark Source Mass Spectrography

With the spark source, matrix effects of the type encountered in emission spectrography are not a serious problem. There may, however, be interferences from multiatom lines or multicharged ions with the same m/e ratio as the analyte. For example, if graphite is used as a diluent and matrix, ions of the type C_n^+ (where n may be from 1 to 20) can appear in the spectrum. The C_4^+ ion, for example, would interfere with ^{48}Ti$^+$. If elements such as Zr and Mo are present in high concentration, the latter line could also have interferences from ^{96}Zr^{2+} and ^{96}Mo^{2+}, which have the same m/e.

Since the area of contact of an individual spark is small, short exposures can result in erroneous data if the sample is not very homogeneous. One way round this problem is to fuse the sample with a flux such as lithium tetraborate and then grind the mixture very finely.

TABLE 15.2 Typical Scheme for Spark-Source Mass Spectrographic Analysis of Geological Materials

Parameter	Procedure
Preparation of electrodes	Sample is ground (1:1) with graphite powder and pressed into the shape of an electrode in a hydraulic press. Two electrodes are required for each sample.
Sparking	Samples are sparked with 20 kV rf spark operating at 1000 cycles/s with spark duration of 50 μs.
Accelerating voltage	A 20 kV potential accelerates the ions through the slit system into the electrostatic analyzer.
Electrostatic analyzer	The electrostatic analyzer filters out all but a narrow band of energy and allows this monoenergetic beam to enter the magnetic analyzer.
Magnetic analyzer	The magnetic field separates the beam into ions of similar m/e ratios. For example at mass 45, it is possible to have $^{45}Sc^+$, $^{90}Zr^{2+}$, and $^{135}Ba^{3+}$.
Detector	Photographic detection with emulsion sensitive to positive ion impact. The normal 25 × 5 cm plate permits the focusing of m/e 7 to 238 at the highest magnetic field strength. Smaller mass ranges can be brought to focus by lowering the magnetic field strength.
Total ion current monitor	This device is used to monitor the accumulation of charge at the photoplate, hence permitting a graded series of exposures.
Vacuum system	The system operates under vacuum with typical pressures of 5×10^{-8} torr in the analyzer and 5×10^{-7} torr in the source.

From Ref. 28.

The spark source spectrographic method gives high sensitivity with low detection limits (commonly about 0.003 μg/g) for most elements in geological materials. Unlike most other methods, spark source mass spectrography gives approximately the same limits of detection for all elements. A permanent record of the spectra is provided by the use of a photographic plate. With careful control of sparking conditions and the use of internal standards, accuracy and precision (2-5%) compare favorably with many other methods.

A major disadvantage of spark-source mass spectrography is the cost of the equipment, which for an instrument with a *photographic* detection system is about the same as for an automatic *direct-reading* emission spectrometer, but without the advantage of the latter instrument's high speed of analysis. The small sample size used in spark source mass spectrography can be disadvantageous if the analyte is not distributed evenly through the sample.

15.4.3 Applications in Geochemical Analysis

Geochemical applications of spark source mass spectrography have been reviewed by Deines[5] and Carver and Johnson.[29] The first application of this technique was in the field of geochronology.[30] Early work was centered on isotopic ratios because of errors involved in photographic measurements of line intensities. Improvements in measurement techniques have now reduced this problem considerably. Earlier work on direct measurement of abundances of trace elements in silicates was confined mainly to standard rocks.[24–27,29,31,32] Later, work was carried out on meteorites,[33–35] andesites,[36] rhyolites[37] and basalts.[38,39]

Water samples were analyzed by Brown et al.[40] using a rotating disk electrode. Concentrations in the nanogram per gram range were measured for several elements.

Morrison et al.[41,42] analyzed eight lunar soils and rocks for 67 elements using a combination of spark source mass spectrography and neutron activation analysis. The data obtained for a lunar soil were compared[43] with those obtained at a number of other laboratories. The methods used by other workers included neutron activation, isotope dilution mass spectrometry, spark source mass spectrography, emission spectroscopy, and X-ray fluorescence. Agreement was in general very good.

In spite of its advantages, the use of spark source mass spectrography is probably limited by competition from the cheaper and faster direct-reading emission spectrometry. For example, in the period 1965–1969, only 1.8% of all analyses for 10 standard reference rocks were performed by spark source mass spectrography compared with 35.7% by emission spectrography.[44]

References

1. W. A. Bassett, in *Encyclopedia of Geochemistry and Environmental Sciences* (ed. R. W. Fairbridge), Van Nostrand Reinhold, New York, 1972, p. 688.
2. A. O. Nier, *Rev. Sci. Instrum.*, **18**, 398 (1947).
3. K. Kunsman, *Science*, **62**, 269 (1925).
4. G. Slodzian, Thesis, Univ. of Paris, Paris, 1964.
5. P. Deines, *Pa. State Univ. Mineral Sci. Exp. Stn. Circ.*, **78**, 1 (1970).
6. B. E. Knox, in *Trace Analysis by Mass Spectrometry* (ed. A. J. Ahearn), Academic Press, New York, 1972, p. 423.
7. R. Herzog, *Z. Phys.*, **89**, 447 (1934).
8. J. Roboz, in *Trace Analysis—Physical Methods* (ed. G. H. Morrsion), Interscience, New York, 1965, p. 435.
9. A. J. Ahearn (ed.), *Trace Analysis by Mass Spectrometry*, Academic Press, New York, 1972.

10. J. Roboz, *Introduction to Mass Spectrometry. Instrumentation and Techniques*, Wiley-Interscience, New York, 1968.
11. J. N. Weber and P. Deines, in *Modern Methods of Geochemical Analysis* (eds. R. E. Wainerdi and E. A. Uken), Plenum, New York, 1971.
12. A. J. B. Robertson, in *M.T.P. International Review of Science, Physical Chemistry Series One*, Vol. 13: Analytical Chemistry, Part 2 (ed. T. S. West), Butterworths, London, 1973.
13. R. K. Webster, in *Methods in Geochemistry* (eds. A. A. Smales and L. R. Wager), Interscience, New York, 1960.
14. E. Rona, L. O. Gilpatrick, and L. M. Jeffrey, *Trans. Am. Geophys. Union*, **37**, 697 (1956).
15. A. A. Smales and R. K. Webster, *Geochim. Cosmochim. Acta*, **11**, 139 (1957).
16. R. K. Webster, J. W. Morgan, and A. A. Smales, *Trans. Am. Geophys. Union*, **38**, 543 (1957).
17. A. A. Smales, T. C. Hughes, D. Mapper, C. A. J. McInnes, and R. K. Webster, *Geochim. Cosmochim. Acta*, **28**, 209 (1964).
18. M. Fleischer, *Geochim. Cosmochim. Acta*, **33**, 65 (1969).
19. B.-M. Jahn, C.-Y. Shih, and V. R. Murthy, *Geochim. Cosmochim. Acta*, **38**, 611 (1974).
20. J. A. Philpotts, C. C. Schnetzler, M. L. Bottino, S. Schumann, and H. H. Thomas, *Earth Planet. Sci. Lett.*, **13**, 429 (1972).
21. M. Tatsumoto and J. N. Rosholt, *Science*, **167**, 461 (1970).
22. D. A. Papanastassiou and G. J. Wasserburg, *Earth Planet. Sci. Lett.*, **5**, 335 (1969); **11**, 37 (1971); **16**, 289 (1972); **17**, 52 (1972).
23. C. M. Gray, D. A. Papanastassiou, and G. J. Wasserburg, *Icarus*, **20**, 213 (1973).
24. R. Brown and W. A. Wolstenholme, *Nature*, **201**, 598 (1964).
25. S. R. Taylor, *Geochim. Cosmochim. Acta*, **29**, 1243 (1965).
26. S. R. Taylor, *Nature*, **205**, 34 (1965).
27. G. D. Nicholls, A. L. Graham, E. Williams, and M. Wood, *Anal. Chem.*, **39**, 584 (1967).
28. P. C. Rankin, *N. Z. Soil News*, **19**, 168 (1971).
29. R. D. Carver and P. G. Johnson, 16th Ann. Conf. Mass Spectrom. Allied Topics, Pap. 116, Pittsburgh, 1968.
30. F. D. Leipziger and W. J. Croft, *Geochim. Cosmochim. Acta*, **28**, 268 (1964).
31. L. F. Herzog, T. J. Eskew, and P. Deines, *Trans. Am. Geophys. Union*, **45**, 112 (1964).
32. L. F. Herzog, T. J. Eskew, and P. Deines, *Pa. State Univ. Mineral Ind. Exp. Stn. Sp. Rep.* **1–65** (1965).
33. E. Berkey and G. H. Morrison, *Int. Ser. Monogr. Earth Sci.*, Pergamon, Oxford, 1968, p. 345.
34. E. Berkey and G. H. Morrison, Mater. Sci. Center Rep. No. 1129, Cornell Univ., New York, 1969.

35. A. A. Smales, *Lab. Practice*, **16**, 701, 713 (1967).
36. S. R. Taylor and A. R. J. White, *Bull. Volcanol.*, **29**, 177 (1966).
37. A. Ewart, S. R. Taylor, and A. C. Capp, *Contr. Mineral. Petrol.*, **18**, 76 (1968).
38. G. H. Morrison and A. T. Kashuba, *Anal. Chem.*, **41**, 1842 (1969).
39. G. D. Nicholls and A. L. Graham, *Trans. Am. Geophys. Union*, **49**, 354 (1968).
40. R. Brown, W. J. Richardson, and P. Swift, 16th Ann. Conf. Mass. Spectrom. Allied Topics, Pap. 110, Pittsburgh, 1968.
41. G. H. Morrison, J. T. Gerard, A. T. Kashuba, E. V. Gangadharam, A. M. Rothenberg, N. M. Potter, and G. B. Miller, *Science*, **167**, 505 (1970).
42. G. H. Morrison, J. T. Gerard, A. T. Kashuba, E. V. Gangadharam, A. M. Rothenberg, N. M. Potter, and G. B. Miller, *Geochim. Cosmochim. Acta*, **34**, Suppl. 1, Vol. 2, 1383 (1970).
43. G. H. Morrison and J. R. Roth, in *Trace Analysis by Mass Spectrometry* (ed. A. J. Ahearn), Academic Press, New York, 1972, p. 297.
44. R. R. Brooks, C. R. Boswell, and R. D. Reeves, *Chem. Geol.*, **21**, 25 (1978).

CHAPTER

16

PHYSICAL AND CHEMICAL FIELD TESTS FOR TRACE ELEMENTS

Field methods used in mineral exploration include physical methods such as panning and magnetic separation, and chemical methods involving simple colorimetric tests or the use of various portable instruments. In recent years, increasing use has been made of mobile geochemical laboratories. These allow moderately sophisticated analytical apparatus to be used in the field, and facilitate the rapid analysis of the large numbers of samples involved in exploration surveys.

16.1 PHYSICAL FIELD TESTS

The gold pan has been found to be useful for examining many heavy metals and minerals[1-3] in stream sediments, in particular. It can be used to separate minerals for identification with a hand lens or small microscope, and can act as a screening tool to select for chemical analysis only those samples containing minerals of interest. For example, Theobald et al.[4] reported the analysis of 291 samples of magnetite from alluvia of the Inner Piedmont belt of North and South Carolina. Lead, zinc and copper were determined by colorimetric field methods[5] and a further nine elements were found by emission spectrography.

A good example of the use of the gold pan in mineral prospecting is the work of Guigues and Devismes[6] who surveyed 50,000 km^2 of the Massif Armoricain of Western France. The survey involved the panning of stream sediments and microscopic examination of 113,520 heavy-metal concentrates over a period of 10 years. Simple physical separations allowed about 50 minerals to be identified and their approximate compositions estimated. Further work by chemical methods at all anomalous sites resulted in the discovery of an appreciable number of economic deposits of tin, titanium, zirconium, rare earths, lead, zinc, and silver.

16.2 CHEMICAL FIELD TESTS

16.2.1 Colorimetric Tests

The majority of chemical field tests are colorimetric because of the simplicity of the apparatus required. Many procedures have been developed by the U. S. Geological Survey[5,7–11] and by the Applied Geochemistry Department, Imperial College, London.[12,13] The field performance of these techniques has been reviewed,[11,12] and the book by Stanton[13] is a good source of information on colorimetric tests for field work.

There is some sacrifice of precision and accuracy by comparison with similar analysis carried out in the laboratory. However, in geochemical exploration, high accuracy is usually less important than good precision. Although many colorimetric tests relying on visual interpretation are only semiquantitative, better precision and sensitivity can be obtained by using a spectrophotometric finish.

16.2.2 Instrumental Techniques

Levinson[14] has described several instruments that are sufficiently portable to be used in the field. Gamma-source and gamma-X, X-ray fluorescence units (Chapter 12) weigh only about 10 kg and can be used to obtain *in situ* measurements of several elemental concentrations in rocks and minerals. These instruments do, however, suffer disadvantages of cost (approximately $5000), matrix effects, and a general lack of sensitivity for trace concentrations.[15,16]

Ion-selective electrodes (Chapter 14) may be used for suitably digested rocks[17,18] or for water samples.[19] They have the advantages of cheapness and portability, but careful control of ionic strength and ion interferences is necessary.

Many instruments, with varying degrees of portability, have been developed to measure natural radioactivity. The use of portable γ-ray spectrometers to measure potassium, uranium, and thorium has been noted briefly in Chapter 13, and has been reviewed more extensively by Rybach[20] and Adams et al.[21] Whitehead and Brooks[22] have compared radioactive methods and chemical procedures for uranium. A portable radioactivation instrument specific for beryllium is the beryllometer.[14] Bombardment of the sample with γ radiation induces the reaction $^9Be(\gamma, n)2^4He$ (see p. 291). For use under field conditions, beryllium concentrations considerably higher than 20 $\mu g/g$ are required.

Because of the importance of mercury as a pathfinder for sulfide minerals, portable units have been developed for the determination of this element.[23] Many instruments developed for this purpose are derived from the earliest analytical atomic absorption apparatus,[24,25] using a mercury vapor lamp and a simple photocell. One of the main problems in the flameless atomic absorption determination of mercury in air or in solid samples (using the various types of vapor-generation apparatus) is the enhancing interference from absorption by molecular species in the light path. Several methods can be used to correct for this effect. Ballard et al.[26] subtracted the signal obtained with a continuous source from that given with the mercury-line source. Modern commercial apparatus employs a hydrogen or deuterium discharge lamp for this purpose (p. 179). Measurements of the nonatomic component of the absorption have also been made using the wings of a pressure- or Lorentz-broadened mercury-line source[27,28] or of Zeeman-split components.[29] The field use of atomic absorption apparatus for mercury has been described by Jonasson et al.[30] An alternative device for mercury determination is a quartz crystal microbalance,[31] in which the frequency shift of a quartz crystal is related to the increase in mass resulting from the amalgamation of mercury on to a gold electrode.

16.3 MOBILE GEOCHEMICAL LABORATORIES

Mobile geochemical laboratories are able to combine the advantages of speed of analysis and sophistication of instrumentation. Significant work in pioneering the development of mobile laboratories has taken place in the Soviet Union[32] and North America.[33,34] The "workhorse" of many mobile laboratories has been a simple optical emission spectrograph. However, there are advantages in using a direct-reading spectrograph[35] now that these instruments have been miniaturized. A mobile spectrographic laboratory designed specifically for biogeochemical work has been described.[36]

The U. S. Geological Survey uses truck-mounted spectrographic units, housed both in large van-type trucks and in smaller step-van type vehicles.[37] Electric power can be obtained from gasoline-powered generators or from a temporary installation of the local electric power company. Operations such as mineral separations, colorimetric tests and atomic absorption analysis, are also carried out in camper laboratories which are hauled to temporary operation sites in the field. In addition to housing a spectrographic laboratory, the step-van type vehicle is used for sample preparation, using such equipment as jaw crushers, pulverizers, and ashing and sieving apparatus. At present, these mobile laboratories are largely used to furnish analytical data to field projects connected with geochemical assessment programs.

Other advances in the use of mobile laboratories have occurred in the area of data handling.[37] At selected field locations, a portable terminal equipped with telephone access to a centralized time-shared computer facility has been set up. Data can be fed into the computer by telephone, and statistical parameters and graphical analysis can be transmitted back to the workers in the field.

Mobile laboratories have recently been of value in the exploration of several areas with difficult access. For example, units equipped with atomic absorption instruments have been shipped to sites in the Solomon Islands and have been flown to within a few hundred kilometers of the North Pole.[14] The use of mobile laboratories will continue to increase, because of the advantage of rapid data acquisition during exploration programs, and because of their convenience in the exploration of remote areas.

References

1. R. P. Fischer and F. S. Fischer, *U. S. Geol. Surv. Circ.*, **592**, 1 (1968).
2. P. K. Theobald, *U. S. Geol. Surv. Bull.*, **1071-A**, 1 (1957).
3. W. C. Overstreet, R. G. Yates, and W. C. Griffitts, *U. S. Geol. Surv. Bull.*, **1162-F**, 1 (1963).
4. P. K. Theobald, W. C. Overstreet, and C. E. Thompson, U. S. Geol. Surv. Prof. Pap. **554-A**, 1 (1967).
5. F. N. Ward, H. W. Lakin, F. C. Canney et al., *U. S. Geol. Surv. Bull.*, **1152**, 1 (1963).
6. J. Guigues and P. Devismes, *Mem. BRGM*, **71**, 1 (1969).
7. H. W. Lakin, H. Almond, and F. N. Ward, *U. S. Geol. Surv. Circ.*, **161**, 1 (1952).
8. L. E. Reichen, *Anal. Chem.*, **23**, 727 (1951).
9. L. E. Reichen and H. W. Lakin, *U. S. Geol. Surv. Circ.*, **41**, 1 (1949).
10. L. E. Reichen and F. N. Ward, *U. S. Geol. Surv. Circ.*, **124**, 1 (1951).
11. J. H. McCarthy, *Proc. 20th Int. Geol. Congr.*, Mexico City, 1959, p. 363.
12. J. S. Tooms, *Proc. 20th Int. Geol. Congr.*, Mexico City, 1959, p. 377.
13. R. E. Stanton, *Rapid Methods of Trace Analysis*, Arnold, London, 1966.
14. A. A. Levinson, *Introduction to Exploration Geochemistry*, Applied Publishing, Calgary, 1974.
15. H. Wollenberg, H. Kunzendorf, and J. Rose-Hansen, *Econ. Geol.*, **66**, 1048 (1971).
16. H. Kunzendorf, *Geochemical Exploration*, Inst. Min. Metall., London, 1972, p. 401.
17. W. H. Ficklin, U. S. Geol. Surv. Prof. Pap. **700-C**, C186 (1970).
18. S. E. Kesler, J. C. Van Loon, and J. H. Bateson, *J. Geochem. Explor.*, **2**, 11 (1973).
19. M. O. Schwartz and G. H. Friedrich, *J. Geochem. Explor.*, **2**, 103 (1973).
20. L. Rybach, in *Modern Methods of Geochemical Analysis* (eds. R. E. Wainerdi and E. A. Uken), Plenum Press, New York, 1971.
21. J. A. S. Adams, G. E. Fryer, and J. J. W. Rogers, *Anal. Chem.*, **41** (6), 22A (1969).
22. N. E. Whitehead and R. R. Brooks, *Econ. Geol.*, **64**, 50 (1969).

23. J. C. Robbins, *Geochemical Exploration*, Inst. Min. Metall., London, 1973, p. 315.
24. T. T. Woodson, *Rev. Sci. Instr.*, **10**, 308 (1939).
25. C. W. Zuehlke and A. E. Ballard, *Anal. Chem.*, **22**, 953 (1950).
26. A. E. Ballard, D. W. Stewart, W. O. Kamm, and C. W. Zuehlke, *Anal. Chem.*, **26**, 921 (1954).
27. A. R. Barringer, *Trans. Inst. Min. Metall.*, B, **75**, 120 (1966).
28. C. Ling, *Anal. Chem.*, **40**, 1876 (1968).
29. T. Hadeishi and R. D. McLaughlin, *Science*, **174**, 404 (1971).
30. I. R. Jonasson, J. J. Lynch, and L. J. Trip, Geol. Surv. Can. Pap., 73-21 (1973).
31. Q. Bristow, *J. Geochem. Explor.*, **1**, 55 (1972).
32. E. A. Ratsbaum, *Razv. Nedr.*, **10**, 38 (1939).
33. F. C. Canney, A. T. Myers, and F. N. Ward, *Econ. Geol.*, **52**, 289 (1957).
34. R. H. C. Holman and C. C. Durham, Geol. Surv. Can. Pap. 66-55, 1 (1966).
35. A. H. Debnam, *Proc. 1st Int. Geochem. Explor. Symp.*, Colorado, 1969, p. 217.
36. J. A. C. Fortescue and E. H. W. Hornbrook, Geol. Surv. Can. Pap. 67-23, 1 (1967).
37. F. N. Ward, personal communication.

CHAPTER

17

USES OF DATA ON TRACE ELEMENTS IN GEOLOGICAL MATERIALS

The development of methods for the analysis of silicates and the evaluation of elemental crustal abundances have been discussed briefly in Chapter 5. Compilations of abundance data in various geological materials can be found in several works.[1-5] Knowledge of the overall crustal abundance of trace elements and of their concentrations in specific materials is important in many fields, including geochemistry and cosmochemistry, mineral exploration, soil science, plant science, animal and human health, environmental chemistry, and archaeology. Several of these applications are outlined briefly in this chapter.

17.1 GEOCHEMICAL STUDIES

17.1.1 Natural Mobilization of the Elements

We first note briefly the mechanisms by which the elements have achieved their present distribution in the earth. Under natural conditions, the elements have first been mobilized as the liquid magma crystallized and have been distributed among the different rock types (*hypogene mobility*). After deposition of the rocks, various weathering processes occur, leading to a further redistribution of the elements (*supergene mobility*).

Hypogene mobility occurs at depth under conditions of high temperature and pressure. The major constituents of the earth's crust form a sequence of minerals dependent on the prevailing temperature and pressure.[6] The minor elements frequently occupy spaces in the lattices of these minerals according to the rules of diadochic substitution. The residual fluids, deposited as pegmatites or hydrothermal veins, are usually rich in many of the elements whose overall crustal abundances are low. In high-temperature metamorphism the last-formed minerals tend to be the first to become liquid again as the temperature rises. The mobilization and transport of elements in the primary environment is known as primary dispersion.

In the course of primary dispersion, the elements become concentrated in certain types of geological formations, leading to localized concentrations (ores). The siderophile elements (including iron, cobalt, nickel, chromium, and the platinum metals) are concentrated in iron deposits and in the iron-nickel core of the earth. Chalcophile elements, associated with sulfur, include antimony, arsenic, cadmium, copper, lead, mercury, selenium, silver, and zinc. The chalcophile elements, although rare in overall abundance, are easy to obtain commercially because they tend to accumulate in accessible sulfide deposits. Lithophile elements, such as the alkali metals, magnesium, calcium, chromium, and vanadium, have an affinity for silicates.

Supergene mobility within the surface environment is of great importance in elemental differentiation in rocks and rock-derived materials, and takes place under conditions of low temperature and pressure. Mobilization is strongly influenced by local conditions of electrode potential and pH and by the stability of the minerals that have to be decomposed. Mechanical, physical, chemical, and biological processes may all be involved. Garrels[7] has calculated the theoretical mobilities of trace elements in drainage waters from consideration of the physical chemistry of the relevant mineral and ionic species. However, equilibrium between the mobile and immobile phases in the natural environment is seldom encountered, and the situation is often complicated by such processes as adsorption on clay materials and humic material. General discussion of chemical weathering can be found in texts such as that of Krauskopf.[1]

17.1.2 Abundance Data and Geochemical Behavior

A considerable amount of effort has been devoted by geochemists to the acquisition of abundance data as an information-gathering exercise and as a test of analytical techniques. In addition, however, there are a number of areas of geochemistry in which trace element data have been very useful. For example, a great deal of interest has centered on the geochemical behavior of the less common elements during such processes as the fractional crystallization of magma of various compositions, and the formation of sediments.

Several different approaches have been used in attempts to classify the behavior of trace elements. Limited success can be achieved by making use of relationships with major elements of the same group in the periodic table. The frequent association of selenium with sulfur, gallium with aluminum, and the rarer alkali metals with potassium are obvious examples.

An alternative way of obtaining a broad general guide to trace-element behavior is based on work by Goldschmidt,[8] who examined the way in which the elements tend to distribute themselves among three kinds of phase: metallic (siderophile elements), sulfide (chalcophile elements), and silicate (lithophile elements). Evidence of general tendencies has been obtained from the analysis of such phases as they occur in the earth's crust and as they are formed in metallurgical operations, and from analysis of different kinds of meteorite.

A more detailed description of trace element distribution must take into account the actual sequence of rock-forming processes and the accompanying changes of major-element composition. During the fractional crystallization of a silicate melt, for example, many trace elements become incorporated in the major silicate structures, often as isomorphous replacements for an abundant element. Pioneering studies in this field were carried out by Goldschmidt,[8] Ahrens,[9] Ramberg,[10] and others. Isomorphous replacement of one element by another occurs if (a) the ions of the two elements have the same charge and similar ionic radii, (b) the ions have charges differing by one unit, provided the radii are very similar. Where two ions can occupy the same position in a crystal lattice, the one that forms stronger bonds is the one with smaller radius, or higher charge, or both. If the two elements form bonds having very different covalent character, substitution of one for the other may be very limited, even if the size criterion is fulfilled.[11,12]

These generalizations have been established from the analysis of a large number of rock samples taken from different stages of magmatic differentiation. They help to explain, for example: the substitution of nickel, cobalt and chromium for magnesium and iron, and their tendency to be found concentrated in ultramafic rocks; the substitution of rubidium, barium, and lead for potassium; the substitution of strontium for calcium, and lithium for magnesium. Elements that do not meet the criteria for ready substitution for one of the major elements, become enriched in the residual liquids and are concentrated in pegmatites or in sulfide veins. Elements concentrated in pegmatites include Be, Sn, Ta, Nb, W, Th, and U. Some elements, such as zirconium, and to some extent, chromium, show a tendency to form separate minerals (zircon, chromite) even when present in very small amounts.

Typical examples of studies on major and trace-element behavior in different rock series can be found elsewhere.[13-16] More detailed accounts of trace element distribution in rocks have been given by Krauskopf,[1] Mason,[2] Shaw,[17] and Taylor.[18] Theoretical (including thermodynamic) approaches to the study of trace element distribution have been discussed by Kretz[19] and others.[20-22]

The distribution of trace elements in sedimentary rocks is influenced by a wide variety of physical and chemical properties. Minerals that are very resistant to mechanical abrasion and to dissolution tend to occur as detrital sediments. In other cases, where the sediments result from precipitation, properties such as solubility, ionic substitution, and the adsorption of traces on particles of fine-grained materials, are of major importance. The activity of organisms and the formation of organometallic compounds may be responsible for certain characteristic accumulations of trace elements in sedimentary materials (e.g., germanium in coal; copper and uranium in black shales; vanadium, nickel, and sometimes molybdenum, in oils, oil shales, and asphalts). The reader is referred to standard works[1,2,8,23,24] for further discussion of sediments and the behavior of trace elements in sedimentation processes. Extensive tabulations of elemental abundances in sedimentary rocks and deep-sea sediments, as well as igneous rocks, have been given by Turekian and Wedepohl.[5]

17.1.3 Trace Elements in Extraterrestrial Material

The chemical composition of meteorites[25-28] has been of interest for many years, particularly in connection with the development of theories on the constitution of the interior of the earth and on the formation of the solar system. Whether meteorites are regarded as fragments of a disintegrated planet or asteroid, or simply as residual material of the same kind as that from which the earth was formed, it has been considered that their average composition should be representative of that of the earth as a whole, at least with respect to the nonvolatile elements.

Extensive studies have been made of the trace element composition of meteorites to determine whether elemental distributions parallel those found in comparable terrestrial materials. In addition, trace element isotopic abundance work has been directed toward the establishment of the ages of meteorites, and the development of theories of the sequence of events involved in the formation of the solar system.

Particular attention has been paid to groups of trace elements that are almost undifferentiated in terrestrial material (e.g., zirconium and hafnium[29]), or have well-established interelement abundance patterns (e.g., the lanthanides[30-32]). Among the lanthanides, the abundances of cerium and europium are of special interest because of the existence of oxidation states additional to the normal trivalent ions of the lanthanide series [Ce(IV), Eu(II)]. These may serve to indicate the red-ox state under which the meteorite (or its precursor) was formed. Where europium exists as Eu^{2+} it may follow strontium and barium, and give an anomaly in the lanthanide distribution pattern. Relative to the other lanthanides, europium is depleted in some achondrites,[33] when the distribution is compared to that in chondrites. A europium anomaly has also been observed in lunar material.[34]

The return to Earth of lunar samples from the Apollo and Luna missions set off an unprecedented volume of analytical activity. Most lunar rock specimens examined to date belong to one of three broad categories: mare basalt, anorthosite, and a norite ("KREEP," for K, rare earth elements, P) containing unusually high concentrations of potassium, barium, lanthanides, thorium, uranium, and phosphorus. Analysis of the lunar basaltic rocks has shown that they are considerably depleted in the more volatile elements (including Ga, In, Tl; Ge, Pb; As, Sb, Bi; Cu, Ag, Au; Zn, Cd, Hg; Br), both relative to inferred cosmic abundances and to terrestrial continental basalts.

Trace element data on lunar rocks and the significance of these data for theories of lunar formation and differentiation processes have been discussed at several conferences[35-40] and elsewhere (see, for example, *Science*, **167**, pp. 449-582; *Geochim. Cosmochim. Acta*, **37**, pp. 719-1109). Recent progress in understanding lunar history, resulting from intensive geochemical work on lunar materials, has been summarized.[41-44]

17.1.4 Trace Analysis in the Solution of Some Geochemical Problems

The literature of geochemistry contains many studies in which trace element analysis (including isotope-ratio measurement on trace elements) has been able to shed light on the origin or geochemical history of a particular type of rock or other geological material. The various techniques of geochronology, using ratios such as $^{238}U/^{206}Pb$, $^{235}U/^{207}Pb$, $^{40}K/^{40}Ar$, and $^{87}Rb/^{87}Sr$, fall into this category. As this field is too large to be covered here, the reader is referred elsewhere for short reviews or more detailed information.[45-49]

Two other geochemical problems in which trace-element analysis has made a valuable contribution are discussed briefly below.

The Origin of Tektites

Tektites[50] consist of small masses of a silica-rich glassy material somewhat resembling obsidian but having characteristic differences of composition and texture. Most have been found in areas that do not suggest a volcanic origin. For many years, several possibilities were argued: that tektites were of lunar origin, or consisted of some other extraterrestrial material, or that they were products of the impact of an extraterrestrial object (e.g., a comet or a giant meteorite) in the vicinity of their occurrence.

Extensive analytical work has been carried out on tektites. The distribution of many elements, including groups of related elements (K, Rb, Cs; Zr, Hf; lanthanides), has been compared for tektites, stony meteorites, and terrestrial rocks such as those of the regions where tektites have been located. Figure 17.1, for example, shows some of the abundance data obtained by Taylor[51,52] using spark-source mass spectrography for 51 elements in tektites compared with values for the subgreywacke rocks surrounding the Henbury impact crater in Australia.

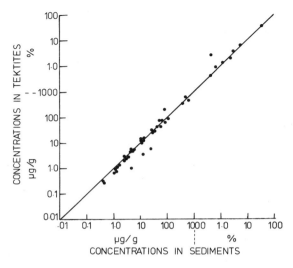

Fig. 17.1. Comparison of the average abundance of 51 elements in tektites compared with the averages for Henbury subgreywacke which surrounds the impact crater. *Source*: Taylor.[51]

A large body of compositional and other evidence has now accumulated,[29,51–56] indicating that tektites consist of terrestrial material, and are impact glasses formed by the melting of local rocks around the impact crater.

A Prehistoric Nuclear Reactor

In 1972 an anomaly was observed in the proportions of uranium isotopes in a uranium hexafluoride sample.[57] The isotopic abundance of ^{235}U was found to be 0.7171 ± 0.0010%, slightly depleted from the normal value of 0.7202 ± 0.0006%. Shortly afterward, much more spectacular depletions were found (0.621%, 0.440%, 0.296% ^{235}U) in uranium samples derived from a deposit at Oklo, Gabon. It was suggested that the loss of ^{235}U had occurred because the deposit had behaved like a giant nuclear fission reactor for a period of perhaps half a million years, between about 1.7 and 1.8 billion years ago. Because of the differing half-lives of ^{235}U and ^{238}U, the proportion of ^{235}U in natural uranium at that time would have been just over 3%. Aided by the presence of ground water to act as a moderator, and by the low concentration of neutron-absorbing elements in the deposit, criticality could have been achieved in the ore deposit.

This explanation was supported by spark-source mass spectrometry on the lanthanides[58,59] (e.g., Nd, Sm, Eu, Dy) in the deposit, which showed modified isotopic abundances, consistent with the addition of isotopes of these elements as fission products.

It is possible that the accumulation of high uranium concentrations in this deposit was aided by the presence of organic matter. If so, the natural occurrence of controlled nuclear fission was the result of an unusual combination of circumstances. At an earlier time (i.e., more than 2.0 billion years ago), the organic matter may not have existed in sufficient quantity. On the other hand had the deposit been formed much later, the ^{235}U abundance of natural uranium would have been too low to sustain a chain reaction.

Studies of this remarkable event have been summarized elsewhere.[59-61] In spite of the coincidences required, it seems possible that the Oklo reactor was not unique.

17.2 MINERAL EXPLORATION

17.2.1 Geochemical Exploration

Geochemical methods of mineral prospecting rely on systematic measurement of the chemical composition of rocks, soils, stream sediments, waters, or vegetation. The purpose of these measurements is to discover and define abnormal chemical patterns, or geochemical anomalies, related to mineralization. The following discussion outlines the various targets of the exploration geochemist. More detailed information can be found in standard works.[62-66]

Analysis of Bedrock Samples

The use of bedrock samples in geochemical exploration is based on the examination of elemental patterns resulting from primary dispersion. Primary dispersion patterns can be divided into two basic types: *syngenetic*, which implies no movement from the original pattern of mineralization, and *epigenetic* patterns formed by subsequent migration of the ore-forming elements into the surrounding *country rock*. Unless outcrops are readily available (which by definition may well represent uncharacteristic weathered material), bedrock samples usually have to be obtained by trenching or drilling. This has tended to make them rather unpopular for geochemical reconnaissance surveys.

Analysis of Soil Samples

Because of the complexity of elemental dispersions in this secondary environment, soil analyses are often difficult to interpret. Trace element distribution patterns are caused by natural chemical and mechanical processes that cause soils to be divided into independent layers known as *horizons*.[67] There are basically three main types of horizon. The uppermost, or *A* horizon, contains the highest content of organic matter and living organisms. Below this, in the *B* horizon, is a region containing the highest concentration of clay minerals. The *A* and *B* horizons comprise the true soil or *solum*. The *C* horizon represents the parent material from which the solum has been formed. In residual soils the *C* horizon, which consists essentially of weathered bedrock fragments, exhibits geochemical anomalies similar to those of the local bedrock.

There is a general tendency for elements to be leached from the *A* horizon into the *B* horizon, and the concentrations of specific elements can be greatly different in the two horizons. It is therefore essential that the same horizon be sampled consistently during a geochemical survey.

There are several other problems inherent in soil-sampling programs. One is the presence of a *transported* soil, which may have been moved some distance by the agency of wind or water and which may not reflect to any extent the composition of bedrock. Another problem that can arise is a shielding of the surface soil from the bedrock by a nonpermeable barrier such as siliceous hardpan. In such a case, the upper horizons of the soil are not necessarily related to the bedrock.

It is not only essential that sampling of horizons be consistent, but also that the same soil fraction be analyzed. Sampling of different fractions during a survey introduces another variable unnecessarily, and may obscure a clear anomaly. In some analytical procedures (e.g., emission spectrography), soil samples are usually ignited at 500°C to remove organic material. During a geochemical soil survey, however, data should be recalculated to a dry weight basis, otherwise the presence of geochemical anomalies can be indicated for ground that is mineralogically barren.

Analysis of Stream Sediments

Stream sediments are the end product of weathering of bedrock by natural waters. They represent a composite of the material upstream from the sampling site and have proved very useful for detecting mineralization. Low-density sampling (i.e., 1 sample per 50–250 km^2) has been used successfully. For example, Armour–Brown and Nichol[68] carried out a stream-sediment survey for copper over 210,000 km^2 in Zambia (density of one sample per 192 km^2) and showed close agreement between known mineralized districts and stream-sediment patterns (see Fig. 17.2). The normal procedure is to carry out a broad stream-sediment program followed by more detailed soil analyses over the more promising areas.

Analysis of Biological Samples

Biogeochemical prospecting (the analysis of vegetation to discover mineralization) has been described.[65,66] The method relies on the uptake of minerals by vegetation. In favorable cases, the concentration of an element in the plant material reflects its concentration in the soil, allowing mineralized areas to be identified. Because of the many factors that may influence the uptake by plants of trace elements from the soil, the biogeochemical method alone is less reliable than soil sampling, and requires considerably more preliminary testing. Nevertheless, it can be useful, especially when the plant's roots penetrate the overburden and where the soil is not representative of the bedrock.

Analysis of Waters

Soil formation processes involve the leaching of soluble products of the overburden by ground waters. A portion of this solubilized material is subsequently reprecipitated in the soil profile, but a further component is transported by ground water to the surface drainage system. In this way, the recognition of an anomalous metal content in waters derived from mineralization may serve as an indication of relatively distant mineralization. When the ground waters reach the surface, changes of the electrode potential (Eh) and pH of the environment often result in the precipitation of the metal, with formation of a stream-sediment anomaly and a resultant decrease in the water anomaly. Although the hydrogeochemical method is quite useful for mobile elements, it is generally possible for the exploration geochemist to trace a mineralized area for a longer distance by stream-sediment anomalies than by water anomalies.

Hydrogeochemistry is one of the less favored techniques for geochemical exploration because samples are bulky and may be chemically unstable. In general, elements must be determined at lower concentrations in water samples than in solid materials. This disadvantage may be only partially offset by the greater simplicity of the aqueous matrix. There are also greater problems of data interpretation, arising from seasonal (or even daily) fluctuations in the water flow rates.

Air Sampling

Mercury vapor from subterranean sulfide deposits ultimately reaches the surface and can be detected easily in samples of air.[69,70] It is possible to monitor mercury directly by an instrument suspended from a helicopter, provided that completely still conditions can be obtained. Air sampling can also be used to detect radon gas emanating from uranium deposits.[64]

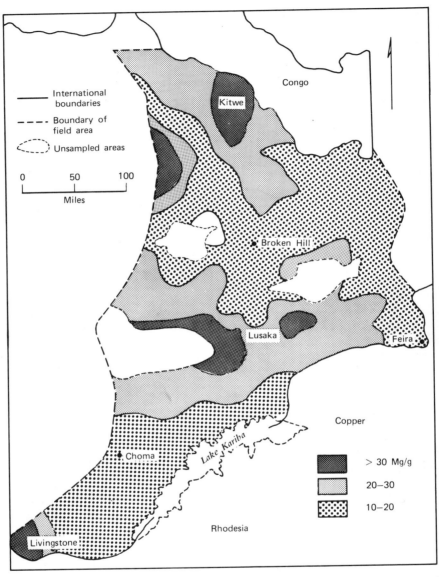

Fig. 17.2a. Copper concentrations in Zambian stream sediments. *Source*: Armour-Brown and Nichol.[68]

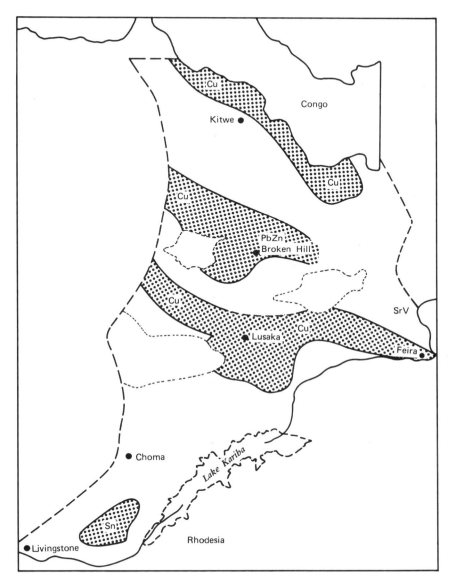

Fig. 17.2b. Copper mineralization (dotted areas) in Zambia. *Source*: Armour-Brown and Nichol.[68]

TABLE 17.1 Examples of Pathfinder Elements Used to Detect Mineralization

Pathfinder(s)	Type of Deposit
As	Au, Ag; vein-type
As	Au-Ag-Cu-Co-Zn; complex sulfide ores
B	W-Be-Zn-Mo-Cu-Pb; skarns
B	Sn-W-Be; veins or greisens
Hg	Pb-Zn-Ag; complex sulfide deposits
Mo	W-Sn; contact metamorphic deposits
Mn	Ba-Ag; vein deposits; porphyry copper
Se, V, Mo	U; sandstone-type
Cu, Bi, As, Co, Mo, Ni	U; vein-type
Mo, Te, Au	Porphyry copper
Pd, Cr, Cu, Ni, Co	Platinum in ultramafic rocks
Zn	Ag-Pb-Zn; sulfide deposits in general
Zn, Cu	Cu-Pb-Zn; sulfide deposits in general
Rn	U; all types of occurrences
SO_4	Sulfide deposits of all types

Source: Levinson.[64]

Pathfinders in Mineral Exploration

Many elements may be found together in easily measurable concentrations in a given geological environment. For example, the chalcophile elements (e.g., As, Cu, Hg, Pb, Sb, Se, and Zn) are found together to some extent in most sulfide deposits. In some cases it is possible to explore for an element of economic significance by analyzing for a second (pathfinder) element.[71,72] The pathfinder may have the advantage of being more mobile than the target element and hence of having a wider dispersion halo. Alternatively, the pathfinder may be more easily and cheaply determined than the target element. Copper, nickel, and chromium, for example, may be used as pathfinders for platinum in ultramafic rocks. Representative pathfinders[64] are listed in Table 17.1.

17.2.2 Abundance Data and Economic Reserves

Apart from the identification of mineral deposits from the analysis of specific samples in localized areas, the exploration geologist is also interested in the general crustal abundances of the elements in igneous rocks and sediments. There is a trend for exploitation of lower grades of economic minerals, and the logical end result would be exploitation of "unmineralized" rocks such as dunites for nickel, pegmatites for uranium, or granites for tin.

However, although the technology exists for ordinary rocks to be exploited in this way, the prerequisite is an enormous supply of cheap energy.[73]

Relationships have been proposed between the crustal abundance of an element and the total exploitable resources. It has been suggested that, except for iron and aluminum, the exploitable reserves R of an element in the earth's crust are related to the percentage crustal abundance A by a formula of the form

$$R = kA$$

For estimating the exploitable reserves of the United States[74,75] at depths to 1 km, it has been calculated that the appropriate value of k is approximately 1 to 3×10^{10} ton. This has been used to suggest that most of the United States reserves of bismuth, copper, lead, and molybdenum have already been discovered, whereas substantial amounts of chromium, cobalt, and nickel remain to be located. Calculations of this kind, however, make an assumption that may be very difficult to justify: that the same proportion of each element in the earth's crust is economically recoverable.

17.2.3 Regional Geochemical Reconnaissance Surveys

Regional geochemical reconnaissance surveys have been used since about 1952 for assessing the mineral potential of large areas by sampling and analysis of soils, rocks, waters, vegetation, and particularly stream sediments.[63,64,76] Earlier surveys were carried out over small areas and for only a few elements because of the limitations of existing analytical methods. The situation was radically transformed by the development of atomic absorption spectrophotometry and direct-reading emission spectrometry, which permit a much greater analytical output than was previously possible.

Regional surveys of stream sediments can yield other types of geological information in addition to evidence of ore deposits.[76,77] The data can be used for geological mapping, stratigraphy, paleogeography, and metamorphic geology. In addition, it may be possible to delineate areas where plant and animal disorders are caused by excesses or deficiencies of certain elements in the geological environment (p. 355).

In the 1950s Webb[78,79] proposed that reconnaissance surveys of stream sediments could be used to prepare regional geochemical atlases showing the distribution of the elements. Work was carried out in Zambia,[80] Sierra Leone,[81,82] and Eire[83,84] in the mid-1960s. A particularly ambitious project involved the collection and analysis of 50,000 stream-sediment samples throughout the United Kingdom. The samples were analyzed for 20 elements and the data are being published in a series of reports, atlases, and bulletins[85–91] that will ultimately cover the whole country. Errors resulting from field sampling, sample preparation, and analysis have been examined.[85] Detailed study was made of emission spectrographic data for 18 elements and atomic absorption data for 5 elements. For routine large-scale surveys analysis of the unground −80 mesh sediment fraction proved satisfactory.

Other regional reconnaissance work has been carried out in Canada[92-95] and the United States.[96-99] The U. S. Geological Survey has been particularly active in analyzing many thousands of samples of soils, rocks, stream sediments, and vegetation collected from across the whole country.

A few case histories must suffice to show how regional reconnaissance surveys have led to the discovery of potentially economic mineral deposits. The usual procedure in such work is to carry out a succession of surveys, beginning with low sampling densities of soils or sediments (e.g., one sample per 100–200 km^2) over very large areas, and continuing with increased density (e.g., one per 12–50 km^2, then one per 2.5–5 km^2) in smaller areas. Precise definition of small deposits is then achieved by surveys involving several samples per square kilometer.

Surveys of stream sediments in Zambia,[68] involving the determination of 16 elements, indicated several metalliferous provinces. Work by Garrett and Nichol[82] over 45,000 km^2 in Sierra Leone identified all previously known chromite deposits and indicated other areas of potential mineralization.

Hawkes et al.[100] have described the results of the collection and analysis of 15,000 stream-sediment samples representing 4900 sites in an area of 65,000 km^2 in Eastern Canada. Several metalliferous geochemical provinces were recognized, and, when these were investigated by high-density sampling surveys, the Charlotte prospect, near Bathurst, New Brunswick, containing over 12% combined lead and zinc, was discovered. The reader is referred to Hawkes[101] for further information on discoveries of this nature.

17.3 PLANT, ANIMAL, AND HUMAN HEALTH

Over the last 50 years, an increasing number of elements has been shown to be present in living tissues of all kinds, and many elements are believed to be essential for plant, animal, and human health. Elements essential for the development of green plants include C, H, O, N, S, P, Ca, Mg, K, Na, Cl, Fe, Mn, Cu, Zn, B, and Mo, and in special cases, Co, Si, Se, and a few others.[102,103] In addition to C, H, O, N, and S, the major constituents of organic compounds, the elements P, Ca, Mg, K, Na, Cl, Fe, Mn, Cu, Zn, Mo, Se, Cr, Co, and I are vital for growth and survival in man and animals,[104] and several other elements (e.g., F, Ni, Sn, and V) produce beneficial effects in many cases. However, excess dietary amounts of F, Mo, Se, and V have detrimental effects, and a number of elements, including As, Be, Cd, Hg, and Pb, are normally regarded as toxic.

The nutritional status of both plants and animals can be strongly affected by the levels of the essential trace elements in the soils on which they are growing. This in turn can have an important bearing on the trace-element intake in the human diet. For this reason, trace element data on geological materials in general, and soils in particular, are playing an important part in plant and animal science and in some studies on human diseases and conditions.

The natural processes involved in soil formation can lead to abnormally low or high concentrations of various trace elements in certain areas. Regions in which adverse effects on animal and plant health can be recognized have been described as biogeochemical provinces.[105] For example, harmful concentrations of toxic elements such as arsenic and mercury can be found in localized regions close to ore bodies or to sources of geothermal activity. Parts of the Northern Plains and Rocky Mountain states of the United States have soils high in selenium, on which grazing animals may show symptoms of selenium toxicity, particularly if they have consumed plant species that accumulate high concentrations of this element. On the other hand, selenium-deficient soils are found in the northeast and northwest of the United States, and in parts of Scotland, New Zealand and elsewhere, leading to a high incidence of selenium-responsive diseases of livestock. The identification of areas where there are deficiencies of elements such as cobalt, copper, iodine, and molybdenum, has also been well documented.[104,106,107] In some cases, the interrelation of two or more trace elements is important, as in the well-established reciprocal antagonism of molybdenum and copper in many animal species.[106]

The ability of regional geochemical reconnaissance surveys to produce results of agricultural significance has been illustrated by Webb et al.[84,89,91,108] Such a survey revealed, for example, areas where hypocuprosis (copper deficiency) in cattle was linked with high molybdenum values in stream sediments.[89,91] The correlation between molybdenum levels and the incidence of hypocuprosis in an area in Derbyshire, England, is illustrated in Fig. 17.3.

A great deal of work has been devoted to the measurement of trace-element levels in soils of many countries, not only to investigate diseases of animals, but also to prepare soil maps. The data are of use in assessing the productive value of farmland and the need for trace-element supplementation in fertilizers. In the United States, for example, nationwide data and maps for trace elements in a variety of surficial materials are available.[97-99] Data of this kind serve as a guide to the existing nutrient status of soils, and also

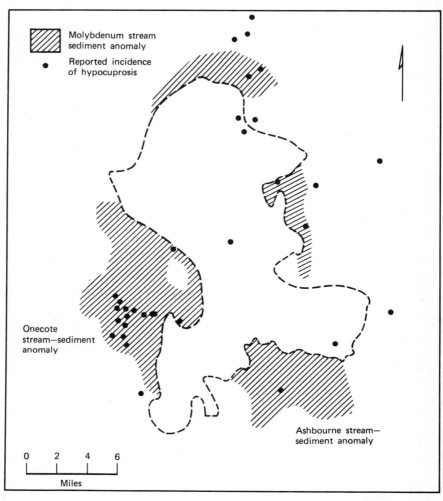

Fig. 17.3. Incidence of hypocuprosis in cattle in Derbyshire, U.K., related to anomalous molybdenum levels in stream sediments. (Central area is an unsampled limestone area.) *Source*: Webb et al.[89]

TABLE 17.2 Diseases Suspected of Being Related to the Geological Environment

Disease	Ref.	Location	Suspected cause	Degree to which cause is established
Balkan nephropathy	119	Yugoslavia	Excess of heavy metals	Poor to fair
Blackfoot disease	119	Taiwan	Excess of arsenic in water	Good
Cardiovascular disease	111–114	Worldwide	Softness of water and deficiency of heavy metals	Fair to good
Dental caries	117, 120	New Zealand, Ohio	Molybdenum deficiency	Good
Goiter	119	Worldwide	Iodine deficiency	Completely
Itai-itai	121	Japan	Excess of cadmium	Good
Kuru	118	New Guinea	Excess of lead	Poor
Minamata disease	122	Japan	Excess of mercury	Completely
Multiple sclerosis	118, 123	Worldwide	Excess of lead	Poor to fair
Parkinson's disease	119	Worldwide	Excess of manganese	Poor

provide a baseline for detecting anomalies due to mineralization or man-made pollution.

Determination of the causes of disease in humans is a complex task, because of the difficulty of proving conclusively a relationship between a particular disease and a specific environmental factor. In recent years there has been an increasing interest in the link between geology and disease. The determination of environmental factors has been assisted by regional reconnaissance surveys and by the use of computers to analyze the data and assist in mapping.[109]

It is difficult to envisage a relationship between local geology and disease in large urban communities with moving populations, varied diets, and bulked and processed foodstuffs. However, in smaller communities and rural populations, especially those where the total trace-element intake is largely derived from locally grown vegetables and animal products and from a local water supply, geochemical factors may have an important influence on human health. A great deal of medically oriented research is therefore being carried out on trace-element levels in town water supplies and in vegetables.[99,110-118]

Table 17.2 lists some of the diseases and conditions believed by some authorities to be associated with trace element excesses or deficiencies. A direct causal relationship between geochemical factors and disease has so far been established in relatively few cases, and the geochemical approach to epidemiology remains controversial.

Diseases and conditions in which trace element analysis may disclose abnormalities or aid diagnosis, have been tabulated by Schroeder and Nason[124] in a review of the content of many trace elements in human tissues. Underwood's book[106] is recommended as a source of information on the role of trace elements in both humans and animals.

17.4 ENVIRONMENTAL CHEMISTRY

An important part of environmental chemistry consists of the analysis of geological and biological materials to determine the presence of abnormally high or low concentrations of various elements. The effects of naturally occurring deficiencies or excesses of some elements have been noted in Section 17.3. Abnormally high concentrations resulting from human activity are potentially even more serious as a public health problem and are being studied worldwide as part of programs directed toward pollution control.

17.4.1 Elemental Redistribution Caused by Man's Activities

Many of man's activities in industry and agriculture depend on the exploitation of economic mineral deposits. These activities create new pathways for the dispersion of the elements, and can be regarded as a tertiary redistribution. In some cases this redistribution causes pollution of the environment, as new biogeochemical provinces develop in which living organisms suffer various kinds of disturbance. Particular attention has been paid recently to the effects of mercury from chlor-alkali plants, lead from the combustion of lead tetra-alkyls in internal combustion engines, and elements such as cadmium, copper, lead, and zinc from mining, smelting, and electroplating operations. An extensive account of industrial aspects of the toxicology of metals and metalloids has been given by Hamilton and Hardy.[125]

Because of the dangerous effects of traces of some elements on human health, limits have been placed by various health authorities on the permissible levels in drinking water. The U. S. Public Health Service, for example, has set limits for cadmium and selenium (0.01 mg/l), arsenic, lead, silver, and hexavalent chromium (0.05 mg/l), barium and copper (1 mg/l), and zinc (5 mg/l). Increasing attention is also being paid to trace element levels in foodstuffs, soils, fertilizers, irrigation waters, and atmospheric air.

17.4.2 Environmental Chemistry of Specific Trace Elements

The environmental chemistry of a number of trace elements has been discussed briefly by Manahan[126] and Brooks.[127] Lead and mercury are of particular concern because of their widespread tertiary distribution in the environment and because their ingestion in hazardous amounts (via foodstuffs, air, and water) is not uncommon. Intensive study of the environmental chemistry of both these elements has been facilitated by recent analytical developments, particularly in flameless atomic absorption.

Aspects of the environmental chemistry of several elements are noted briefly below:

Boron occurs at low levels in most freshwater, but is present at a concentration of 4.6 mg/l in seawater, where it exists chiefly as undissociated boric acid. Boric acid is also a major constituent of some volcanic steam discharges. Borax (sodium tetraborate, $Na_2B_4O_7 \cdot 10H_2O$) is widely used in detergent preparations and food preservatives. Boron is an essential element for plant growth, but is toxic to some plants at concentrations of approximately 1 mg/l in solution. A recommended maximum boron level of 1 mg/l in drinking water has been set by some public health authorities.

Cadmium occurs at an average concentration of about 0.2 μg/g in the earth's crust, and is normally present in soils at about 0.5 μg/g. Seawater contains 0.11 μg/l. Natural levels of cadmium in freshwater are generally very low, except in waters draining regions of sulfide mineralization. Cadmium may therefore be used as a pathfinder in prospecting for zinc.[63] Dangerously high cadmium levels may be found in drainage waters from the mining of sulfide ores and near industrial plants where cadmium is involved (e.g., smelting, electroplating, and the manufacture of pigments and batteries). Cadmium is extremely toxic, its ingestion leading to bone fracture, as illustrated by the occurrence of "itai-itai" in Japan, where water polluted by cadmium from mining operations was used for crop irrigation and for drinking.[128,129] Cadmium has been found in certain shellfish organs[130] at concentrations higher than that of the environmental seawater by factors of 10^5 to 10^6.

Chromium is an essential element, which occurs naturally as Cr(VI) (chromate ion in neutral or basic solution, dichromate ion in strongly acidic solution) or as Cr(III). Soluble Cr(III) levels are normally low, but Cr(VI), which is more toxic, occurs as a pollutant in waters from electroplating works. Maximum levels of hexavalent chromium in drinking waters have been set at 0.05 mg/l.

Copper is an essential element in both plants and animals, being a component of several metalloenzymes. It occurs in the earth's crust with an average abundance of about 60 μg/g. Natural copper levels in rivers and lakes are commonly in the range 1 to 10 μg/l, and concentrations of 0.5 to 3.5 μg/l have been reported in seawater. Mining, smelting, and various other industrial operations contribute to copper pollution.

Fluorine, as fluoride in drinking water at a concentration of about 1 mg/l, helps to reduce the incidence of dental caries. This concentration occurs naturally in some regions, but elsewhere the artificial supplementing of naturally low fluoride levels is widely practised. The long-term use of water containing excess fluoride (e.g. >5 mg/l) causes mottling of tooth enamel, and higher concentrations interfere with collagen formation.[131] Fluoride and fluorine pollution is a problem near fluorine plants and in the vicinity of aluminum smelters which use cryolite (Na_3AlF_6) as a flux.

Lead has been studied more extensively than any other trace-element pollutant in recent years. In the period 1950–1975, approximately 10,000 scientific papers were published on the biological effects of lead and on lead pollution.

Lead occurs in the earth's crust with an average abundance of about 15 μg/g. Soil lead levels are commonly 5 to 50 μg/g, although values exceeding 100 μg/g are found in some areas. Fresh waters often contain approximately 0.02 mg/l. The major uses of lead are in storage batteries and in the manufacture of lead tetra-alkyls for use as gasoline additives. Lead and lead alloys are used in building construction, cable coverings, ammunition, solder and type metal, and various lead compounds are used in pigments and glazes and as additives in the plastics industry.

The most widespread form of tertiary redistribution of lead occurs as a result of gasoline combustion by motor vehicles. Much smaller amounts are discharged into the atmosphere from combustion of coal and oil, from smelting, and from lead alkyl manufacture. Significant amounts of lead are ingested by humans from air, foodstuffs, and drinking water, the relative importance of these sources depending on the locality. Urban air often contains lead at concentrations of 1–5 μg/m^3, and lead levels above 1000 μg/g have been found in dust from urban roads in many parts of the world. The effects of gasoline combustion[132–138] and lead smelting[139–141] on lead levels of air, soil, vegetation, and the ocean surface, have been studied very extensively. The ubiquitous nature of man's dispersion of lead is indicated by the extent to which this element has accumulated in the Greenland ice-cap during the last century.[142]

The toxicology of lead has been summarized.[143] Inorganic lead is a general metabolic poison and enzyme inhibitor. Lead interferes, for example, with the action of enzymes involved in heme synthesis, thereby causing anemia. The average lead content of human blood is 0.2 to 0.3 μg/ml, not far below the levels at which metabolic disturbances have been demonstrated.[143] Young children seem particularly vulnerable and can suffer brain damage and mental retardation from exposure to excessive amounts of lead. Apart from the general public health problem arising from lead in air, food, and water, there are several other sources from which inordinately large amounts of lead may be ingested accidentally. These include flakes of lead-containing paints, and lead leached by acidic drinks from lead-glazed pottery.

Alleviation of lead pollution is likely to come about gradually with reductions in the use of leaded gasoline and with better control of emissions from smelters. Further discussion of aspects of lead in the environment can be found in several reviews.[127,143,144]

Mercury has an extremely low abundance in the earth's crust: various estimates range from about 0.05 to 0.17 μg/g. Soil levels are often about 0.1 μg/g, and values of 0.15 to 0.27 μg/l have been reported in seawater. Each year about 5000 tons of mercury are added to the oceans by natural processes, and a similar amount is added as a result of human activities.[127] The relatively high vapor pressure of mercury leads to the presence of this element in the atmosphere at concentrations of approximately 0.003 to 0.009 μg/m^3. Much higher values are found in the air near mercury deposits, enabling prospecting for mercury to be carried out by air sampling (p. 349). Mercury is concentrated as deposits of cinnabar (HgS) from which the metal is readily obtained.

Major uses of mercury are in the chlor-alkali industry (electrolytic production of chlorine and sodium hydroxide), in batteries, lamps, switchgear, and other electrical apparatus, in fungicides for agricultural seeds, in marine paints, and as a catalyst for some chemical manufacturing processes, especially in the plastics industry. Natural weathering processes contribute traces of mercury to fresh waters, and elevated mercury levels have been found in waters near geothermal areas.[145,146] Extensive losses of mercury into the environment have occurred from chlor-alkali and other chemical manufacturing plants. Other man-made contributions come from the burning of fossil fuels (anthracites may contain 1–3 μg/g, and crude oils 2–20 μg/g), and from the sowing of mercury-treated seed.

The toxicity of high concentrations of mercury vapor and inorganic mercury compounds has been known for many centuries. However, the effects of prolonged exposure to traces of mercury have become appreciated only since the episode of poisonings among residents of fishing villages on Minamata Bay, Japan,[128,147] during 1956–1971. In this case the mercury came from chemical plant wastes draining into the bay and was accumulated by fish, which formed an important part of the local diet. Other serious incidents involving human fatalities from consumption of mercury-contaminated fish or grain have occurred at Niigata, Japan (1964–1965), in Iraq (1961, 1972), and in Guatemala (1968).[128]

Different mercury compounds differ in their toxicity and in their metabolic behavior in living organisms. All mercury compounds are toxic in sufficient quantity, but the greatest general health hazard appears to come from alkylmercury compounds. These are manufactured as fungicides, but are also produced from inorganic and arylmercury compounds by the action of anaerobic microorganisms in river and lake sediments. Biological processes[148] produce methylmercury ions, CH_3Hg^+, and dimethylmercury, which are accumulated by fish lipid tissue. Mercury levels of 0.2 to 4 μg/g (wet weight basis) have been found in edible fish living in areas of man-made or natural mercury pollution.[146,149–151] More detailed accounts of mercury use, pollution and toxicology can be found elsewhere.[122,127,144,152]

Selenium is an essential element, which must occur within a remarkably narrow concentration range for normal human and animal health. It is a chalcophile element, with a crustal abundance of about 0.1 μg/g. Selenium is chiefly obtained as a byproduct of electrolytic copper refining. The toxic effects of selenium-accumulating plants (e.g., species of *Astragalus*, which may contain up to 1% Se) on cattle are well known. A very low limit of 0.01 mg/l has been set for selenium in drinking water. The routine analysis of low selenium concentrations, both for pollution studies and for investigation of deficiency diseases, has been assisted by the development of hydride-generation atomic absorption techniques[153,154] and of automated fluorimetric methods.[155]

Zinc is an essential element in both plants and animals, being involved in the functioning of a large number of enzymes. It occurs in the earth's crust with an average abundance of about 70 μg/g. Soils typically contain about 50 μg/g, and values of 1.5 to 10 μg/l have been reported for zinc in seawater.

The toxicity of zinc to man is low, although some forms of aquatic life[156] are disturbed by concentrations as low as 0.25 mg/l. Concentrations up to 5 mg/l are permitted in drinking water.

Major industrial uses are in galvanizing and the manufacture of brass and other alloys. Important sources of environmental zinc pollution are mining, smelting and galvanizing operations. The effects of smelting on zinc levels in soil and vegetation have been studied extensively.[139–141] High levels of zinc are found in various organs of some shellfish, even in unpolluted waters.[130] Where the concentrations have been elevated by pollution from nearby smelting,[151,157] shellfish and other marine life can accumulate zinc to concentrations of 800 $\mu g/g$ to more than 10,000 $\mu g/g$ (dry weight basis). In an extreme case[151] consumption of only six shellfish provided an emetic dose (150 mg) of zinc.

Other elements that present localized pollution problems or occupational health hazards include vanadium (emitted as a result of the burning of fossil fuels), nickel (from fossil fuel combustion, nickel carbonyl plants), cobalt, manganese, and tin. Low limits for manganese (0.05 mg/l) and iron (0.3 mg/l) in drinking waters have been recommended because of undesirable effects when the water is used for domestic purposes such as laundering, rather than for any demonstrated threat to human health. Elements with well-known toxicity but that are used in small amounts and are regarded as hazardous only to those using them include beryllium, thallium, antimony, and tellurium.

17.5 ARCHAEOLOGY

The characterization of sources of raw materials of prehistoric civilizations is a field of major interst in modern archaeology. Lithic materials such as chert (flint, jasper) and obsidian have been widely used for the manufacture of tools and ornaments, and in many instances were transported large distances from their natural sources. Important insights into traveling and trading patterns of many people in prehistoric times have recently been obtained as a result of source identification of these materials found at archaeological sites.

Chemical analysis is one of the most valuable means of characterization. Distinction between materials from different sources can sometimes be made on the basis of major element composition, but greater relative differences can often be found in trace-element concentrations. Potassium, for example, is generally present in obsidian at a level of 2.5% to 4.5%, but concentrations ranging over at least one order of magnitude may be found in trace elements such as manganese (typically 120–1600 $\mu g/g$), barium (2–1300 $\mu g/g$), strontium (1–250 $\mu g/g$), rubidium (50–800 $\mu g/g$), zirconium (20–1600 $\mu g/g$), and niobium (2–60 $\mu g/g$).

Successful use of chemical analysis in sourcing archaeological raw materials usually involves the following steps: (a) location of natural sources of the material, defined by occurrence within a restricted geographical area; (b) collection and analysis of a representative range of samples from each source; (c) demonstration that, at least for the concentrations of some elements, intersource variability is considerably greater than intrasource variability; (d) establishment of a suitable set of parameters, based on the analytical data, which will allow material from different sources to be distinguished with a high degree of confidence; (e) analysis of artifactual material and assignment to sources in accordance with criteria established in (d).

The most widespread success in source identification of archaeological lithic materials has been achieved with obsidian. This is a volcanic glass found in many parts of the world where volcanic activity has occurred, such as the Mediterranean, western parts of North America, Central America, Iceland, Japan, New Zealand, and other islands throughout the Pacific. Obsidian is formed when silica-rich lava has cooled too rapidly to allow crystallization to occur. By the nature of its formation, it is a relatively homogeneous material; the small variation of composition within a single source or flow has been a key factor in allowing obsidian to be characterized with a high degree of confidence. Although some chemical analysis of obsidian for major elements was carried out before 1930, the first comprehensive trace-element investigation was reported by Cann and Renfrew[158] in 1964. This work, on obsidian from the Mediterranean and Near East, has been followed by studies from nearly every area where prehistoric people exploited obsidian.

Cann and Renfrew used emission spectrography to determine 16 trace elements, of which barium and zirconium were the most useful for discrimination. Obsidian from the Near East[159] was classified into eight groups on the basis of these two elements, and Aegean obsidians were also characterized.[160] An account of recent work in the Mediterranean and Near East has been given.[161]

Seven New Zealand sources of obsidian were characterized by emission spectrography,[162] using intensity ratios of emission lines of manganese, zirconium, calcium, and beryllium. The source of a New Guinea artifact was also established by emission spectroscopy.[163]

The potential of X-ray fluorescence for this work was demonstrated by Parks and Tieh,[164] who characterized obsidian from three Californian sources, using strontium/rubidium intensity ratios. X-ray analysis has been used extensively in later work on Californian,[165] Mexican,[166] and New Zealand[167] obsidians. Where the X-ray fluorescence is carried out by powdering the sample, the technique is nondestructive in the sense that repeated measurements can be made on the same sample. Direct rapid-scan analysis of artifact material for Rb, Sr, Y, Zr, and Nb has been used[165] where it is desirable to preserve the sample intact.

Analysis has also been carried out by neutron activation, the instrumental form of this technique being preferred to preserve the samples.[168–171] The sodium/manganese ratio, accessible with short irradiation times, can be used for preliminary characterization, but a more comprehensive subdivision of sources requires the use of additional parameters, such as concentrations of Sc, Fe, Rb, La, and Sm. Several other elements can be determined if necessary. The computerized reduction of neutron activation γ-ray data obtained with a high-resolution detector has been described elsewhere.[172]

The success of trace-element measurements as a basis for identifying sources of archaeological obsidian (or other lithic materials) depends largely on two factors: (a) each source should be relatively homogeneous; (b) variations of composition between different sources should be greater than the variations within each source. Careful studies using analytical methods capable of high precision, such as flame emission and atomic absorption spectroscopy,[173] have illustrated a high degree of homogeneity in some obsidian flows. In other cases, however, systematic variations of obsidian composition within a single flow have been observed.[174] More extensive discussion of recent work on obsidian analysis and source characterization can be found elsewhere.[161,165,175,176]

Methods of data treatment used by different workers have varied in their complexity. Characterization can occasionally be achieved in terms of a single-element concentration or a two-element concentration ratio. More often, pairs of concentrations or ratios have been taken[158,161,162,173] to give a two-dimensional plot in which fields corresponding to different sources are defined. Triangular diagrams have been used to illustrate data on groups of three elements.[161,165] Where such plots fail to discriminate between sources, further element ratios can be investigated. With a limited number of homogeneous sources and with analytical data of high precision, these methods may be used successfully. However, when the number of sources is large, the analytical data imprecise, or the source compositions rather inhomogeneous, it becomes desirable to use a multivariate data-analysis technique. In its simplest form, this consists of visual inspection of multielement data. A graphical method was illustrated by Key,[163] who plotted the element concentrations in an artifact on the y axis and the mean source concentrations on the x axis, identification being made by finding the source giving the smallest overall deviation from the line $y = x$. More sophisticated methods of data analysis include discriminant analysis (Section 18.4.4) as used by Sieveking et al.[177] for European flints and by Ward[178] for New Zealand obsidians, and computer-assisted pattern-recognition techniques.[179–181]

Much less work has been done on other archaeological lithic materials such as chert[177,179-83] (including flint, jasper, etc.) and marble[184] than on obsidian. Results have not always been encouraging. As might be expected from their conditions of formation, chert and marble from any given source generally show greater variations in trace-element composition than is found in single obsidian sources. For this reason, the characterization of material from prehistoric flint mines and classical marble quarries has usually required that a large number of trace elements be determined and that statistical techniques be used in data analysis.

References

1. K. B. Krauskopf, *Introduction to Geochemistry*, McGraw-Hill, New York, 1967.
2. B. Mason, *Principles of Geochemistry*, 3rd ed., Wiley, New York, 1966.
3. A. B. Ronov and A. A. Yaroshevskii, *Geophys. Monogr. Am. Geophys. Union*, **13**, 37 (1969).
4. H. C. Urey, *Ann. N. Y. Acad. Sci.*, **194**, 35 (1972).
5. K. K. Turekian and K. H. Wedepohl, *Bull. Geol. Soc. Am.*, **72**, 175 (1961).
6. N. L. Bowen, *The Evolution of the Igneous Rocks*, Princeton Univ. Press, Princeton, N. J., 1972.
7. R. M. Garrels, *Mineral Equilibria at Low Temperature and Pressure*, Harper and Row, New York, 1960.
8. V. M. Goldschmidt, *Geochemistry*, Clarendon Press, Oxford, England, 1954.
9. L. H. Ahrens, *Geochim. Cosmochim. Acta*, **2**, 168 (1952); **3**, 1 (1952).
10. H. Ramberg, *J. Geol.*, **61**, 318 (1953).
11. A. E. Ringwood, *Geochim. Cosmochim. Acta*, **7**, 189, 242 (1955).
12. L. H. Ahrens, *Phys. Chem. Earth*, **5**, 1 (1964).
13. L. R. Wager and R. L. Mitchell, *Geochim. Cosmochim. Acta*, **1**, 129 (1951).
14. E. M. Patterson, *Geochim. Cosmochim. Acta*, **2**, 283 (1952).
15. L. R. Wager and R. L. Mitchell, *Geochim. Cosmochim. Acta*, **3**, 217 (1953).
16. S. R. Nockolds and R. Allen, *Geochim. Cosmochim. Acta*, **4**, 105 (1953); **5**, 245 (1954); **9**, 34 (1956).
17. D. M. Shaw, *Interprétation Geochimique des Éléments en Traces dans les Roches Cristallines*, Masson, Paris (1964).
18. S. R. Taylor, *Phys. Chem. Earth*, **6**, 133 (1965).
19. R. Kretz, *J. Geol.*, **67**, 371 (1959).
20. W. L. McIntire, *Geochim. Cosmochim. Acta*, **27**, 1209 (1963).
21. C. D. Curtis, *Geochim. Cosmochim. Acta*, **28**, 389 (1964).
22. R. F. Mueller, in *The Encyclopedia of Geochemistry and Environmental Sciences* (ed. R. W. Fairbridge), Van Nostrand Reinhold, New York, 1972, p. 1194.

23. K. Rankama and T. G. Sahama, *Geochemistry*, University of Chicago Press, Chicago, 1950.
24. K. Krejci-Graf, in *The Encyclopedia of Geochemistry and Environmental Sciences* (ed. R. W. Fairbridge), Van Nostrand Reinhold, New York, 1972, p. 1201.
25. B. Mason, *Meteorites*, Wiley, New York, 1962.
26. J. A. Wood, *Meteorites and the Origin of the Planets*, McGraw-Hill, New York, 1968.
27. B. Mason (ed.), *Handbook of Elemental Abundances in Meteorites*, Gordon and Breach, New York, 1971.
28. G. J. H. McCall, *Meteorites and their Origins*, Halsted, New York, 1973.
29. J. L. Setser and W. D. Ehmann, *Geochim. Cosmochim. Acta*, **28**, 769 (1964).
30. L. A. Haskin, F. A. Frey, R. A. Schmitt, and R. H. Smith, *Phys. Chem. Earth*, **7**, 167 (1966).
31. S. R. Taylor, in *The Encyclopedia of Geochemistry and Environmental Sciences* (ed. R. W. Fairbridge), Van Nostrand Reinhold, New York, 1972, p. 1020.
32. J.-G. Schilling, in *The Encyclopedia of Geochemistry and Environmental Sciences* (ed. R. W. Fairbridge), Van Nostrand Reinhold, New York, 1972, p. 1029.
33. R. A. Schmitt, R. H. Smith, J. H. Lasch, A. W. Mosen, D. A. Olehy, and J. Vasilevskis, *Geochim. Cosmochim. Acta*, **27**, 577 (1963).
34. H. Wakita, R. A. Schmitt, and P. Rey, in *Proc. Apollo 11 Lunar Sci. Conf.*, Vol. 2 (ed. A. A. Levinson), Pergamon, New York, 1970, p. 1685.
35. A. A. Levinson (ed.), *Proc. Apollo 11 Lunar Sci. Conf.*, Vol. 2, Pergamon, New York, 1970.
36. A. A. Levinson (ed.), *Proc. 2nd Lunar Sci. Conf.* 1971, MIT, Cambridge, Mass., 1971.
37. E. A. King (ed.), *Proc. 3rd Lunar Sci. Conf.* 1972, Vol. 1, MIT, Cambridge, Mass., 1972.
38. D. Heymann (ed.), *Proc. 3rd Lunar Sci. Conf.* 1972, Vol. 2, MIT, Cambridge, Mass., 1972.
39. W. A. Gose (ed.), *Proc. 4th Lunar Sci. Conf.* 1973, Vol. 2, Pergamon, New York, 1974.
40. W. A. Gose (ed.), *Proc. 5th Lunar Sci. Conf.* 1974, Vol. 2, Pergamon, New York, 1975.
41. A. A. Levinson and S. R. Taylor, *Moon Rocks and Minerals*, Pergamon, New York, 1971.
42. J. A. Wood, *Sci. Am.*, **233** (3), 92 (1975).
43. R. A. Pacer and W. D. Ehmann, *J. Chem. Ed.*, **52**, 350 (1975).
44. S. R. Taylor, *Lunar Science: A Post-Apollo View*, Pergamon, New York, 1975.
45. J. L. Kulp, in *The Encyclopedia of Geochemistry and Environmental Sciences* (ed. R. W. Fairbridge), Van Nostrand Reinhold, New York, 1972, p. 446.

46. J. L. Kulp and J. Engels, *Radioactive Dating*, International Atomic Energy Agency, Vienna, 1963.
47. D. York and R. M. Farquhar, *The Earth's Age and Geochronology*, Pergamon, New York, 1972.
48. G. Faure and J. L. Powell, *Strontium Isotope Geology*, Springer, New York, 1972.
49. G. B. Dalrymple and M. A. Lanphere, *Potassium-Argon Dating*, Freeman, San Francisco, 1969.
50. J. A. O'Keefe (ed.), *Tektites*, University of Chicago Press, Chicago, 1963.
51. S. R. Taylor, *Geochim. Cosmochim. Acta*, **30**, 1121 (1966).
52. S. R. Taylor, *Geochim. Cosmochim. Acta*, **31**, 961 (1967).
53. J. W. Chase, C. C. Schnetzler, G. K. Czamanske, and J. W. Winchester, *J. Geophys. Res.*, **68**, 577 (1963).
54. L. A. Haskin and M. Gehl, *Science*, **139**, 1056 (1963).
55. S. R. Taylor and M. Sachs, *Geochim. Cosmochim. Acta*, **28**, 235 (1964).
56. S. R. Taylor and M. Kaye, *Geochim. Cosmochim. Acta*, **33**, 1083 (1969).
57. R. Bodu, H. Bouzigues, N. Morin, and J. P. Pfiffelmann, *C. R. Acad. Sci., Ser. D*, **275**, 1731 (1972).
58. M. Neuilly, J. Bussac, C. Frejacques, G. Nief, G. Vendryes, and J. Yvon, *C. R. Acad. Sci., Ser. D*, **275**, 1847 (1972).
59. Commissariat a l'Énergie Atomique, *Énerg. Nucl.*, **14**, 376 (1972).
60. J. R. Lancelot, A. Vitrac, and C. J. Allegre, *Earth Planet. Sci. Lett.*, **25**, 189 (1975).
61. R. Naudet, *Recherche*, **6**, 508 (1975).
62. P. M. D. Bradshaw, D. R. Clews, and J. L. Walker, *Exploration Geochemistry*, Barringer Research, Rexdale, Ont., 1972.
63. H. E. Hawkes and J. S. Webb, *Geochemistry in Mineral Exploration*, Harper and Row, New York, 1962.
64. A. A. Levinson, *Introduction to Exploration Geochemistry*, Applied Publishing, Ltd., Wilmette, Ill., 1974.
65. R. R. Brooks, *Geobotany and Biogeochemistry in Mineral Exploration*, Harper and Row, New York, 1972.
66. D. P. Malyuga, *Biogeochemical Methods of Prospecting*, Consultants Bureau, New York, 1964.
67. J. R. Corbett, *The Living Soil*, Martindale Press, Sydney, 1969.
68. A. Armour-Brown and I. Nichol, *Econ. Geol.*, **65**, 312 (1970).
69. Q. Bristow and I. R. Jonasson, *Can. Min. J.*, **93**, 39 (1972).
70. J. C. Robbins, in *Geochemical Exploration*, Inst. Min. Metall., London, 1972, p. 315.
71. H. V. Warren and R. E. Delavault, *Min. Eng.*, **5**, 980 (1953).
72. H. V. Warren and R. E. Delavault, *Proc. 20th Int. Geol. Congr.*, Mexico City, 1956, p. 359.

73. A. A. Saukov, *Ocherki Geokhim. Otd. Element.*, **1973**, 23 (1973).
74. V. E. McKelvey, *Am. J. Sci.*, **258A**, 234 (1960).
75. R. L. Erickson, U. S. Geol. Surv. Prof. Pap. 820, 21 (1973).
76. J. S. Webb, I. Nichol, and I. Thornton, *Proc. 23rd Int. Geol. Congr.*, Prague, **6**, 131 (1968).
77. J. S. Webb, *Proc. Geol. Assoc.*, **81** (Part 3), 585 (1970).
78. J. S. Webb, in *The Future of Non-Ferrous Mining in Great Britain and Ireland*, Inst. Min. Metall., London, 1959, p. 419.
79. J. S. Webb, *R. School Mines Res. Rep.* **1954-7**, 19 (1957).
80. J. S. Webb, J. A. C. Fortescue, I. Nichol, and J. S. Tooms, *Regional Geochemical Maps of the Namwala Concession Area, Zambia*, Nos. 7-X, Geol. Surv., Lusaka, 1964.
81. I. Nichol, L. D. James, and K. A. Viewing, *Trans. Inst. Min. Metall. Sec. B*, **75**, B146 (1966).
82. R. G. Garrett and I. Nichol, *Trans. Inst. Min. Metall. Sec. B*, **76**, B97 (1967).
83. I. Thornton, W. J. Atkinson, and J. S. Webb, *Irish J. Agric. Res.*, **5**, 280 (1966).
84. J. S. Webb and W. J. Atkinson, *Nature*, **208**, 1056 (1965).
85. R. J. Howarth and P. L. Lowenstein, *Trans. Inst. Min. Metall., Sec. B*, **80**, B363 (1971).
86. I. Nichol, I. Thornton, and J. S. Webb, *G. B. Inst. Geol. Sci. Rep.* No. 70/8, 1 (1970).
87. I. Nichol, I. Thornton, J. S. Webb, W. K. Fletcher, R. F. Horsnail, and J. Khaleelee, *G. B. Inst. Geol. Sci. Rep.* No. 71/2, 1 (1971).
88. I. Nichol, I. Thornton, J. S. Webb, W. K. Fletcher, R. F. Horsnail, J. Khaleelee, and D. Taylor, *G. B. Inst. Geol. Sci. Rep.* No. 70/2, 1 (1970).
89. J. S. Webb, I. Thornton, and W. K. Fletcher, *Nature*, **217**, 1010 (1968).
90. D. Taylor, I. Nichol, and J. S. Webb, *Trans. Inst. Min. Metall. Sec. B*, **76**, B214 (1967).
91. I. Thornton, R. N. B. Moon, and J. S. Webb, *Nature*, **221**, 457 (1969).
92. R. H. C. Holman, in *Liaison Symposium on Regional Geochemistry in the British Commonwealth*, Brit. Commonwealth Geol. Liaison Office, London, 1963.
93. R. W. Boyle et al., Geol. Surv. Canada Pap. 65-42, 1 (1966).
94. E. M. Cameron, *Can. J. Earth Sci.*, **6**, 247 (1969).
95. W. K. Fletcher and P. J. Doyle, *Can. Inst. Metall. Trans.*, **74**, 1 (1971).
96. *U. S. Geol. Surv. Bulletins* 1230-A–1230-J (10), 1261-A–1261-G (7), 1319-A–1319-F (6), 1325, 1353-A, 1353-B.
97. H. T. Shacklette, J. C. Hamilton, J. G. Boerngen, and J. M. Bowles, U. S. Geol. Surv. Prof. Pap. 574-D, 1 (1971).
98. H. T. Shacklette, J. G. Boerngen, J. P. Cahill, and R. N. Rahill, *U. S. Geol. Surv. Circ.*, **673**, 1 (1973).
99. J. J. Connor and H. T. Shacklette, U. S. Geol. Surv. Prof. Pap. 574-F, 1 (1975).

100. H. E. Hawkes, H. Bloom, J. E. Riddell, and J. S. Webb, *Proc. 20th Int. Geol. Congr.*, Mexico City, **3**, 607 (1960).
101. H. E. Hawkes, *Exploration Geochemistry Bibliography Special Vol. No. 1*, Association of Exploration Geochemists, Toronto, 1972.
102. V. Sauchelli, *Trace Elements in Agriculture*, Van Nostrand Reinhold, New York, 1969.
103. L. P. Miller and F. Flemion, in *Phytochemistry*, Vol. III (ed. L. P. Miller), Van Nostrand Reinhold, New York, 1973.
104. M. L. Scott, in *Micronutrients in Agriculture* (eds. J. J. Mortvedt, P. M. Giordano, and W. L. Lindsay), Soil Sci. Soc. Amer., Madison, Wisc., 1972.
105. A. P. Vinogradov, *Int. Monogr. Earth Sci.*, **15**, 317 (1964).
106. E. J. Underwood, *Trace Elements in Human and Animal Nutrition*, 3rd ed., Academic Press, New York, 1971.
107. J. Kubota and W. H. Allaway, in *Micronutrients in Agriculture* (eds. J. J. Mortvedt, P. M. Giordano, and W. L. Lindsay), Soil Sci. Soc. Amer., Madison, Wisc, 1972.
108. I. Thomson, I. Thornton, and J. S. Webb, *J. Sci. Food Agric.*, **23**, 879 (1972).
109. R. W. Armstrong, in *Medical Geography, Techniques and Field Studies* (ed. N. D. McGlashan), Methuen, London, 1972, p. 69.
110. J. S. Webb, in *Environmental Geochemistry in Health and Disease* (eds. H. L. Cannon and H. C. Hopps), Geological Society of America, Boulder, Col., 1971, p. 31.
111. H. A. Schroeder, *J. Am. Med. Assoc.*, **172**, 1902 (1960); **195**, 81 (1966).
112. G. Biorck, H. Bostrom, and A. Widstrom, *Acta Med. Scand.*, **178**, 239 (1965).
113. M. D. Crawford, M. J. Gardner, and I. N. Morris, *Lancet* I, 827 (1968).
114. H. T. Shacklette, H. L. Sauer, and A. T. Miesch, U. S. Geol. Surv. Prof. Pap. 574-C (1970).
115. H. V. Warren and R. E. Delavault, in *Environmental Geochemistry in Health and Disease* (eds. H. L. Cannon and H. C. Hopps), Geological Society of America, Boulder, Col. 1971, p. 97.
116. H. V. Warren, R. E. Delavault, K. Fletcher, and E. Wilks, in *Trace Substances in Environmental Health IV* (ed. D. D. Hemphill), Univ. of Missouri, 1971, p. 94.
117. T. G. Ludwig, W. B. Healy, and F. L. Losee, *Nature*, **186**, 695 (1960).
118. H. V. Warren, *J. Biosoc. Sci.*, **6**, 223 (1974).
119. H. C. Hopps, in *Environmental Geochemistry in Health and Disease* (eds. H. L. Cannon and H. C. Hopps), Geological Society of America, Boulder, Col., 1971, p. 1.
120. F. L. Losee and B. D. Adkins, in *Environmental Geochemistry in Health and Disease* (eds. H. L. Cannon and H. C. Hopps), Geological Society of America, Boulder, Col., 1971, p. 203.
121. K. Tsuchiya, *Keijo J. Med.*, **18**, 181, 195 (1969).

REFERENCES

122. L. Friberg and J. Vostal (eds.), *Mercury in the Environment*, Chemical Rubber Co., Cleveland, Ohio, 1972.
123. H. V. Warren, R. E. Delavault, and C. H. Cross, *Ann. N. Y. Acad. Sci.*, **136**, 657 (1967).
124. H. A. Schroeder and A. P. Nason, *Clin. Chem.*, **17**, 461 (1971).
125. A. Hamilton and H. L. Hardy, *Industrial Toxicology*, 3rd ed., Publishing Sciences Group, Acton, Mass,. 1974.
126. S. E. Manahan, *Environmental Chemistry*, Willard Grant, Boston, 1972.
127. R. R. Brooks, in *Environmental Chemistry* (ed. J. O'M. Bockris), Plenum Press, New York, 1976.
128. World Health Organization, *Health Hazards of the Human Environment*, Geneva, 1972.
129. L. Friberg, M. Piscator, and G. Nordberg, *Cadmium in the Environment*, Karolinska Institute, Stockholm, 1971.
130. R. R. Brooks and M. G. Rumsby, *Limnol. Oceanog.*, **10**, 521 (1965).
131. D. Hunter, *The Diseases of Occupation*, 4th ed., Eng. Univ. Press, London, 1969.
132. H. V. Warren and R. E. Delavault, *Trans. R. Soc. Can.*, **54**, 11 (1960).
133. H. L. Cannon and J. M. Bowles, *Science*, **137**, 765 (1962).
134. R. H. Daines, H. Motto, and D. M. Chilko, *Environ. Sci. Technol.*, **4**, 318 (1970).
135. J. V. Lagerwerff and A. W. Specht, *Environ. Sci. Technol.*, **4**, 583 (1970).
136. A. L. Page and T. J. Ganje, *Environ. Sci. Technol.*, **4**, 140 (1970).
137. H. L. Motto, R. H. Daines, D. M. Chilko, and C. K. Motto, *Environ. Sci. Technol.*, **4**, 231 (1970).
138. R. R. Brooks, B. J. Presley, and I. R. Kaplan, *Talanta*, **14**, 809 (1967).
139. G. T. Goodman and T. M. Roberts, *Nature*, **231**, 287 (1971).
140. A. Burkitt, P. Lester, and G. Nickless, *Nature*, **238**, 327 (1972).
141. P. Little and M. H. Martin, *Environ. Pollut.*, **3**, 241 (1972).
142. M. Murozumi, T. J. Chow, and C. C. Patterson, *Geochim. Cosmochim. Acta*, **33**, 1247 (1969).
143. D. Bryce-Smith, *Chem. Brit.*, **7**, 54 (1971).
144. H. S. Stoker and S. L. Seager, *Environmental Chemistry: Air and Water Pollution*, Scott Foresman, Glenview, Ill., 1972.
145. D. E. White, H. E. Hinkle, and I. Barnes, U. S. Geol. Surv. Prof. Pap. 713, 25 (1970).
146. B. G. Weissberg and M. G. R. Zobel, *Bull. Environ. Contam. Toxicol.*, **9**, 148 (1973).
147. J. Ui, *Nord. Hyg. Tidskr.*, **50**, 139 (1969).
148. S. Jensen and A. Jernelöv, *Nature*, **223**, 753 (1969).
149. G. Grimstone, *Chem. Brit.*, **7**, 244 (1971).

150. R. J. Evans, J. D. Bails, and F. M. D'Itri, *Environ. Contam. Toxicol.*, **6**, 901 (1972).
151. S. J. Thrower and I. J. Eustace, *Food Technol. Aust.*, **25**, 546 (1973).
152. N. Fimreite, *Environ. Pollut.*, **1**, 119 (1970).
153. K. C. Thompson and D. R. Thomerson, *Analyst*, **99**, 595 (1974).
154. P. D. Goulden and P. Brooksbank, *Anal. Chem.*, **46**, 1431 (1974).
155. M. W. Brown and J. H. Watkinson, *Anal. Chim. Acta*, **89**, 29 (1977).
156. J. Cairns and R. E. Sparks, *Water Pollut. Control Res. Ser.*, No. 18050 EDQ12/71 (1971).
157. J. Butterworth, P. Lester, and G. Nickless, *Mar. Poll. Bull.*, **3**, 72 (1972).
158. J. R. Cann and C. Renfrew, *Proc. Prehist. Soc.*, **30**, 111 (1964).
159. C. Renfrew, J. E. Dixon, and J. R. Cann, *Proc. Prehist. Soc.*, **32**, 30 (1966).
160. C. Renfrew, J. R. Cann, and J. E. Dixon, *Ann. Brit. Sch. Archaeol. Athens*, **60**, 225 (1965).
161. J. E. Dixon, in *Advances in Obsidian Glass Studies* (ed. R. E. Taylor), Noyes Press, Park Ridge, N. J., 1976.
162. R. C. Green, R. R. Brooks, and R. D. Reeves, *N. Z. J. Sci.*, **10**, 675 (1967).
163. C. A. Key, *Nature*, **219**, 360 (1968).
164. G. A. Parks and T. T. Tieh, *Nature*, **211**, 289 (1966).
165. R. N. Jack, in *Advances in Obsidian Glass Studies* (ed. R. E. Taylor), Noyes Press, Park Ridge, N. J., 1976.
166. R. H. Cobean, M. D. Coe, E. A. Perry, K. K. Turekian, and D. P. Kharkar, *Science*, **174**, 666 (1971).
167. G. K. Ward, *J. R. Soc. N. Z.*, **4**, 47 (1974).
168. A. A. Gordus, W. C. Fink, M. E. Hill, J. C. Purdy, and T. R. Wilcox, *Archaeometry*, **10**, 87 (1967).
169. A. A. Gordus, G. A. Wright, and J. B. Griffin, *Science*, **161**, 382 (1968).
170. J. B. Griffin, G. A. Wright, and A. A. Gordus, *Arctic*, **22**, 152 (1969).
171. W. W. Patton and T. P. Millar, *Science*, **169**, 760 (1970).
172. P. A. Baedecker, in *Advances in Obsidian Glass Studies* (ed. R. E. Taylor), Noyes Press, Park Ridge, N. J., 1976.
173. G. C. Armitage, R. D. Reeves, and P. Bellwood, *N. Z. J. Sci.*, **15**, 408 (1972).
174. H. R. Bowman, F. Asaro, and I. Perlman, *J. Geol.*, **81**, 312 (1973).
175. F. H. Stross, T. R. Hester, R. F. Heizer, and R. N. Jack, in *Advances in Obsidian Glass Studies* (ed. R. E. Taylor), Noyes Press, Park Ridge, N. J., 1976.
176. R. D. Reeves and G. K. Ward, in *Advances in Obsidian Glass Studies* (ed. R. E. Taylor), Noyes Press, Park Ridge, N. J., 1976.
177. G. de G. Sieveking, P. Bush, J. Ferguson, P. T. Craddock, M. J. Hughes and M. R. Cowell, *Archaeometry*, **14**, 151 (1972).
178. G. K. Ward, *Archaeometry*, **16**, 41 (1974).

REFERENCES

179. A. Aspinall and S. W. Feather, *Archaeometry*, **14**, 41 (1972).
180. M. de Bruin, P. J. M. Korthoven, C. C. Bakels, and F. C. A. Groen, *Archaeometry*, **14**, 55 (1972).
181. M. de Bruin, P. J. M. Korthoven, R. P. W. Duin, F. C. A. Groen, and C. C. Bakels, *J. Radioanal. Chem.*, **15**, 181 (1973).
182. G. K. Ward and I. E. Smith, *Mankind*, **9**, 281 (1974).
183. N. E. Whitehead, *Mem. Nat. Mus. Vic.*, **34**, 225 (1973).
184. L. Conforto, M. Felici, D. Monna, L. Servi, and A. Taddeucci, *Archaeometry*, **17**, 201 (1975).

CHAPTER

18

STATISTICAL INTERPRETATION OF GEOCHEMICAL DATA

18.1 INTRODUCTION

A sound statistical interpretation of data is required if maximum information is to be obtained from many geological investigations. The advent of high-speed computer facilities has allowed increasingly sophisticated statistical calculations to be carried out. Such facilities have also been an important factor in statistical manipulation of the very large amounts of data that can now be provided rapidly by various instrumental methods of analysis.

The exploration geochemist of today is faced with the need to detect ore bodies that are becoming increasingly difficult to find, and with the need to identify mineralized anomalies that are less and less differentiated from the background. The following summary of statistical procedures is formulated largely with the needs of the exploration geochemist in mind, but the principles are applicable in many other fields in which large amounts of trace element data are acquired. Definitions and symbols for the statistical terms used in the discussion are set out in Table 18.1.

18.1.1 Populations and Distributions

An existing distribution of data is known as a *population*. It is assumed that all components of this population are linked by some generic factor. They may be, for example, geochemical abundance data for samples drawn from a particular geological formation. Such samples may represent one geological population only, or they may represent several populations within a particular geological province. For example, if copper is measured in soils in the vicinity of an ore body, some values may belong to the *background* population and others to a population representing unweathered fragments of the ore body; there may well be a third population representing weathered components of the ore itself.

A knowledge of the *distribution* of the population is of importance in interpreting geochemical data as the reliability of many statistical procedures is dependent on the nature of the distribution.

TABLE 18.1 Statistical Notation

Symbol	Interpretation
n	Number of observations
μ	Population mean $= \dfrac{1}{N} \sum_{i=1}^{N} x_i$
$\bar{x}$	Sample mean $= \dfrac{1}{n} \sum_{i=1}^{n} x_i$
σ^2	Population variance $= \dfrac{1}{N} \sum_{i=1}^{N} (x_i - \mu)^2$
$\hat{\sigma}^2$	Best estimate of population variance calculated from sample $= \dfrac{1}{n-1} \sum_{i=1}^{n} (x_i - \bar{x})^2$
σ or $\hat{\sigma}$	Standard deviation $= (\sigma^2)^{1/2}$ or $(\hat{\sigma}^2)^{1/2}$
ϕ	Number of degrees of freedom on which statistic is based
H_0	Null hypothesis
P	Probability

From Ref. 1.

18.1.2 Normal and Log-Normal Distributions

Over 20 years ago Ahrens[2] suggested that the concentrations of most elements in felsic and mafic rocks are log-normally distributed. This paper initiated a lively controversy which still persists.[3-7] Assumptions about the nature of the distribution are important in expressing the "average" composition of rocks, soils, and biological samples. If all the data on a particular element are plotted as a histogram with linear coordinates, a symmetrical or skewed distribution arises if only one population is present. If the histogram is symmetrical the distribution is *normal* and the median, arithmetic mean, and mode of the data coincide. If the histogram is asymmetric with a pronounced skew, it is possible that a normal distribution will be obtained if the logarithms of the elemental concentrations are used. The original distribution is then termed *log-normal*. In such a case, the median of the data corresponds to the geometric mean.

Figure 18.1 shows two histograms from a single log-normal population, plotted on linear and logarithmic scales. Alternatively, a simple calculation can replace this comparison of histograms. Agreement between the median and geometric mean suggests log-normality; agreement between the median

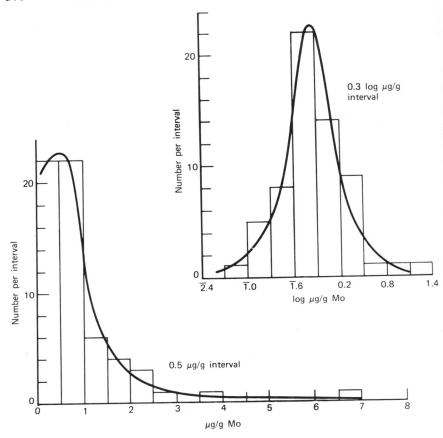

Fig. 18.1. Molybdenum concentrations in granites illustrating a log-normal distribution. *Source*: Ahrens.[2] Courtesy of *Geochimica et Cosmochimica Acta*.

and arithmetic mean indicates normality. For the data shown in Figure 18.1, the median is 0.70 µg/g and the arithmetic and geometric means are 1.2 µg/g and 0.95 µg/g respectively. It is clear that the data approximate more closely to a log-normal distribution.

The most important reason for deciding whether data are normally or log-normally distributed, is that the most sensitive statistical procedures (such as those for deciding whether two sets of samples are taken from the same or different populations) assume that the data are normally distributed. If the data are log-normally distributed, the statistical procedures should be applied to the logarithms of the elemental concentrations.

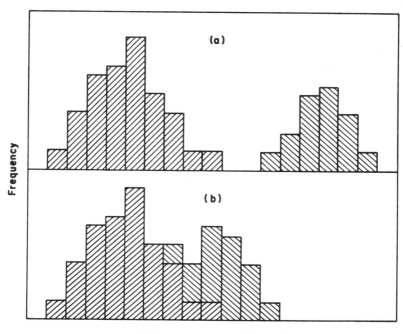

Fig. 18.2. (a) Histograms for separate populations representing anomalous and background values for concentrations of an element in soils or vegetation with both distributions well separated. (b) The same as (a) but with considerable overlap of the two populations.

18.1.3 Mixed Populations

Homogeneous populations are somewhat rare in geochemical data collected in the vicinity of an anomaly. In a mineralized area, there may be two or more different populations of an elemental concentration in rocks, soils, or plants. There will probably be a population representing the normal background values and another representing anomalous levels from an ore body or dispersion halo.

It is common practice to use histograms for soil or rock data to indicate the concentration that should be taken as anomalous for the area. This procedure is straightforward when the two populations are quite separate, as shown in Figure 18.2a. In Figure 18.2b, however, the two populations are nearly coincident and there is some difficulty in deciding which concentration levels should be taken as representing mineralization and which

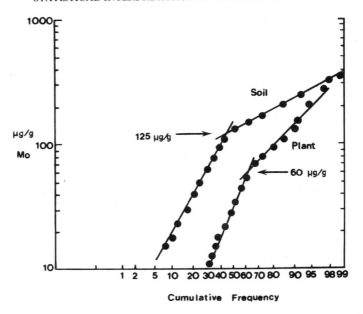

Fig. 18.3. Threshold values for plant and soil anomalies for molybdenum (Takaka, New Zealand), obtained from a plot of cumulative frequencies on logarithmic probability paper. *Source*: Brooks.[9]

should represent the threshold values between the anomaly and the background.

It has been shown[8] that if the cumulative frequency of the concentrations is plotted on probability paper, each population will be represented by a straight line or curve. The intersection of the lines or curves represents a level that may be taken, somewhat arbitrarily, as being the threshold level between background and anomaly. On this plot, a single distribution appears as a straight line if its type (normal or log-normal) is matched with the type of scale (linear or logarithmic) used as the ordinate of the probability paper. Figure 18.3 shows a cumulative frequency plot of molybdenum concentrations found in a study of soils and plants in a mineralized area. Two separate log-normal distributions are evident in both cases.

The intersection of two cumulative frequency lines or curves does not necessarily indicate a sharp discontinuity between two populations and there can be considerable overlap on each side of the intersection. The two distributions may be separated[10] and the degree of overlap calculated. Williams[11] has given worked examples of this method of separation. A more detailed interpretation of cumulative frequency diagrams has been given by Herdan.[12]

In using the cumulative frequency plot, the data are arranged in increasing order, and points are plotted with the cumulative frequency (expressed as a percentage of the total number of samples) along the x axis and the concentration values along the y axis. If there is a large number of points, the data may be grouped and the highest value in each group recorded.

18.1.4 Parametric and Nonparametric Methods

Statistical methods may be classified into two groups, parametric and nonparametric, according to the nature of the data used. Parametric methods use raw data (e.g., elemental concentrations, masses, particle sizes, etc.), whereas nonparametric methods use the ranks or other classifications of the raw data. Univariate and multivariate tests, discussed in the following sections, may be carried out on data of either type.

Parametric methods of statistical analysis are often only valid if the data are normally distributed. It is therefore necessary to know the type of distribution beforehand and to transform to logarithms if necessary. Nonparametric methods do not depend on the data being normally distributed but seldom have the same predictive power at a given confidence limit, necessitating the use of a greater number of samples to give the same degree of significance as a parametric procedure.

The various measurement scales that may be used for parametric and nonparametric data have been described by Marshall[13] and Siegel.[14]

18.2 SIMPLE TESTS OF SIGNIFICANCE

18.2.1 The Null Hypothesis and the Concept of Probability

A frequent problem in geology and geochemistry is the decision as to whether a given number of samples originates from a common source or from several sources. This problem arises because it is hardly ever possible to measure an entire population. Instead, reliance has to be made on samples from this population. Element concentrations measured in these samples vary partly because of the process of sampling from an inhomogeneous material, and partly because of experimental error involved in the methods of measurement.

As an example we may consider a case where two groups of granite samples were analyzed for copper. The means of the two groups were 23.9 and 27.0 μg/g respectively. One person might judge that the groups represent

TABLE 18.2 Levels of Significance

Probability (P)	Conclusion	Symbol
>0.1	Not significant	NS
0.1–0.05	Possibly significant, some doubt cast on H_0	PS
0.05–0.01	Significant, H_0 is rejected though with reservations	S
0.01–0.001	Highly significant, H_0 confidently rejected	S*
<0.001	Very highly significant, H_0 very confidently rejected	S**

Source: Brookes et al.[1]

different populations, whereas another might decide that they represent one population only. By using simple tests of significance, judgements can be placed on a quantitative basis and their reliability stated.

The first step in applying a test of significance is to adopt a null hypothesis H_0. With the null hypothesis, the assumption is made that the parameter being considered does not differ from some specified value. A determination is then made of the probability of obtaining the observed value (or more extreme values) if this hypothesis is true. Taking the above example of copper levels in granites, the null hypothesis would be that there is no significant difference between the means and that both groups are derived from the same population.

We can now calculate the probability P that the above statement is true; that is, there is no significant difference and H_0 is accepted. Suppose the probability is very small (e.g., 0.001). Two conclusions may be drawn: either (a) the 1 in 1000 chance has occurred and H_0 is true, or (b) H_0 is false.

Before the data analysis is begun, the risk we are prepared to take of wrongly asserting a difference should be decided. The level of this risk will vary according to the nature of the experiment being carried out.

As a general rule in geology and geochemistry, the null hypothesis is rejected only if there is less than a 10% probability ($P \leq 0.1$) of its being true. Brookes et al.[1] have listed the generally accepted levels and divisions for tests of significance (Table 18.2).

18.2.2 The *F* Test

The *F* test is used to make a comparison of variability. For example, a geologist may wish to establish which of the two sets of granite samples mentioned above, has the more variable chromium content, or he may wish to compare the reliability of two different service laboratories to decide which of them should perform the analyses. In the latter case, a single sample of a rock could be prepared, divided into 20 replicates, and 10 of each sent to laboratories A and B. An F test can be carried out on the results, as shown below.

The F test is based on the ratio of the variances ($\hat{\sigma}^2$) of two sets of data. The variance estimate is given by

$$\hat{\sigma}^2 = \frac{1}{n-1} \sum_{i=1}^{n} (x_i - \bar{x})^2 \qquad (18.1)$$

although the equivalent expression

$$\hat{\sigma}^2 = \frac{1}{n-1} \left[\sum_{i=1}^{n} x_i^2 - \frac{1}{n} \left(\sum_{i=1}^{n} x_i \right)^2 \right] \qquad (18.2)$$

is often more convenient to calculate.

Suppose we have two variance estimates $\hat{\sigma}_1^2$ and $\hat{\sigma}_2^2$, based on ϕ_1 and ϕ_2 degrees of freedom, respectively. We adopt the null hypothesis H_0 that $\hat{\sigma}_1^2$ and $\hat{\sigma}_2^2$ estimate the same population variance, and then test the probability that $\hat{\sigma}_1^2 > \hat{\sigma}_2^2$, where $\hat{\sigma}_1^2$ is the larger estimate of variance. The greater the ratio of the two variance estimates the less likely it is that H_0 is true.

Suppose the results of the chromium analyses (in $\mu g/g$) for the two laboratories mentioned above are as follows.

Laboratory A: 20.8, 19.5, 22.1 21.0, 21.8, 17.4, 21.5, 18.2, 19.0, 18.0
Laboratory B: 17.5, 18.8, 19.3, 18.9, 19.2, 20.0, 18.8, 19.5, 19.2, 19.6

The variances are $\hat{\sigma}_A^2 = 2.971$ and $\hat{\sigma}_B^2 = 0.451$. Hence

$$F = \frac{\hat{\sigma}_A^2}{\hat{\sigma}_B^2} (\phi_A, \phi_B) = \frac{2.971}{0.451} (9, 9) = 6.58(9, 9)$$

where ϕ_A refers to the set of data with the greater variance.

Reference to Appendix I shows that the experimental value for F (6.58) exceeds the value (5.35) that corresponds to the 1% probability level ($P = 0.01$) and the null hypothesis is confidently rejected. There is clearly a highly significant (S*) difference in the variability of the results produced by the two laboratories.

It should be noted that the above example is a *single-sided test*, i.e., the question asked was whether data from laboratory A were more variable than those from laboratory B. In some cases, however, the question is framed differently: which laboratory gives more variable results? There are now two possibilities: either $\hat{\sigma}_1^2 > \hat{\sigma}_2^2$ or $\hat{\sigma}_2^2 > \hat{\sigma}_1^2$. The test now would be for *any* difference in variability rather than for a difference in a particular direction. To make a *double-sided* test, the same calculation as before is carried out and the probability values obtained from the tables are then doubled. For the above example there is still less than a 2% probability ($P = 0.02$) that there is no difference in the variability of the two sets of analytical data, and H_0 is still confidently rejected.

18.2.3 Student's t Test

We return to the case of the two groups of granites analyzed for copper (Sec. 18.2.1) and ask whether there is a difference between the means, and if so with what degree of confidence can we assert that a difference exists. This is answered by using Student's t test. The following layout is based on an example given by Hinchen.[15]

The concentrations of copper (in micrograms per gram) in granites sampled from two different areas were as follows.

Set 1:   25.4, 26.1, 23.7, 20.9, 18.4, 27.1, 30.9, 21.2, 23.4, 22.4
Set 2:   27.1, 26.3, 28.9, 27.7, 29.3, 20.7, 21.6, 28.9, 29.8, 30.1

1. Calculate the respective means ($\bar{x}_1$ and $\bar{x}_2$). These are 23.94 and 27.04, respectively.

2. Calculate the standard deviations ($\hat{\sigma}_1$ and $\hat{\sigma}_2$) for each set from the formula:

$$\hat{\sigma} = \left[\sum \frac{(x_i - \bar{x})^2}{n - 1}\right]^{1/2} \qquad (18.3)$$

where n is the number of samples in a set and $x_i - \bar{x}$ is the deviation of each sample value from the appropriate mean. In this case the standard deviations are 3.58 and 3.33, respectively.

3. Use the F test to verify that the two sets of data have the same variance. Only if the F test shows the variances to be insignificantly different can the t test be used to examine the significance of any difference in the means. In the present example, the variances are not significantly different.

4. Obtain a better estimate of the variance, σ_p^2 by pooling the information from the two sets of data.

$$\sigma_p^2 = \frac{(n_1 - 1)\hat{\sigma}_1^2 + (n_2 - 1)\hat{\sigma}_2^2}{n_1 + n_2 - 2} = 11.97$$

5. Calculate Student's t for the comparison of the two means, using

$$t = \frac{\bar{x}_1 - \bar{x}_2}{\sigma_p(1/n_1 + 1/n_2)^{1/2}} = -1.99$$

This value of t is based on $(n_1 + n_2 - 2)$ degrees of freedom.

SIMPLE TESTS OF SIGNIFICANCE 383

6. Refer to Appendix II and note that a value of $t = -1.99$ occurs at a probability between 10% and 5% for 18 degrees of freedom (ϕ). We may now assert that H_0 is rejected at the 10% level and that there is a possibly significant (PS) difference between the means of the two sets of data. However, there remains the possibility (>5%) that the observed difference in the means has occurred by chance from populations with the same mean.

The significance of a given value of t depends on the number of degrees of freedom involved. A value of t of -1.99 arising from a situation where $\phi = 120$ (e.g., two sets of 61 analyses) implies that the means are significantly different (S), and the probability of H_0 being true is less than 0.05.

This test is only valid if the raw data are normally distributed. If the raw data have a log-normal distribution, values should be converted to logarithms before making the calculations. The t test is nevertheless comparatively "robust"; that is, small deviations from normality can be permitted without greatly affecting the validity of the conclusions.

18.2.4 The χ^2 Test

The χ^2 test compares the variation of a sample of n random values (normal distribution) with the variance of the distribution.

$$\chi^2 = \sum_{i=1}^{n} \frac{(x_i - \mu)^2}{\sigma^2} \qquad (18.4)$$

Where data have been classified in groups we may compare frequencies observed in these groups (f_o) with the expected frequencies (f_e) based on the normal distribution. To do this, an alternative version of the test is used:

$$\chi^2 = \sum \frac{(f_o - f_e)^2}{f_e} \qquad (18.5)$$

where the summation is carried out over all groups.

The use of the χ^2 test is illustrated by the following example, in which the activity of a radioactive sample is being measured. Observations of a rate of radioactive decay follow the Poisson distribution, which approximates to the normal distribution when the number of observations is large. We may be interested to know whether a set of successive observations of the same activity follows the expected distribution, or whether some fault in the counting apparatus (e.g., voltage drift) is causing a large discrepancy between the observed and theoretical distributions.

In a test of this kind, 30 one-minute observations of the activity were made, the number of counts recorded being as follows: 1020, 969, 994, 1007, 944, 977, 1033, 1001, 1066, 995, 1010, 956, 988, 1037, 1036, 980, 1005, 1048,

960, 1000, 1033, 1010, 1008, 970, 942, 1039, 1029, 990, 1026, 987. The mean activity is 1002 counts/min, and the standard deviation σ is 31.4 counts/min. We examine the frequency of observations f_o in six ranges, each covering one standard deviation, compare each f_o with the theoretical frequency f_e, and calculate χ^2 as tabulated below. There are five degrees of freedom, as the frequency for the sixth range is fixed when the other five are known.

χ^2 Test

Range	f_o	f_e	$(f_o - f_e)^2/f_e$
> 1064.8	1	0.69	0.140
1033.4–1064.8	4	4.08	0.002
1002.0–1033.4	10	10.23	0.005
970.6–1002.0	9	10.23	0.148
939.2–970.6	6	4.08	0.902
< 939.2	0	0.69	0.690

$$\chi^2 = 1.887, \phi = 5$$

By consulting χ^2 tables (given in abbreviated form in Appendix III) we find that $P = 0.86$. In other words there is an 86% probability that the discrepancy between the observed and theoretical distributions has arisen purely by chance. The null hypothesis H_0, that the discrepancy is not significant, is therefore accepted. Had the observations led to a value of χ^2 greater than 11.1 ($P < 0.05$), some fault in the counting apparatus would have been suspected.

The χ^2 test can be used in other ways, for example to test whether or not a given distribution is normal or log-normal. The data are divided into class (i.e., concentration) intervals. The observed number is then compared with the theoretical number for a normal distribution (obtained from standard tables) and χ^2 is calculated as above. If a significant difference is shown to exist, the data may then be transformed logarithmically and the calculation repeated to see if a better fit results.

More comprehensive tables of the normal distribution, and of the F, t, and χ^2 distributions, can be found in standard statistical compilations.[16–19]

18.3 BIVARIATE ANALYSIS

Relationships between two variables are studied by the techniques of *regression* and *correlation*. In regression we study the relationship by expressing one variable (y) in terms of a linear (or more complex) function of the other (x). This functional relationship finds particular uses in (a) supporting hypotheses about the possible causation of changes in y by changes in x, (b) predicting values of y for given values of x, and (c) using changes in x to explain some of the observed variation in y. In correlation analysis we study principally the extent to which two variables are associated, or vary together, without any suggestion that one of the variables is dependent on the other.

18.3.1 Regression Analysis

If the assumption can be made that one variable (x) is independent and is measured without error and the other variable (y) is dependent and subject to error, it is possible to seek a relationship between the two by calculating a *least squares line* through the data points. In the simplest case, that of a linear relationship, this is done by comparing the data points (x_i, y_i) with a line of the form

$$y = a + bx \tag{18.6}$$

and minimizing the sum of the squared deviations of the dependent variable. This quantity, $\sum (y_i - y)^2$, is minimized by partially differentiating it with respect to a and b, and equating each derivative separately to zero.

The parameters a and b are found to be

$$b = \frac{n \sum x_i y_i - \sum x_i \sum y_i}{n \sum x_i^2 - (\sum x_i)^2} \tag{18.7}$$

$$a = \frac{1}{n}(\sum y_i - b \sum x_i) = \bar{y} - b\bar{x} \tag{18.8}$$

The slope of the line, b, is described as the *regression coefficient* for the regression of y on x, and the function 18.6 with the least squares values of a and b inserted is the regression equation.

A worked example shows the calculation of the least squares line for the absorbance (y) of a set of solutions as a function of the concentration (x) of the absorbing species. For the purpose of this example it is assumed that the latter quantity is without error. The data and relevant calculations are shown in Table 18.3.

The equation of the regression line is found to be

$$y = 0.04567x + 0.0049$$

The setting of confidence limits on a and b and the use of the regression line for prediction of a value of y corresponding to a given value of x are covered in specialized statistical texts, such as that of Acton.[20] This work

TABLE 18.3 Calculation of Regression Line for a Plot of
Solution Absorbance as a Function of Concentration

Concentration (mmol/l)(x)	Absorbance (y)	x^2	xy
1.0	0.052	1.00	0.052
2.5	0.118	6.25	0.295
5.0	0.233	25.00	1.165
7.5	0.346	56.25	2.595
10.0	0.463	100.00	4.630
$\sum x = 26.0$	$\sum y = 1.212$	$\sum x^2 = 188.5$	$\sum xy = 8.737$

also discusses the problem, common in analytical chemistry where calibration curves are used extensively, of predicting x from a value of y, measured with error.

Many attempts have been made to fit straight lines to sets of data when both variates are measured with error. The regression line for the regression of y on x, calculated as above, is in general different from the line for the regression of x on y. The latter line is obtained by minimizing the sum of the squared deviations of x_i from the line, the deviations being measured parallel to the x-axis, as shown in Fig. 18.4.

Each of these regression lines gives a biased estimate of the functional relation between the two variables, and often in this situation, neither line corresponds to the one that would be drawn by visual methods. One approach to this problem involves minimizing the sum of the squares of the perpendicular deviations (Fig. 18.4). However, this line is not invariant under simple transformations, such as a change of scale along one of the axes. This may be remedied by plotting the data in units of their respective standard deviations. The maximum-likelihood methods of straight-line fitting[20] used for this rather complex situation require some knowledge of the standard deviations of the errors and of the correlation between them. These methods are appropriate if it is desired to establish and estimate the *dependence* of one random variable upon another. However, a more common problem in geochemical work involves the establishment and estimation of the degree of *association* between two random variables, such as concentrations of two elements in a set of samples. For this purpose a calculation of the *correlation coefficient* is appropriate. A careful discussion of the distinction between problems properly treated as regression and those properly treated by correlation methods has been given by Sokal and Rohlf.[21]

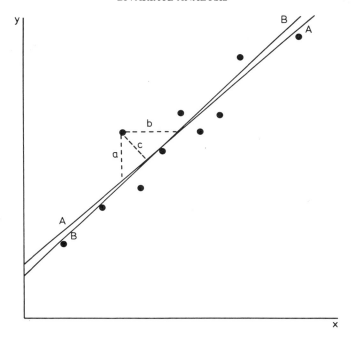

Fig. 18.4. Possible criteria in the minimization of deviations from fitted lines. *a*, Regression of *y* on *x*, minimum sum of squares of *y*-deviations; *b*, regression of *x* on *y*, minimum sum of squares of *x*-deviations; *c*, minimization of perpendicular distances. "The two regression lines are for *y* on *x* (line AA); *x* on *y* (line BB)."

18.3.2 Correlation Analysis

Correlation analysis is particularly applicable to the problem of testing the strength of the association between two random variables. It is neither known nor assumed that there is any cause-and-effect relationship between them. Where a strong association is shown to exist, it is sometimes (but not always) reasonable to regard the two variables as effects of a common cause. In geochemical work this often applies in cases where the variables are trace-element concentrations in materials such as rocks and soils.

The most common approach to estimating the degree of association is through the calculation of the Pearson product moment correlation coefficient r. For a set of n paired data points (x_i, y_i), the sample correlation coefficient is calculated conveniently from the expression

$$r = \frac{n \sum x_i y_i - \sum x_i \sum y_i}{\{[n \sum x_i^2 - (\sum x_i)^2][n \sum y_i^2 - (\sum y_i)^2]\}^{1/2}} \qquad (18.9)$$

TABLE 18.4 Calculation of the Correlation Coefficient for Nickel and Cobalt Concentrations (µg/g) in Soils

x_i(Ni)	y_i(Co)		
50	6.2	$\sum x_i$	$= 764$
53	6.5	$\sum y_i$	$= 101.3$
57	6.8	$\sum x_i^2$	$= 49{,}380$
60	7.2	$\sum y_i^2$	$= 890.47$
61	8.0	$\sum x_i y_i$	$= 6592.8$
62	7.0		
64	8.6		
68	9.8	$r = 0.89$	
68	11.8		
70	8.9		
73	10.5		
78	10.0		

The correlation coefficient can vary between $+1$ and -1, positive values indicating a direct relationship and negative values implying an inverse relationship between the variables. The closer the value of r is to $+1$ or -1, the closer the linear relationship. A value of r near zero implies that the variables are not significantly associated.

Table 18.4 shows the nickel and cobalt concentrations of a set of soil samples and gives the summations required for the calculation of r. The value of 0.89 obtained for r implies that a direct relationship exists. To establish the confidence with which this assertion is made, the probability can be obtained from Appendix IV for $n - 2$ degrees of freedom. For 10 degrees of freedom in the above example, the value lies below 0.001. This is the probability that the relationship has occurred by chance. The linear relationship bwtween the two variables can therefore be judged as very highly significant. Examination of Appendix IV shows that the significance of a given value of r depends on the number of degrees of freedom.

The Pearson correlation coefficient provides a valid estimate of the association between two variables only if the data are approximately normally distributed. If either variable has a log-normal distribution it should be transformed logarithmically to preserve the validity of the correlation.

An awkward situation arises when a few points have much higher values of the variables than the majority. For example, Figure 18.5a shows a graph

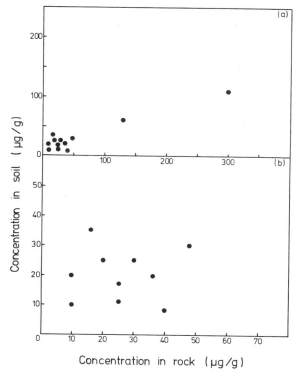

Fig. 18.5. Hypothetical values for concentrations of an element in a rock and its related soils. (a) Includes two abnormally high values and gives a highly significant value for the correlation coefficient (r) between the two variables. (b) Ignores the two high values and shows that the resultant value of r implies a nonsignificant relationship.

of some hypothetical values of an elemental concentration in some soils compared with the concentration of the same element in the associated rocks. There are 10 points clustered near the origin and 2 points with much higher values of both variables. Statistically, this distribution almost behaves as 3 points, with the cluster acting as a heavily weighted single point. The value for r is very high ($r = 0.95$). If the 2 high points are ignored and the remaining 10 points examined, the distribution of Figure 18.5b results. The value of r for these remaining points is 0.06, which is not significant.

The effect of outlying points of this kind is reduced by logarithmic transformation. Some guidance as to whether such points are outliers of a normal distribution or are part of a log-normal distribution can often be obtained by constructing a histogram or cumulative frequency diagram. Even under a logarithmic transformation, however, a few points may appear as outliers, and standard tests for rejection of these extreme values should be considered. The above example illustrates the danger of inferring a strong correlation from distributions containing one or two outlying points.

If an approximately normal distribution of the variables is not found, the nonparametric Spearman rank correlation coefficient (r_s) can be used to evaluate the correlation, as follows. The pairs of data are arranged in two columns. The first column is ranked with the highest values at the top and the lowest values at the bottom (or vice versa). The values in the second column are then arranged so that all pairs are matched horizontally. If there is a complete correlation, then the ranks of the second column will be exactly as in the first and the Spearman coefficient will be 1.0. The extent of deviation from perfect ranking in the second column will determine the value of r_s which ranges from -1 to $+1$.

In its simplest form r_s is given by

$$r_s = 1 - \frac{6 \sum d^2}{n^3 - n} \qquad (18.10)$$

where n is the number of pairs of values and d is the difference of rank of each horizontal pair in the two columns. When there are tied values in either column, a modified formula must be used:

$$r_s = \frac{A + B - \sum d^2}{2(AB)^{1/2}} \qquad (18.11)$$

where

$$A = \frac{n^3 - n}{12} - \sum T_a \quad \text{and} \quad B = \frac{n^3 - n}{12} - \sum T_b \qquad (18.12)$$

The terms $\sum T_a$ and $\sum T_b$ are corrections for the presence of ties in columns a and b respectively. If t is the number of values tied for any given ranking,

$$T = \frac{t^3 - t}{12} \qquad (18.13)$$

The summations are carried out for all the ties in each column.

As an example, the data of Table 18.4 may be considered. The data of the first column are ranked in ascending order, the ranks being: 1, 2, 3, 4, 5, 6, 7, 8.5, 8.5, 10, 11, 12. The tied values are assigned the means of the ranks that would have been assigned if no tie had occurred. The ranking of the second column as a consequence of this arrangement of the first, gives ranks of 1, 2, 3, 5, 6, 4, 7, 9, 12, 8, 11, 10. The rank differences taken across the columns are: 0, 0, 0, -1, -1, $+2$, 0, -0.5, -3.5, $+2$, 0, $+2$. From this, $\sum d^2 = 26.5$ and successive application of equations 18.13, 18.12, and 18.11 gives $r_s = 0.91$. This value compares with 0.89 for the Pearson coefficient.

The Spearman rank correlation test is less efficient than the parametric Pearson coefficient,[13] but is generally more satisfactory where a small number of samples is involved, and where it is difficult to ascertain that the data are normally (or log-normally) distributed. In terms of computer time, there is some disadvantage in using the Spearman coefficient because of the need to rank the data.

18.4 MULTIVARIATE ANALYSIS

It is usually quite laborious to do statistical calculations manually, even for simple tests of significance and correlation coefficients. Until the development of computers it was hardly ever practicable to do calculations involving more than two variables, but such operations are now widely used in analyzing geochemical data.

18.4.1 Multiple-Regression Analysis

In many cases it is appropriate to attempt to express a dependent variable y in terms of more than one independent variable, $x_1, x_2, \ldots, x_m$. One general model used frequently is that in which the data are fitted to an equation of the form

$$y = a_0 + a_1 x_1 + a_2 x_2 + \cdots + a_m x_m \qquad (18.14)$$

An estimate of a coefficient a_i in the regression model can be found from the rate of change of the dependent variable with respect to variable x_i, with all other independent variables being held constant. These estimates, the sample *partial regression coefficients*, are generally different from the simple regression coefficients of y on x_i. The multiple regression can be expected to account for more of the variation in y than any simple regression does. It should be noted that, because the variance of only one variable (y) is subjected to analysis, multiple regression is multivariate only in the sense that several variables are measured on each sample.

As in the two-variable case, the method of least squares can be used to provide the best estimates of the parameters $a_0, a_1, a_2, \ldots, a_m$. By minimizing the sum of the squared deviations of the observed values of y from the line, a set of $m + 1$ simultaneous equations is obtained. Computational procedures for dealing with situations where m is large are discussed in several works on applications of statistics.[22-24]

One such procedure is described as stepwise regression.[24,25] At each stage of the computation, this procedure selects the independent variable that increases by the most significant amount, the variance of y that is explained by the regression equation. If, and only if, this amount is more significant than a predetermined probability level, this independent variable is inserted into the regression equation. Following the insertion of a variable into the regression equation, the other variables already in the equation are rechecked to find if they are still significant at the prescribed level. If not, they are deleted from the equation. On completion of the analysis, the percentage of variance in the dependent variable which has been explained by the regression equation is computed.

Examples of the use of multiple-regression analysis, multivariate analysis of variance, and multivariate extensions of simple statistical tests, have been given by several authors.[23,26-28]

18.4.2 Trend Surface Analysis

Trend surface analysis is a mathematical method of analyzing map data, using techniques similar to those of multiple regression. In geochemistry it can be used to separate the values of a variable (such as a metal concentration in soil or rock) into various components to facilitate interpretation of the geological environment. Trend analysis leads to an equation from which an estimated value of the variable may be obtained for each point on the survey area.

Geochemical map data may be considered as having three main components: (a) regional trends due to large-scale geological processes (such as changes in the nature of the bedrock) that affect a major part of the sampling area; (b) local deviations from the trend, extending over more than one sample site and generally arising from small-scale geological anomalies such as mineralization; (c) random variation over the sampling area, usually called "noise," caused by local geological effects and by sampling and analytical error.

It is necessary to postulate a suitable function to describe the regional trend. For example, a linear trend surface is an equation of the form

$$y = a_0 + a_1 x_1 + a_2 x_2 \qquad (18.15)$$

where the variables x_1 and x_2 are geographical coordinates. In some cases it is necessary to try fitting the regional trend to a more complex function, for example, a second-degree trend surface:

$$y = a_0 + a_1 x_1 + a_2 x_2 + a_3 x_1^2 + a_4 x_2^2 + a_5 x_1 x_2 \qquad (18.16)$$

The trend surface is found by a least-squares procedure, minimizing the sum of the squared deviations of observed values of y from the surface. This best-fit surface is considered to be the regional component of the data, and the local component and noise are contained in the deviations. The deviations, or *residuals*, may be separated into local and noise components by observing the signs of adjacent residuals. Where a group of adjacent residuals all have the same sign they are said to be *autocorrelated*. Positive autocorrelated residuals reflect regions of anomalous local influence. If the dependent variable is a metal concentration in soil or rock, these anomalous features may be mineralized zones. Residuals whose signs vary in a random fashion contain the noise component of the data.

In practice, it is often difficult to decide at what stage of refinement the best-fit surface begins to fit small-scale geological effects as well as the regional trend. If the geology of the sampling area is known, refinement can be made until the surface-of-best-fit agrees with the known geology. If, however, the geology of the sampling area is unknown, one technique is to retain in the trend surface equation only those terms which result in a significant reduction in the residual sum of squares. The degree of significance is specified for a particular analysis. The method is usually designed to give a regular surface with few maxima. In general, the higher the level of significance specified, the fewer maxima the surface will contain. Because the prescribed significance level is controlled by the operator, the surface-of-best-fit that is obtained will not necessarily correspond to the actual regional trend over the sampling area. Similarly, the trend residuals may not arise solely from small-scale geological effects but may include some of the regional trend. It is the operator who must decide what the components of the trend analysis represent.

A simple example of the use of trend analysis is illustrated in Fig. 18.6, where the curves represent data obtained in a transect across a suspected mineralized area.[29] Data sets *a* and *b* represent copper and molybdenum values, respectively, in bedrock.

It can be seen from Figures 18.6A and 18.6B that the data sets *a* and *b* are each influenced by different regional processes and that apparently they do not indicate the same anomalous areas, if only isoconcentration contours are used (Fig. 18.6B). If, however, estimates of the trends due to the regional processes can be computed (Fig. 18.6C) and the residuals plotted, then it becomes clear that, in fact, the data sets do indicate the same anomalies (Fig. 18.6D).

A recent example of the application of trend surface analysis to a more complex three-dimensional situation can be found in the work of Putman.[30] For a more comprehensive discussion of the technique, and for other applications, the reader is referred to the literature.[23,31–36]

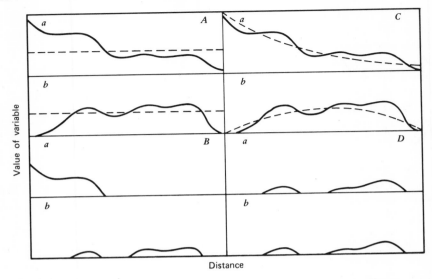

Fig. 18.6. Trend analysis of hypothetical data for concentrations of two trace elements in bedrock (elements *a* and *b*) from a transect across a mineralized zone. *A*, Raw data; *B*, anomalies indicated by isoconcentration contours only; *C*, estimated trends (broken lines); *D*, residuals. *Souce*: Timperley et al.[29]

18.4.3 Factor Analysis

Factor analysis is a technique used in the interpretation of relationships between the variables in a multivariate collection of data. It effectively allows the data to be expressed in terms of a number of factors that is fewer than the number of original variables. The factors created are new variables, having the form of linear combinations of the original variables. Each factor extracted successively accounts for a decreasing proportion of the data variance.

The analysis may be carried out by examining the relationship between the *samples* (Q mode) or the relationship between the *variables* (R mode). R-mode factor analysis is particularly useful in clarifying the relationship between variables in a complex array of data. Where the variables are elemental concentrations in a set of geological samples, the analysis can lead to explanations of the data variability in terms of geochemical interelement associations. If factor analysis is appropriate for a given data set (i.e., if the variances observed are the result of correlations between variables and underlying factors), only a few factors should be needed to account for a high proportion of the variance.

Space does not permit the mathematical techniques of factor analysis to be discussed here. The reader is referred to the literature for accounts of basic principles[23,37-40] and applications to geochemical problems.[35,41-44]

18.4.4 Discriminant Analysis

Discriminant analysis is essentially a method of *classifying* individuals into one of several groups on the basis of a number of measured variables. The discriminant function is a linear combination of the variables x_i:

$$y = a_0 + a_1 x_1 + a_2 x_2 + \cdots + a_m x_m$$

We seek values of a_i such that the differences in the average values of y for the various groups are maximized.

The degree of discrimination can be tested by computing the D^2 statistic.[45-47] This statistic is calculated from the means, variances, and covariances of the variables measured in samples from the various groups. In effect, the discriminant function combines and weights the variables in such a way that the D^2 scores of samples within each group have a minimum spread, whereas the separation of the groups is maximized.

The D^2 statistic can be used to test the hypothesis that the means are the same in all populations for each of the m variables. The null hypothesis probability values indicate the confidence with which the population groups can be distinguished. The adequacy of the discriminant function should be tested by investigating its ability to classify members of a fresh sample of data, not used in determining the function itself.

A simple recent example of the use of discriminant analysis of trace element data on two population groups can be found in the work of Whitehead and Govett.[48] It was desired to discriminate between two groups of acid volcanic rocks above and below a massive sulfide deposit. Concentrations of lead and zinc were found to be the most useful variables in separating the two populations. The discriminant functions were used to identify a trace element halo extending about 400 m above the sulfides and 1200 m along the same stratigraphic horizon.

As applied to discrimination between two populations groups, the data analysis involved calculating a function f_1 to describe group g_1, in terms of its measured variables (such as Cu, Pb, Zn, ..., concentrations)

$$f_1 = a_1 \text{Cu} + b_1 \text{Pb} + c_1 \text{Zn} + \cdots + C_1$$

and a function f_2 to describe group g_2

$$f_2 = a_2 \text{Cu} + b_2 \text{Pb} + c_2 \text{Zn} + \cdots + C_2$$

TABLE 18.5 Discriminant Analysis Classification of Defined Halo and Background Samples By Functions Based on Different Variables

Test no.	Elements used as variables	Samples misclassified (no.)		D^2 statistic
		Halo group ($N = 20$)	Background group ($N = 18$)	
1	Cu, Co, Ni	5	7	13.3
2	Co, Ni, Mn	4	6	22.2
3	Co, Ni, Mn, Fe	1	6	30.3
4	Cu, Co, Ni, Mn, Fe	1	7	34.2
5	Pb, Zn	1	2	68.2
6	Cu, Pb, Zn	1	2	68.3
7	Cu, Pb, Zn, Co	1	2	71.9
8	Cu, Pb, Zn, Co, Ni	1	2	73.1
9	Cu, Pb, Zn, Co, Ni, Fe, Mn	0	2	78.8

From Ref. 48.

where the chemical symbols represent concentrations of the elements concerned, $a_i, b_i, c_i, \ldots$, are coefficients, and C_i are constants.

The functions were determined using different combinations of the measured variables. Some of the functions were shown to classify correctly most of the samples of known origin, as indicated in Table 18.5. Although the best classification (in terms of number of samples correctly classified and the value of the D^2 statistic) was given by seven variables (Test 9, Table 18.5), all of the tests between 5 and 9 were regarded as giving satisfactory results. The simplicity of test 5, which required determination of lead and zinc concentrations only, made this a suitable choice for the subsequent classification of "unknown" samples. Application of this test to a suite of 133 quartz-feldspar porphyry samples, made it possible to define an extensive halo region surrounding the massive sulfide deposit. Use of the lead or zinc concentrations alone would have been much less successful, because of considerable overlap of the elemental concentrations found in the halo and background groups.

Other applications of discriminant analysis of trace element data to geochemical problems include those of Cameron,[43] Govett,[49] and Ward.[50] General discussion of discriminant analysis is given in several works.[23,47,51]

18.4.5 Cluster Analysis

Cluster analysis[52-54] is a technique for classifying objects into related groups and subgroups (taxonomic classification). Many methods of computation and presentation are available and the data can be correlated by

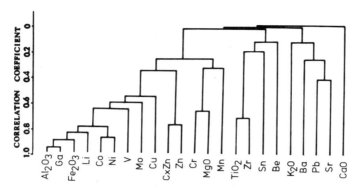

Fig. 18.7. Dendrogram of cluster analysis of the elemental content of stream sediments in the United Kingdom. *Source*: Obial.[55]

comparing variables (R mode) or samples (Q mode). As with most other multivariate techniques, the method is most practicable with a computer.

The simplest way of correlating multielement data is to generate a matrix that contains the correlations between all possible pairs of elements considered. Such a matrix is, however, difficult to interpret and impossible to plot meaningfully upon a map. With use of cluster analysis, however, it is possible to group variables according to their mutual correlations, that is, to choose mutually correlated variables that exhibit the greatest *within-group* correlation relative to the *between-group* correlations, taking into account all possible combinations of the given elements. Obial[55] has used cluster analysis to interpret the major, minor, and trace-element contents of 170 stream sediments collected in the United Kingdom. The data are shown as a dendrogram in Fig. 18.7.

Clearly the best correlation ($r = 0.95$) is for Al_2O_3 and Ga and both are correlated well with Fe_2O_3 ($r = 0.90$). In this particular case six elemental groupings were recognized: (*a*) Al_2O_3, Ga, Fe_2O_3, Li, Co, Ni, V, Mo, Cu; (*b*) extractable Zn and Zn; (*c*) Cr, MgO, Mn; (*d*) TiO_2, Zr, Sn, Be; (*e*) K_2O, Ba, Pb, Sr; (*f*) CaO.

Maps of individual elements with high correlations within a group, showed close similarity in their dispersion patterns, whereas elements from different groups showed different patterns. It should be noted that attempts to produce larger groupings invariably result in lower correlation coefficients and consequently lesser significance.

For further discussion of multivariate statistical techniques useful in dealing with geochemical problems, the reader is referred to standard texts.[23,26–28,37,47,51,54,56]

18.5 ACCURACY AND PRECISION IN TRACE-ELEMENT ANALYSIS

18.5.1 Definitions

In an attempt to standardize the usage of the terms *accuracy* and *precision* the Advisory Board of the journal *Analytical Chemistry*[57] has published the following guide. These recommendations are applicable to sets of results based on five or more determinations.

Series. A number of test results that possess common properties that identify them uniquely.

Mean. The sum of a series of test results divided by the number in the series. Arithmetic mean is understood.

Precision Data. Measurements that relate to the variation among the test results themselves, that is, the scatter or dispersion of a series of tests results, without assumption of any prior information. The following measures apply: (a) *Variance.* The sum of squares of deviations of the test results from the mean of the series after division by one less than the total number of test results. (b) *Standard Deviation.* The square root of the variance. (c) *Relative Standard Deviation.* The standard deviation of a series of test results as a percentage of the mean of this series. This term is preferred over "*coefficient of variation.*" (d) *Range.* The difference in magnitude between the highest test result and the lowest test result in a series.

Accuracy Data. Measurements which relate to the difference between the average test results and the true results when the latter are known or assumed. The following measures apply: *Mean Error.* The average difference with regard to sign of the test results from the true results. Also equal to the difference between the average of a series of test results and the true result.

18.5.2 Assessment of Acceptability of Analytical Methods

Both precision and accuracy must be taken into account when judging the acceptability of a given analytical technique. The relative importance of these factors differs in different situations. For example, in a geochemical exploration survey of trace elements in stream sediments, the exploration geochemist is less concerned with accuracy than with precision, because he will select his anomalies by comparison with background values subject to the same sources of systematic error. On the other hand, a smelting works buying raw materials from a mining company will be greatly concerned with accuracy because the amount paid for the raw materials is governed by the analyzed content of the ore metal.

The precision of an analysis is readily found from a series of test results (Sec. 18.5.1). The accuracy, however, is much more difficult to assess, because a "true" result must be known or assumed.

In some cases where the analytical method itself is absolute (i.e., the results are not dependent on a comparison with synthetic standards), it is possible to estimate the accuracy of the method by making measurements on samples of known composition prepared from standard materials of high purity. More often, however, the accuracy of an instrumental method of trace analysis cannot be assessed in this way. In the first place, most of the widely used techniques are comparative: the analytical result is obtained by referring a signal given by the unknown sample to a working curve resulting from signals given by a set of standards. (This comparative approach is used whenever the atomic or molecular properties or instumental parameters that determine the signal are not known or cannot be measured accurately enough.) Second, in dealing with geological materials in particular, the overall composition of the sample matrix is often not well known, and it may be difficult to be sure that the standards are subject to exactly the same matrix effects as the sample.

One procedure that can be adopted in testing analytical accuracy is to carry out determinations on standard reference materials, such as those noted in Chapter 5. Many trace elements in these materials have been measured by a wide variety of techniques, and "recommended" values have been given where the data are sufficiently concordant.

The reliability and accuracy of analytical procedures are often tested by adding known amounts of the analyte to the sample, and showing that these amounts are quantitatively recoverable (i.e., that the "spiking" produces exactly the change expected in the analytical signal.) It should be noted, however, that fulfilment of this test is a necessary, but not sufficient, condition for the method to be considered accurate. This test cannot demonstrate with certainty the absence of a systematic error introduced by some component present in the unspiked and spiked samples.

Criteria for judging the "acceptability" of analytical methods, based on a consideration of both accuracy and precision, have been discussed by McFarren et al.[58] The total error T is regarded as consisting of two components, the relative error (a measure of accuracy), and the standard deviation (a measure of precision). T is defined by:

$$T = \frac{\text{(absolute value of mean error)} + 2\,\text{(standard deviation)}}{\text{true value}} \times 100\%$$

(18.17)

By taking twice the standard deviation, all but about 2.3% of the area under either end of the normal distribution curve is included.

McFarren et al. observed that interlaboratory studies on the analysis of trace inorganic and organic substances in waters seldom yield a relative standard deviation less than 2.5%, and that total error as defined by equation 18.17 often exceeds 25%. In the studies summarized by McFarren et al. a method was judged "excellent" if the total error did not exceed 25%, and "acceptable" if the total error was below 50% and better procedures were not available. Although this classification of acceptability of analytical methods for trace elements has been used extensively, it must be remembered that the choice of acceptable levels of T is somewhat arbitrary and that the relative importance of accuracy and precision varies in different situations. Ultimately, the decisions reached by the analyst on these matters are dependent on the use he wishes to make of the analytical measurements.

References

1. C. J. Brookes, I. G. Betteley, and S. M. Loxston, *Mathematics and Statistics for Chemists*, Wiley, New York, 1966.
2. L. H. Ahrens, *Geochim. Cosmochim. Acta*, **5**, 49 (1954).
3. K. V. Aubrey, *Geochim. Cosmochim. Acta*, **9**, 83 (1956).
4. F. Chayes, *Geochim. Cosmochim. Acta*, **6**, 119 (1954).
5. R. L. Miller and E. D. Goldberg, *Geochim. Cosmochim. Acta*, **8**, 53 (1955).
6. A. B. Vistelius, *J. Geol.*, **68**, 1 (1960).
7. D. M. Shaw, *Geochim. Cosmochim. Acta*, **23**, 116 (1961).
8. C. B. Tennant and M. L. White, *Econ. Geol.*, **54**, 1281 (1959).
9. R. R. Brooks, *Geobotany and Biogeochemistry in Mineral Exploration*, Harper and Row, New York, 1972.
10. R. M. Cassie, *Austr. J. Mar. Freshw. Res.*, **5**, 513 (1954).
11. X. K. Williams, *N. Z. J. Geol. Geophys.*, **10**, 771 (1967).
12. G. Herdan, *Quality Control by Statistical Methods*, Nelson, London, 1948.
13. N. J. Marshall, *Proc. 2nd UNESCO Sem. Prospect. Methods Tech.*, United Nations, New York, 1970, p. 314.
14. S. Siegel, *Non-Parametric Statistics for the Behavioral Sciences*, McGraw-Hill, Tokyo, 1956.
15. J. D. Hinchen, *Practical Statistics for Chemical Research*, Methuen, London, 1969.
16. R. A. Fisher and F. Yates, *Statistical Tables for Biological, Agricultural and Medical Research*, 6th ed., Longman, London, 1974.
17. E. S. Pearson and H. O. Hartley (eds.), *Biometrika Tables for Statisticians*, Vol. 1, 3rd ed., Cambridge University Press, Cambridge, England, 1966.
18. W. H. Beyer (ed.), *Handbook of Tables for Probability and Statistics*, 2nd ed., Chemical Rubber Co., Cleveland, Ohio, 1968.

19. D. B. Owen, *Handbook of Statistical Tables*, Addison-Wesley, Reading, Mass., 1962.
20. F. S. Acton, *Analysis of Straight-Line Data*, Wiley, New York, 1959.
21. R. R. Sokal and F. J. Rohlf, *Introduction to Biostatistics*, Freeman, San Francisco, 1973.
22. N. R. Draper and H. Smith, *Applied Regression Analysis*, Wiley, New York, 1966.
23. J. C. Davis, *Statistics and Data Analysis in Geology*, Wiley, New York, 1973.
24. M. A. Efroymson, in *Mathematical Methods for Digital Computers* (eds. A. Ralston and H. S. Wilf), Wiley, New York, 1960, p. 191.
25. A. T. Miesch and J. J. Connor, *Computer Contr.* No. 27, University of Kansas, 1968.
26. G. S. Koch and R. F. Link, *Statistical Analysis of Geological Data*, Vol. 2, Wiley, New York, 1971.
27. W. W. Cooley and P. R. Lohnes, *Multivariate Data Analysis*, Wiley, New York, 1971.
28. D. F. Morrison, *Multivariate Statistical Methods*, McGraw-Hill, New York, 1967.
29. M. H. Timperley, R. R. Brooks, and P. J. Peterson, *Econ. Geol.*, **67**, 669 (1972).
30. G. W. Putman, *Econ. Geol.*, **70**, 1225 (1975).
31. F. P. Agterberg, *Col. School Mines Quart.*, **59**, 111 (1964).
32. E. M. Cameron, *Can. J. Earth Sci.*, **5**, 287 (1968).
33. J. J. Connor and A. T. Miesch, *Stanford Univ. Pub. Geol. Sci.*, **9**, 110 (1964).
34. W. C. Krumbein, *J. Geophys. Res.*, **64**, 823 (1959).
35. I. Nichol, R. G. Garrett, and J. S. Webb, *Econ. Geol.*, **64**, 204 (1969).
36. E. C. Dahlberg, *Econ. Geol.*, **63**, 409 (1968).
37. H. Harman, *Modern Factor Analysis*, University of Chicago Press, Chicago, 1967.
38. J. E. Klovan, *J. Sed. Petrol.*, **36**, 115 (1966).
39. R. B. Cattell, *Biometrics*, **21**, 190, 405 (1965).
40. D. N. Lawley and A. E. Maxwell, *Factor Analysis as a Statistical Method*, Butterworth, London, 1963.
41. R. G. Garrett and I. Nichol, *Col. School Mines Quart.*, **64**, 245 (1969).
42. R. Saagar and P. A. Esselaar, *Econ. Geol.*, **64**, 445 (1969).
43. E. M. Cameron, *Can. J. Earth Sci.*, **6**, 247 (1969).
44. K. M. Dawson and A. J. Sinclair, *Econ. Geol.*, **69**, 404 (1974).
45. P. C. Mahalanobis, *J. Asiatic Soc. Bengal*, **26**, 541 (1930).
46. P. C. Mahalanobis, *Proc. Nat. Inst. Sci. India*, **12**, 49 (1936).
47. C. R. Rao, *Advanced Statistical Methods in Biometric Research*, Wiley, New York, 1952.
48. R. E. S. Whitehead and G. J. S. Govett, *J. Geochem. Explor.*, **3**, 371 (1974).
49. G. J. S. Govett, *J. Geochem. Explor.*, **1**, 77 (1972).
50. G. K. Ward, *Archaeometry*, **16**, 41 (1974).

51. W. C. Krumbein and F. A. Graybill, *An Introduction to Statistical Models in Geology*, McGraw-Hill, New York, 1965.
52. J. C. Gower, *Biometrika*, **55**, 588 (1966).
53. J. M. Parks, *J. Geol.*, **74**, 703 (1966).
54. R. R. Sokal and P. H. A. Sneath, *Principles of Numerical Taxonomy*, Freeman, London, 1963.
55. R. C. Obial, *Trans. Inst. Min. Metall. (Lond.) Sec. B*, **79**, 175 (1970).
56. R. L. Miller and J. S. Kahn, *Statistical Analysis in the Geological Sciences*, Wiley, New York, 1962.
57. Anon., *Anal. Chem.*, **40**, 2271 (1968).
58. E. F. McFarren, R. J. Lishka, and J. Parker, *Anal. Chem.*, **42**, 358 (1970).

APPENDIX

I

TABLE OF VALUES FOR *F* TEST (SINGLE-SIDED)

Table of Values for F Test (Single-sided)

ϕ_1 (corresponding to greater mean square)

Probability (P)	ϕ_2	1	2	3	4	5	6	7	8	9	10	15	∞
0.10	1	39.9	49.5	53.6	55.8	57.2	58.2	58.9	59.4	59.9	60.2	61.2	63.3
0.05		161	199	216	225	230	234	237	239	241	242	246	254
0.01		4,052	4,999	5,403	5,625	5,764	5,859	5,928	5,982	6,022	6,056	6,157	6,366
0.10	2	8.53	9.00	9.16	9.24	9.29	9.33	9.35	9.38	9.39	9.39	9.42	9.49
0.05		18.5	19.0	19.2	19.2	19.3	19.3	19.4	19.4	19.4	19.4	19.4	19.5
0.01		98.5	99.0	99.2	99.2	99.3	99.3	99.4	99.4	99.4	99.4	99.4	99.5
0.10	3	5.54	5.46	5.39	5.34	5.31	5.28	5.27	5.25	5.24	5.23	5.20	5.13
0.05		10.1	9.55	9.28	9.12	9.01	8.94	8.89	8.85	8.81	8.79	8.70	8.53
0.01		34.1	30.8	29.5	28.7	28.2	27.9	27.7	27.5	27.3	27.2	26.9	26.1
0.10	4	4.54	4.32	4.19	4.11	4.05	4.01	3.98	3.95	3.94	3.92	3.87	3.76
0.05		7.71	6.94	6.59	6.39	6.26	6.16	6.09	6.04	6.00	5.96	5.86	5.63
0.01		21.2	18.0	16.7	16.0	15.5	15.2	15.0	14.8	14.7	14.5	14.2	13.5
0.10	5	4.06	3.78	3.62	3.52	3.45	3.40	3.37	3.34	3.32	3.30	3.24	3.10
0.05		6.61	5.79	5.41	5.19	5.05	4.95	4.88	4.82	4.77	4.74	4.62	4.36
0.01		16.3	13.3	12.1	11.4	11.0	10.7	10.5	10.3	10.2	10.1	9.72	9.02
0.10	6	3.78	3.46	3.29	3.18	3.11	3.05	3.01	2.98	2.96	2.94	2.87	2.72
0.05		5.99	5.14	4.76	4.53	4.39	4.28	4.21	4.15	4.10	4.06	3.94	3.67
0.01		13.7	10.9	9.78	9.15	8.75	8.47	8.26	8.10	7.98	7.87	7.56	6.88
0.10	7	3.59	3.26	3.07	2.96	2.88	2.83	2.78	2.75	2.72	2.70	2.63	2.47
0.05		5.59	4.74	4.35	4.12	3.97	3.87	3.79	3.73	3.68	3.64	3.51	3.23
0.01		12.2	9.55	8.45	7.85	7.46	7.19	6.99	6.84	6.72	6.62	6.31	5.65

ν_2	α													
8	0.10	3.46	3.11	2.92	2.81	2.73	2.67	2.62	2.59	2.56	2.54		2.46	2.29
	0.05	5.32	4.46	4.07	3.84	3.69	3.58	3.50	3.44	3.39	3.35		3.22	2.93
	0.01	11.3	8.65	7.59	7.01	6.63	6.37	6.18	6.03	5.91	5.81		5.52	4.86
9	0.10	3.36	3.01	2.81	2.69	2.61	2.55	2.51	2.47	2.44	2.42		2.34	2.16
	0.05	5.12	4.26	3.86	3.63	3.48	3.37	3.29	3.23	3.18	3.14		3.01	2.71
	0.01	10.6	8.02	6.99	6.42	6.06	5.80	5.61	5.47	5.35	5.26		4.96	4.31
10	0.10	3.29	2.92	2.73	2.61	2.52	2.46	2.41	2.38	2.35	2.32		2.24	2.06
	0.05	4.96	4.10	3.71	3.48	3.33	3.22	3.14	3.07	3.02	2.98		2.85	2.54
	0.01	10.0	7.56	6.55	5.99	5.64	5.39	5.20	5.06	4.94	4.85		4.56	3.91
12	0.10	3.18	2.81	2.61	2.48	2.39	2.33	2.28	2.24	2.21	2.19		2.10	1.90
	0.05	4.75	3.89	3.49	3.26	3.11	3.00	2.91	2.85	2.80	2.75		2.62	2.30
	0.01	9.33	6.93	5.95	5.41	5.06	4.82	4.64	4.50	4.39	4.30		4.01	3.36
15	0.10	3.07	2.70	2.49	2.36	2.27	2.21	2.16	2.12	2.09	2.06		1.97	1.76
	0.05	4.54	3.68	3.29	3.06	2.90	2.79	2.71	2.64	2.59	2.54		2.40	2.07
	0.01	8.68	6.36	5.42	4.89	4.56	4.32	4.14	4.00	3.89	3.80		3.52	2.87
16	0.10	3.05	2.67	2.46	2.33	2.24	2.18	2.13	2.09	2.06	2.03		1.94	1.72
	0.05	4.49	3.63	3.24	3.01	2.85	2.74	2.66	2.59	2.54	2.49		2.35	2.01
	0.01	8.53	6.23	5.29	4.77	4.44	4.20	4.03	3.89	3.78	3.69		3.41	2.75
24	0.10	2.93	2.54	2.33	2.19	2.10	2.04	1.98	1.94	1.91	1.88		1.78	1.53
	0.05	4.26	3.40	3.01	2.78	2.62	2.51	2.42	2.36	2.30	2.25		2.11	1.73
	0.01	7.82	5.61	4.72	4.22	3.90	3.67	3.50	3.36	3.26	3.17		2.89	2.21
60	0.10	2.79	2.39	2.18	2.04	1.95	1.87	1.82	1.77	1.74	1.71		1.60	1.29
	0.05	4.00	3.15	2.76	2.53	2.37	2.25	2.17	2.10	2.04	1.99		1.84	1.39
	0.01	7.08	4.98	4.13	3.65	3.34	3.12	2.95	2.82	2.72	2.63		2.35	1.60
∞	0.10	2.71	2.30	2.08	1.94	1.85	1.77	1.72	1.67	1.63	1.60		1.49	1.00
	0.05	3.84	3.00	2.60	2.37	2.21	2.10	2.01	1.94	1.88	1.83		1.67	1.00
	0.01	6.63	4.61	3.78	3.32	3.02	2.80	2.64	2.51	2.41	2.32		2.04	1.00

APPENDIX

II

TABLE OF VALUES FOR t TEST[a]

ϕ	Probability (P)							
	0.5	0.2	0.1	0.05	0.02	0.01	0.005	0.001
1	1.00	3.08	6.31	12.7	31.8	63.7	127	637
2	0.816	1.89	2.92	4.30	6.97	9.92	14.1	31.6
3	0.765	1.64	2.35	3.18	4.54	5.84	7.45	12.9
4	0.741	1.53	2.13	2.78	3.75	4.60	5.60	8.61
5	0.727	1.48	2.01	2.57	3.37	4.03	4.77	6.87
6	0.718	1.44	1.94	2.45	3.14	3.71	4.32	5.96
7	0.711	1.42	1.89	2.36	3.00	3.50	4.03	5.41
8	0.706	1.40	1.86	2.31	2.90	3.36	3.83	5.04
9	0.703	1.38	1.83	2.26	2.82	3.25	3.69	4.78
10	0.700	1.37	1.81	2.23	2.76	3.17	3.58	4.59
11	0.697	1.36	1.80	2.20	2.72	3.11	3.50	4.44
12	0.695	1.36	1.78	2.18	2.68	3.05	3.43	4.32
13	0.694	1.35	1.77	2.16	2.65	3.01	3.37	4.22
14	0.692	1.35	1.76	2.14	2.62	2.98	3.33	4.14
15	0.691	1.34	1.75	2.13	2.60	2.95	3.29	4.07
16	0.690	1.34	1.75	2.12	2.58	2.92	3.25	4.01
17	0.689	1.33	1.74	2.11	2.57	2.90	3.22	3.96
18	0.688	1.33	1.73	2.10	2.55	2.88	3.20	3.92
19	0.688	1.33	1.73	2.09	2.54	2.86	3.17	3.88
20	0.687	1.33	1.72	2.09	2.53	2.85	3.15	3.85
21	0.686	1.32	1.72	2.08	2.52	2.83	3.14	3.82
22	0.686	1.32	1.72	2.07	2.51	2.82	3.12	3.79
23	0.685	1.32	1.71	2.07	2.50	2.81	3.10	3.77
24	0.685	1.32	1.71	2.06	2.49	2.80	3.09	3.74
25	0.684	1.32	1.71	2.06	2.49	2.79	3.08	3.72
26	0.684	1.32	1.71	2.06	2.48	2.78	3.07	3.71
27	0.684	1.31	1.70	2.05	2.47	2.77	3.06	3.69
28	0.683	1.31	1.70	2.05	2.47	2.76	3.05	3.67
29	0.683	1.31	1.70	2.05	2.46	2.76	3.04	3.66
30	0.683	1.31	1.70	2.04	2.46	2.75	3.03	3.65
40	0.681	1.30	1.68	2.02	2.42	2.70	2.97	3.55
60	0.679	1.30	1.67	2.00	2.39	2.66	2.91	3.46
120	0.677	1.29	1.66	1.98	2.36	2.62	2.86	3.37
∞	0.674	1.28	1.64	1.96	2.33	2.58	2.81	3.29

[a] The table gives the probability that the absolute value of t will exceed the tabulated value.

APPENDIX

III

TABLE OF VALUES FOR χ^2 TEST

	Probability of validity of H_0									
ϕ	0.99	0.975	0.95	0.90	0.50	0.10	0.05	0.025	0.02	0.001
1	0.000	0.001	0.004	0.016	0.455	2.71	3.84	5.02	6.63	10.83
2	0.020	0.051	0.103	0.211	1.39	4.61	5.99	7.38	9.21	13.82
3	0.115	0.216	0.352	0.584	2.37	6.25	7.81	9.35	11.34	16.27
4	0.297	0.484	0.711	1.06	3.36	7.78	9.49	11.14	13.28	18.47
5	0.554	0.831	1.15	1.61	4.35	9.24	11.07	12.83	15.09	20.52
6	0.872	1.24	1.64	2.20	5.35	10.64	12.59	14.45	16.81	22.46
7	1.24	1.69	2.17	2.83	6.35	12.02	14.07	16.01	18.48	24.32
8	1.65	2.18	2.73	3.49	7.34	13.36	15.51	17.53	20.09	26.13
9	2.09	2.70	3.33	4.17	8.34	14.68	16.92	19.02	21.67	27.88
10	2.56	3.25	3.94	4.87	9.34	15.99	18.31	20.48	23.21	29.59
11	3.05	3.82	4.57	5.58	10.34	17.28	19.68	21.92	24.73	31.26
12	3.57	4.40	5.23	6.30	11.34	18.55	21.03	23.34	26.22	32.91
13	4.11	5.01	5.89	7.04	12.34	19.81	22.36	24.74	27.69	34.53
14	4.66	5.63	6.57	7.79	13.34	21.06	23.68	26.12	29.14	36.12
15	5.23	6.26	7.26	8.55	14.34	22.31	25.00	27.49	30.58	37.70
16	5.81	6.91	7.96	9.31	15.34	23.54	26.30	28.85	32.00	39.25
17	6.41	7.56	8.67	10.09	16.34	24.77	27.59	30.19	33.41	40.79
18	7.01	8.23	9.39	10.86	17.34	25.99	28.87	31.53	34.81	42.31
19	7.63	8.91	10.12	11.65	18.34	27.20	30.14	32.85	36.19	43.82
20	8.26	9.59	10.85	12.44	19.34	28.41	31.41	34.17	37.57	45.32
21	8.90	10.28	11.59	13.24	20.34	29.62	32.67	35.48	38.93	46.80
22	9.54	10.98	12.34	14.04	21.34	30.81	33.92	36.78	40.29	48.27
23	10.20	11.69	13.09	14.85	22.34	32.01	35.17	38.08	41.64	49.73
24	10.86	12.40	13.85	15.66	23.34	33.20	36.42	39.36	42.98	51.18
25	11.52	13.12	14.61	16.47	24.34	34.38	37.65	40.65	44.31	52.62
26	12.20	13.84	15.38	17.29	25.34	35.56	38.89	41.92	45.64	54.05
27	12.88	14.57	16.15	18.11	26.34	36.74	40.11	43.19	46.96	55.48
28	13.56	15.31	16.93	18.94	27.34	37.92	41.34	44.46	48.28	56.89
29	14.26	16.05	17.71	19.77	28.34	39.09	42.56	45.72	49.59	58.30
30	14.95	16.79	18.49	20.60	29.34	40.26	43.77	46.98	50.89	59.70

APPENDIX

IV

TABLE OF SIGNIFICANCE VALUES FOR r

ϕ	Probability of validity of H_0				
	0.10	0.05	0.02	0.01	0.001
1	0.988	0.997	0.999	1.000	1.000
2	0.900	0.950	0.980	0.990	0.999
3	0.805	0.878	0.934	0.959	0.992
4	0.729	0.811	0.882	0.917	0.974
5	0.669	0.754	0.833	0.874	0.951
6	0.621	0.707	0.789	0.834	0.925
7	0.582	0.666	0.750	0.798	0.898
8	0.549	0.632	0.716	0.765	0.872
9	0.521	0.602	0.685	0.735	0.847
10	0.497	0.576	0.658	0.708	0.823
11	0.476	0.553	0.634	0.684	0.801
12	0.457	0.532	0.612	0.661	0.780
13	0.441	0.514	0.592	0.641	0.760
14	0.426	0.497	0.574	0.623	0.742
15	0.412	0.482	0.558	0.606	0.725
16	0.400	0.468	0.543	0.590	0.708
17	0.389	0.456	0.528	0.575	0.693
18	0.378	0.444	0.516	0.561	0.679
19	0.369	0.433	0.503	0.549	0.665
20	0.360	0.423	0.492	0.537	0.652
25	0.323	0.381	0.445	0.487	0.597
30	0.296	0.349	0.409	0.449	0.554
35	0.275	0.325	0.381	0.418	0.519
40	0.257	0.304	0.358	0.393	0.490
45	0.243	0.287	0.338	0.372	0.465
50	0.231	0.273	0.322	0.354	0.443
60	0.211	0.250	0.295	0.325	0.408
70	0.195	0.232	0.274	0.302	0.380
80	0.183	0.217	0.256	0.283	0.357
90	0.173	0.205	0.242	0.267	0.337
100	0.164	0.195	0.230	0.254	0.321

INDEX

Abbreviations for Separation and Analytical Methods

anodic stripping voltammetry ASV, atomic absorption AA, atomic fluorescence AF, chromatography CHR, colorimetry C, coprecipitation COP, emission spectrography (arc, spark, plasma) ES, fire assay FA, flame emission FE, fluorimetry FL, gravimetry G, ion exchange IE, ion-selective electrodes ISE, mass-spectrometric isotope dilution IDM, neutron activation NAA, polarography POL, potentiometry POT, radioactive isotope dilution IDR, radiometric RAD, solvent extraction SE, titrimetry T, volatilization VAP

Absorbance, definition of, 102
Absorptiometry of solutions, 99-119
  instrumentation for, 104-108
    detectors, 106, 107
    light sources, 105
    sample cells, 106
    wavelength selection devices, 105, 106
  multicomponent analysis by, 115
  precision, 114
  reagents for, 111
  selection of methods for, 109, 110
  table of selected methods for, 118, 119
  typical applications of, 116, 117
  typical instruments for, 107, 108
  see also Spectrophotometry
Absorption of radiation, by atomic vapors, 161-165
  by solutions, 99, 100
Absorption spectra, molecular, 99, 100
Abundances, elemental, 76, 79-81
  compilations of, 80, 81, 342-344
  data of Clarke and Washington for, 76
  data of Fairbairn *et al.*, 76
  economic reserves and, 352, 353
  of lanthanides, 344, 345
  in lunar material, 344, 345
  in meteorites, 344
  relation with relative standard deviation, 76, 77
Acceptability of analytical methods, 2, 3, 398-400
Accuracy in trace element analysis, 398-400
Actinides, separation and analytical methods for, 56(SE), 64-66(CHR), 270(RAD)
Activation analysis, 270-293
  by charged particles, 292, 293
  by gamma-ray photons, 291, 292
  by neutrons, 270-291
Adsorption of ions from solution, 7-10, 278
Aerial prospecting, radiometric, 267, 269
Aerial sampling, 11
Alkali metals, separation and analytical methods for, 46, 48, 49(IE), 214, 215, 224, 225(FE)
Alpha particle, measurement of, 264
  spectrometers for, 270
Alpha-particle decay, 252
Alpha-particle emitters, determination of, 269, 270
Aluminosilicates, decomposition of, 30, 31
Aluminum, separation and analytical methods for, 59(CHR), 60(SE), 91(G), 118(C), 133(FL), 151(ES), 188, 193, 196(AA), 222, 225(FE), 283, 287 (NAA)
Ammonium pyrrolidine-1-carbodithioate, solvent extraction with, 57, 187, 193
Amplification procedure, 96, 97
Anion exchange resins, 38, 39, 46

Anionic complexes, solvent extraction of, 49, 51
Annihilation radiation, 256, 263
Anodic stripping voltammetry, 311, 315, 316
  applications of, 318
  detection limits, 318
  with differential pulse stripping, 316
Antimony, separation and analytical methods for, 48, 48(IE), 58, 60(SE), 62, 119(C), 134(FL), 151, 155(ES), 188, 193, 198, 198(VAP), 200(AA), 279, 283, 288(NAA), 318(ASV)
Archaeology, trace element analysis in, 363-366
Arsenic, separation and analytical methods for, 35, 58, 62(VAP), 118(C), 151, 155(ES), 188, 191, 193, 196(AA), 279, 283, 287(NAA)
Atomic absorption coefficients, 164
Atomic absorption lines, 161-165, 188, 189
  broadening of, 162, 163
  table of, 188, 189
Atomic absorption spectrophotometry, 77, 138, 157, 160-213
  analytical curves in, 184, 186
  applications of, 191-202
    to analysis of solid geological materials, 195-201
    to brine analysis, 194, 195
    to sea-water analysis, 192, 194
    to water analysis, 192-195
  aspiration and nebulization in, 175, 176
    effect of organic solvents on, 176, 187
    effect of sample macroconstituents on, 176, 177
    effect of solution temperature on, 177
  atom-forming processes in, 175-183
  detection limits in, 186, 187
  development of, 160, 161
  effect of ionization in, 177, 178
  flame atomization in, 168-171, 177, 178
  flameless atomization in, 171, 172
  indirect methods of analysis by, 190, 191
  instrumentation for, 165-175
    amplifiers and readout systems, 173, 174
    atomizers, 168-172
    detectors, 173
    light sources, 166, 167
    monochromators, 172, 173
    nebulizers, 168-171
    optical systems, 174, 175, 186
    source modulation, 165, 173
  interferences in, 178-183
    atomic spectral line, 178
    chemical, 179-183
    molecular absorption, 178, 179
  sensitivity of, 186, 187
    methods of decreasing, 187
    methods of increasing, 187
  theory of, 161-165
Atomic fluorescence spectroscopy, 225-231
  applications of, 227, 228
  atomizers for, 226, 227
  detection limits in, 225
  instrumentation for, 226, 227
  light sources for, 226
  multielement analysis by, 228
  optical systems for, 226, 227
  quenching of, 226

Background correction, in atomic absorption, 179, 338
Background emission, in flame emission spectrometry, 221
Band spectra, in flame emission background, 221
Barium, analytical methods for, 151, 155 (ES), 188, 196(AA), 222(FE), 279, 283, 288(NAA), 330(IDM)
Beer's law, 103, 104
  deviations from, 104, 110, 112-114, 184, 186
Beryllium, determination of, by low-energy photon activation, 291, 337
  separation and analytical methods for, 39(COP), 60(SE), 118(C), 133, 134 (FL), 151, 155(ES), 188, 193(AA), 222(FE)
Beta particle, absorption by matter, 255
Beta-particle counting, in neutron activation analysis, 285-291
Beta-particle decay, 252-256
Biological material, sampling of, 11
Bismuth, separation and analytical methods for, 35, 48(IE), 58, 62(VAP), 118(C), 151, 155(ES), 188, 196(AA), 279, 283, 289(NAA), 318(POL), 318(ASV)

INDEX

Bisulfates, for sample fusion, 32
Bivariate analysis, 384-391
Boltzmann equation, 161
Bomb digestion, of silicates, 31
Borates, in sample fusion, 33, 34
Boron, analytical methods for, 117, 118(C), 134(FL), 151, 155(ES), 188(AA)
  environmental chemistry of, 359
Bremsstrahlung, 255, 292
Bromide, analytical methods for, 279, 283, 287(NAA)
Burners, for atomic absorption spectrophotometry, 168, 169
  laminar flow pre-mix, 168
  turbulent flow, 168
  for flame emission photometry, 214

Cadmium, environmental chemistry of, 359, 360
  separation and analytical methods for, 39(COP), 48(IE), 60(SE), 96(T), 151, 155(ES), 188, 193-196(AA), 222(FE), 279, 283, 288(NAA), 316, 317(POL), 318(ASV)
  toxicology of, 360
Calcium, separation and analytical methods for, 48(IE), 96(T), 151, 155(ES), 188, 193-196(AA), 222, 225(FE), 227(AF), 287(NAA), 308(ISE)
Carbonates, decomposition of, 30
  for sample fusion, 32
Carriers in radiochemical separations, 278, 279
Cassiterite, decomposition of, 32, 33
Cathodic stripping voltammetry, 315, 316
Cation exchange resins, 40, 46
Cationic complexes, solvent extraction of, 51
Cerium, separation and analytical methods for, 39(COP), 151, 155(ES), 188(AA), 283, 288(NAA)
Cesium, separation and analytical methods for, 46, 48(IE), 151(ES), 188, 196(AA), 224, 225(FE), 279, 283, 288(NAA), 330(IDM)
Chalcophile elements, 342, 352
Charged particles, use in activation analysis, 292, 293
Charging current, in polarography, 311
Chelate complexes, in solvent extraction, 47, 49

Chelating agents, in solvent extraction, 50
Chemical separations, following neutron activation, 278-281
Chemiluminescence, suprathermal, 215, 216, 223
Chi-square test of significance, 383, 384
Chloride, separation and analytical methods for, 118(C), 191(AA), 279, 283, 287(NAA)
Chlorination, for liberation of volatile chlorides, 34, 35
Chromite, decomposition of, 32, 33
Chromium, environmental chemistry of, 360
  separation and analytical methods for, 48(IE), 116, 118(C), 151, 155(ES), 188, 193-196(AA), 222(FE), 227(AF), 279, 283, 287(NAA)
Cluster analysis, 396, 397
Coal ash, atomic absorption analysis of, 196-199, 201
Cobalt, decomposition of ores of, 30
  separation and analytical methods for, 39(COP), 60(SE), 91(G), 117, 118(C), 151, 155(ES), 188, 194-196, 201(AA), 222(FE), 227(AF), 279, 283, 287(NAA), 317(POL)
Colorimetric methods, see Absorptiometry of solutions
Colorimetry, 99. See also Absorptiometry of solutions
Complexometric titrations, 93-96
  indicators for, 94-96
  stability constants for metal-EDTA complexes in, 94, 95
  titrants for, 93, 95, 96
Compton effect, 257
  correction of γ-ray spectra for, 269
Computers, use in emission spectrometry, 152, 153
  use in X-ray fluorescence, 246
Concentration of traces, by coprecipitation, 37, 38
  by electrodeposition, 59, 270
  by ion exchange, 44-47, 187
  by solvent extraction, 56-58, 187
Contamination during grinding, 14, 15
Copper, decomposition of ores of, 30
  environmental chemistry of, 360

separation and analytical methods for, 48(IE), 60(SE), 91(G), 118(C), 134 (FL), 151, 155(ES), 188, 193-196, 201, 202(AA), 222(FE), 227(AF), 279, 283, 287(NAA), 308(ISE), 316, 317(POL), 318(ASV)
Coprecipitation, 37, 38
   precipitates suitable for use in, 37, 38
Correlation analysis, 387, 391
   Pearson correlation coefficient in, 387-389
   Spearman rank correlation coefficient in, 390, 391
Corundum, decomposition of, 31, 34
Country rock, 347
Crude oils, *see* Oils
Crushing and pulverizing methods, 13-16
Cumulative frequency curves, 378, 379

Decay constant, radioactive, 252
Decomposition of solid samples, 29-36
   by acid attack, 29-31
   by fusion mixtures, 31-34
   by other methods, 34-35
Derivative spectrophotometry, 115
Detection limits, in atomic absorption, 186, 187
   in atomic fluorescence, 226
   in colorimetry, 109
   in emission spectrochemical analysis, 151, 152, 155
   in flame emission, 223, 224
   in fluorimetry, 132
   in mass spectrometric isotope dilution, 328, 329
   in neutron activation analysis, 285, 286
   in radiometric methods, 269, 270
   in spark source mass spectrography, 332
   in voltammetric techniques, 312, 313, 318
   in X-ray emission spectrometry, 240, 243
Diethyldithiocarbamates, solvent extraction of, 57
Differential pulse anodic stripping voltammetry, 316
Diffusion current, polarographic, 310
Diffusion flames, 170
Discriminant analysis, 395, 396
   definition of, 395
   D^2 statistic in, 395, 396

Diseases, caused by trace element deficiency or excess, 355-358, 360-363
Distribution ratio, ion exchange, 43
   solvent extraction, 52, 53, 55
Drill sampling of rocks, 4, 347
Dropping mercury electrode, 309, 310
Dry ashing, of organic matter, 35, 36
Dye cations, use in solvent extraction, 51, 52
Dysprosium, analytical methods for, 151, 155(ES), 189(AA), 222(FE), 283, 289(NAA)

Electrochemical methods of analysis, 303
Electrochromatography, 63
Electrodeless discharge lamps, 167, 225
Electrodeposition, for concentrating traces, 59
   of hydrous oxides of α-emitters, 59, 270
   for removing interfering ions, 59
Electrode potential, of ion-selective electrodes, 304-307
Electron capture, 256
Electron microprobe, 246, 247
Electrothermal atomizers, in atomic absorption, 171, 172, 187, 192, 193
   in atomic fluorescence, 226
Emission spectra, 138, 139
   ground state and excited atom ratios in, 139
   persistent lines in, 151
   population of excited atoms, fluctuations of, 139
   spectral order in, 145
Emission spectrochemical analysis, 138-159
   atomic emission spectra in, 138, 139
   comparison with other techniques, 156, 157
   detection limits in, 151, 152
   development of, 138
   direct reading spectrometers for, 152, 153
   instrumentation for, 140-146, 152, 153
   dc arc, 140, 141
   electrodes for liquid samples, 140, 153, 154
   electrodes for solid samples, 140, 141
   high-voltage ac spark, 141
   inductively coupled rf plasma, 142, 143
   laser microprobe, 155, 156
   plasma jet, 154

matrix effect in, 149
optical spectrographs for, 143-146
photographic measurement of line intensities in, 146-149
  characteristic curve in, 146, 147
  internal standards in, 148-150
  plate calibration for, 147, 148
  step sector in, 147
procedures for qualitative and quantitative analysis by, 150, 152
spectral interferences in, 141, 150
spectroscopic buffers in, 150
tape machine for, 153
Environmental chemistry, 358-363
of selected trace elements, 359-363
Epidemiology, 355-358
Epithermal neutrons, 276
activation by, 276
Erbium, analytical methods for, 151, 155 (ES), 189(AA), 222(FE), 283, 289 (NAA)
Europium, separation and analytical methods for, 39(COP), 134(FL), 151, 155 (ES), 189(AA), 222(FE), 283, 289 (NAA)
Exchange constant, in ion exchange, 42, 43
Extraction equilibria, 52-55
Extraction selectivity, 55, 56

Factor analysis, 394, 395
  Q-mode, 394
  R-mode, 394
Fast neutrons, 275
Field tests, beryllometer for, 337
  colorimetric, 337
  determination of mercury in, 337, 338
  gamma-ray spectrometric, 337
  ion selective electrodes for, 337
  X-ray fluorimetric, 337
Filter fluorimeters, 131, 132
Filter photometers, 99, 217, 218
Filtration of natural waters, 25
Fire assay technique, 35, 36, 90, 91
Fission products, concentration by coprecipitation, 38
  determination of, 270
Fission tracks, determination of uranium from, 285
  etching of, 285

from neutron activation, 285
from spontaneous U-238 fission, 285
Flame, for atomic absorption, 168-171
  for atomic fluorescence, 225, 226
  characteristics of, 170, 171
  for flame emission, 217, 218
Flame emission line intensities, effect of excitation energy on, 216
  effect of temperature on, 216
  relation to atom concentration, 216, 217
Flame emission spectrometry, 214
  analytical curves in, 220
  applications of, 224, 225
  detection limits in, 222-224
  development of, 214, 215
  excitation processes in, 215-217
  instrumentation for, 214, 215, 217-219
  interferences in, 221
  sensitivity of, 217
  theory of, 215-217
Flameless atomization, for atomic absorption, 171, 172
Fluorescence, atomic, see Atomic fluorescence spectroscopy
  molecular, 126-130
Fluorescence efficiency, 128
Fluorescence intensity, relationship with concentration, 128-130
Fluorescence quenching, 130
Fluorescence spectra, 126-128
  corrected, 132
Fluoride, environmental chemistry of, 360
  separation and analytical methods for, 35, 48, 49, 49(IE), 60(SE), 62(VAP), 91 (G), 117, 118(C), 133, 134(FL), 306, 308(POT)
Fluorides, in sample fusion, 34
Fluorimetry, advantages over absorptiometry, 130
  applications of, 132, 133
  disadvantages of, 130
  indirect methods of analysis by, 133
  instrumentation for, 130-132
  light sources, 130, 131
  sample cells, 131
  wavelength selection, 131
  on solid samples, 131, 132
Fraser tube, for mineral separation, 23
F-test of significance, 380-382
Fused salts, for mineral separation, 22, 23

INDEX

Fusion mixtures, for sample decomposition, 31-34

Gadolinium, analytical methods for, 151, 155(ES), 189(AA), 222(FE), 283, 288(NAA)
Gallium, separation and analytical methods for, 39(COP), 48(IE), 57, 60(SE), 118(C), 134(FL), 151, 155(ES), 189, 197, 200(AA), 222, 225(FE), 279, 283, 287(NAA)
Gamma decay, 256, 257
Gamma radiation, absorption of, 257
  interaction with matter, 257
  measurement of, 256-266
  use of, for activation analysis, 291, 292
Gamma-ray photopeaks, 260-262
Gamma-ray spectra, with scintillation detectors, 259
  with semiconductor detectors, 263-266
Gamma-ray spectrometry, in determination of thorium and uranium, 267-269
  in neutron activation analysis, 281-283
  in photon activation analysis, 291, 292
Garnet, borate fusion of, 34
Gas-liquid chromatography, 59-61
Geiger-Müller counter, 258, 259
Geochemical exploration, 336-339, 347-354
  air sampling in, 349
  bedrock analysis in, 347
  biological sample analysis in, 349
  soil analysis, 347, 348
  stream sediment analysis in, 348
  water analysis in, 349
Germanium, separation and analytical methods for, 35, 58, 60(SE), 62(VAP), 117, 118(C), 134(FL), 151, 155(ES), 189, 197, 200(AA), 222(FE), 283, 287(NAA)
Germanium-lithium detector, for γ-radiation, 264-266, 269, 284
  resolution of, 265, 266
Gold, separation and analytical methods for, 35, 39(COP), 48(IE), 57, 60(SE), 64(CHR), 91(FA), 118(C), 134(FL), 151, 155(ES), 188, 193, 196(AA), 222(FE), 227(AF), 279, 283, 289 (NAA)
Gold pan in mineral exploration, 336

Gravimetric methods of analysis, amplification procedures in, 90, 92
  cascade processes in, 92
  fire assay methods, 35, 90-92
  for noble element determination, 90-92
  precipitate-analytic mass ratios in, 91

Hafnium, separation and analytical methods for, 48(IE), 58(SE), 151, 155(ES), 189(AA), 279, 283, 289(NAA)
Half-lives, of isotopes produced by neutron activation, 287-289
  of natural radioisotopes, 253
Half-wave potentials, polarographic, 309, 310
Halides, determination of, with ion-selective electrodes, 308
  extraction of covalent, 47
  liberation of volatile, 58
  separation of, by ion exchange, 46
Halogens, determination of, by neutron activation analysis, 279, 283, 287, 288
  by X-ray emission, 246
Heavy liquids, in mineral separation, 22
Heteropolyacids, solvent extraction of, 49, 57, 191
Hollow cathode lamps, 166, 167, 225
  in atomic absorption, 166, 167
  in atomic fluorescence, 225
  multielement, 166, 178
Holmium, analytical methods for, 151, 155 (ES), 189(AA), 222(FE), 283, 289 (NAA)
Hydrides, volatilization of, 35, 58, 170
Hydrochloric acid, in sample decomposition, 30
Hydrofluoric acid, in sample decomposition, 30, 31, 195, 199
Hydrogeochemistry, 349
Hydroxides, for sample fusion, 32, 33

Ilkovic equation, 310
Ilmenite, decomposition of, 32
Indium, separation and analytical methods for, 60(SE), 134(FL), 151, 155(ES), 189(AA), 222(FE), 279, 283, 288 (NAA), 294(IDR), 318(POL), 318 (ASV)
Insoluble salts, dissolution of, by ion

exchange, 44, 45
Instrumental neutron activation analysis, 284
　applications of, 284
　computation procedure in, 284
Interfering ions, removal of, by electrolysis, 59
　by ion exchange, 44, 46
Internal conversion electrons, 256, 257
Internal standards, in emission spectrochemical analysis, 148
　in flame emission, 220, 221
　in spark source mass spectrometry, 332
　in X-ray emission spectroscopy, 245
Iodide, separation and analytical methods for, 49(IE), 96, 97(T), 118(C), 191 (AA), 279, 288(NAA)
Iodine, solvent extraction of, 60
Ion-association complexes, in solvent extraction, 49
Ion exchange, 38-47
　applications of, 44-47
Ion exchange equilibrium coefficient, 42, 43
Ion exchangers, chelating, 42, 46-52
　inorganic, 40, 46, 48, 49
　liquid, 40, 41
　naturally occurring, 39, 41, 42
　organic resins, 39, 40, 48, 49
Ion exchange selectivity, 42-44
Ion exchange selectivity coefficient, 42, 43
Ionic strength, control for electrode potential measurements, 306
Ionization buffer, in atomic absorption, 183, 186
　in flame emission, 219, 220
Ionization effects, in atomic absorption, 183, 186
Ion-selective electrodes, 303-308
　applications of, 307, 308
　concentration measurement with, 306, 307
　errors in measurements with, 306, 307
　glass, 304
　heterogeneous, 305
　liquid, 305, 306
　solid state, 304
　specificity of, 306
Iridium, separation and analytical methods for, 48(IE), 151(ES), 189(AA), 222(FE), 279, 283, 289(NAA)

Iron, separation and analytical methods for, 47, 49(SE), 90(T), 118(C), 151, 155 (ES), 189, 193-195, 197, 201, 202 (AA), 222, 225(FE), 279, 283, 287 (NAA)
Isodynamic mineral separator, 20
Isomeric transitions, 256
Isotope dilution, mass spectrometric, 326-330
　accuracy, precision and sensitivity of, 328-330
　theory of, 326-328
　radiochemical, 293-294
　substoichiometric procedure in, 294
Isotopic neutron sources, 277

K-electron capture, 256

Lanthanides, separation and analytical methods for, 39(COP), 46, 48(IE), 64, 65 (CHR), 134(FL), 151, 155(ES), 198 (AA), 222, 225(FE), 279, 283, 288, 289(NAA), 330(IDM)
Lanthanum, analytical methods for, 151, 155(ES), 189(AA), 222(FE), 279, 283, 288(NAA)
Laser microprobe, 155, 156
Lead, decomposition of ores of, 30
　environmental chemistry of, 360, 361
　separation and analytical methods for, 48 (IE), 60(SE), 62(VAP), 119(C), 151, 155(ES), 188, 193-195, 198(AA), 222 (FE), 279, 289(NAA), 293, 294(IDR), 316-318(POL), 318(ASV)
Limestone, atomic absorption analysis of, 196-199
　decomposition of, 200
Limiting current, voltammetric, 309, 310
Lithium, separation and analytical methods for, 48(IE), 134(FL), 151(ES), 189, 193-195, 197(AA), 222, 224(FE), 327, 328(IDM)
Lithium metaborate, in sample fusion, 34
Lithium tetraborate, in sample fusion, 33
Lithophile elements, 342
Lunar material, analysis of, 333, 344, 345
Lutetium, analytical methods for, 151, 155 (ES), 189(AA), 222(FE), 283, 289 (NAA)

Magnesium, separation and analytical methods for, 48(IE), 91(G), 96(T), 118(C), 134(FL), 151, 155(ES), 189, 193, 194, 197, 201(NAA), 222(FE), 227 (AF), 287(NAA)
Manganese, analytical methods for, 119(C), 151, 155(ES), 188, 193-195, 197, 201(AA), 222(FE), 227(AF), 283, 287(NAA)
Mass spectrometry, 322, 333
  instruments for, 322-326
    ion detectors in, 324, 326
    ion sources in, 324, 325
    mass analyzers in, 324-326
  isotope dilution with, 326-330
  spark source, see Spark source mass spectrography
  theory of, 322, 323
Matrix effects, 81, 85
  in atomic absorption, 175-183
  in emission spectrochemical analysis, 149, 153
  in neutron activation analysis, 286
  in X-ray fluorescence spectrometry, 244
Membrane electrodes, 305, 306. See also Ion-selective electrodes
Mercury, environmental chemistry of, 361, 362
  separation and analytical methods for, 34, 35, 48(IE), 58, 60(SE), 151, 155(ES), 171, 172, 172(VAP), 189, 193, 194, 197, 337, 338(AA), 226, 227(AF), 283, 289(NAA), 294(IDR)
Mineral separation, 19-25
  electrostatic, 20
  by heavy liquids, 21-23
  magnetic, 20
  by planning, 21
Mobile geochemical laboratories, 336, 338, 339
Mobilization of elements, effect of electrode potential on, 342
  effect of pH on, 342
  epigenetic dispersion in, 347
  hypogene mobility in, 341, 342
  supergene mobility in, 341, 342
  syngenetic dispersion in, 347
Molar absorption coefficient, 109, 110
  definition of, 102, 103
Molybdenum, separation and analytical methods for, 39(COP), 48(IE), 60 (SE), 118(C), 155(ES), 188, 193-195, 198(AA), 222(FE), 279, 283, 288 (NAA)
Multichannel analyzer, use of, with Ge(Li) detector, 265, 284
Multiple regression analysis, 391, 392
  partial regression coefficients in, 391, 392
Multivariate analysis, 391-397

Nansen bottle, 7
Natural radioactivity, trace analysis using, 266-270
Natural waters, filtration of, 25
  sampling and storage of, 7-10
Neodymium, analytical methods for, 151, 155(ES), 188, 198(AA), 222(FE), 283, 288(NAA)
Neutron activation analysis, 250, 270-291
  applications to geological materials, 277-284
  development of, 270, 271
  errors in, 286, 290, 291
  growth and decay of induced radioactivity, 271-275
  instrumental, 284
  interfering reactions in, 276
  multielement, with chemical separations, 281-283
  precision of, 291
  procedure for, 277-284
    chemical separations, 278-283
    irradiation, 278
    sample preparation, 277, 278
  sensitivity of, 285-289
  standards for, 274
  statistical errors in, 290, 291
  systematic errors in, 286, 290
Neutron activation cross section, 272
Neutron capture, 275
Neutron generators, 14 (MeV), 176
Neutron sources, 275-277
  isotopic, 277
  neutron generators, 275, 276
  nuclear reactors, 275, 276
Nickel, separation and analytical methods for, 39(COP), 48(IE), 49, 60(SE), 91 (G), 119(C), 151, 155(ES), 188, 193-195, 198, 201, 202(AA), 222(FE), 279, 283, 287(NAA), 317(POL)

INDEX

Niobium, separation and analytical methods for, 60(SE), 64(CHR), 119(C), 151, 155(ES), 188, 198(AA), 222(FE), 279, 288(NAA)
Niskin bottle, 7
Nitric acid, in sample decomposition, 30
Non-metals, flame spectroscopy of, 224
Non-parametric statistical methods, 379
Nuclear isomers, 256
Nuclear radiation, 250-257
  interaction with matter, 255-257
  measurement of, 257-266
    by gas ionization devices, 258, 259
    by Geiger-Müller counter, 258, 259
    by proportional counter, 259
    by scintillation detectors, 259-263
    by semiconductor detectors, 263-266
Nuclear reactor, prehistoric, 346, 347
Null hypothesis in statistics, 379, 380

Obsidian, source identification of, 363-366
Oils, analysis of, by atomic absorption, 201, 202
Organic matter, destruction of, 35, 36
Organic phosphors, in β-radiation measurement, 259
Organic solvents, in atomic absorption, 176
  in flame emission, 223
Osmiridium, analysis of, 200
Osmium, separation and analytical methods for 60(SE), 62(VAP), 188(AA), 279, 283, 289(NAA)
Oxides, volatilization of, 58
Oxygen, excited, as ashing agent, 36

Pair production, 257
Palladium, separation and analytical methods for, 60(SE), 119(C), 151, 155(ES), 188(AA), 222(FE), 279, 283, 288(NAA), 294(IDR)
Paper chromatography, 62, 63
Parametric statistical methods, 379
Partition chromatography, 61-66
  reversed phase, 63, 64
Pathfinding element(s), cadmium as, 359
  mercury as, 337
  in mineral exploration, 352
Perchloric acid, in sample digestion, 30, 31
Phosphates, colorimetric determination of, 119
  dissolution of, by ion exchangers, 44, 45
Phosphomolybdate, solvent extraction of, 60
Photoelectric effect, 257
Photoelectron spectroscopy, 232, 233
Photographic measurement of spectral line intensities, 146-148, 330, 331
  in emission spectrochemical analysis, 146-148
  in spark source mass spectrometry, 330, 331
Photometric errors, 112-113
Photometric titrations, 116
Photon activation analysis, 291, 292
  sensitivity of, 292
Plasmas as atom sources, plasma jet, 140, 154
  plasma torch, 140
  rf plasma, 140, 142, 143
Platinum, separation and analytical methods for, 35, 48(IE), 60(SE), 91(FA), 119 (C), 151, 155(ES), 188, 198(AA), 222 (FE), 279, 283, 289(NAA)
  vessels of, in acid digestion, 309, 310
  in fusion, 32, 33
Polarography, 309
  applications of, 316-318
  differential pulse, 316
  modified forms of, 310, 311
  selectivity and sensitivity of, 311
  oscilloscopic, 312, 313
  pulse, 312, 313
Polypropylene, vessels of, in acid digestion, 31
Populations, geological, 4
Positron emission, 256
Potassium, separation and analytical methods for, 48(IE), 151(ES), 189, 193, 194, 197, 201(AA), 225(FE), 266 (RAD), 283, 287(NAA), 330(IDM)
Potassium-40, radiation from, 266
Potentiometric titrations, with ion-selective electrodes, 304
Potentiometry, with ion-selective electrodes, 304-307
Praseodymium, analytical methods for, 151, 155(ES), 188(AA), 222(FE), 283, 288(NAA)
Precipitation, 37, 38

Precision of trace element analysis, 398-400
Probability concept in statistics, 379, 380
Proportional counter, 259
Proton activation analysis, 292, 293
Pulse height analysis, 259
Pyrosulfates, for sample fusion, 32

Quartz crystal microbalance, for mercury determination, 338

Radiation, nuclear, see Nuclear radiation
Radioactivation methods, 250, 270-294
Radioactive decay, 250-255
  equations for, 252, 254, 255
  modes of, 254, 255
  of natural radioisotopes, 250-252
Radioactive equilibrium, 254, 255, 267
Radioactivity, artificially induced, 250, 270-293
  natural, 250-255
  application to trace analysis, 266-270
Radiochemical isotope dilution, 293, 294
Radioisotopes, natural, decay data for, 252, 253
Radiometric measurements, determination of thorium and uranium by, 266-270
  field use of, 269
Radium, analytical methods for, 270
Rare earths, see Lanthanides
Reconnaissance surveys, geochemical atlas from, 353
  regional geochemical, 353, 354
  for soil maps, 355
Regression analysis, 385-387
  least squares lines in, 385
  regression coefficients in, 385
Residual current, polarographic, 311
Resonance lines, atomic absorption, 161
Rhenium, separation and analytical methods for, 60(SE), 62(VAP), 134(FL), 151(ES), 188, 198(AA), 222(FE), 279, 283, 289(NAA)
Rhodium, analytical methods for, 151, 155(ES), 188, 198(AA), 222(FE), 279, 288(NAA)
Ring-oven separations, 63
Rocks, crushing and pulverizing of, 13-16
  sampling of, 4-6

Rubidium, separation and analytical methods for, 48(IE), 151(ES), 188, 193, 194, 198(AA), 222, 225(FE), 279, 283, 288(NAA), 329, 330(IDM)
Ruthenium, separation and analytical methods for, 60(SE), 62(VAP), 151(ES), 188(AA), 222(FE), 279, 283, 288(NAA)
Rutile, decomposition of, 31-34

Saha equation, 182
Samarium, analytical methods for, 151, 155(ES), 189(AA), 222(FE), 283, 288(NAA)
Sample pretreatment by chemical methods, 29-66
  decomposition by acid attack, 29-31
  decomposition by fusion mixtures, 31-34
  decomposition of organic matter, 35, 36
  miscellaneous decomposition techniques, 34, 35
  separation and concentration techniques, 36-66
    chromatographic techniques, 59-66
    coprecipitation, 37, 38
    electrodeposition, 59
    ion exchange, 38-47
    precipitation, 37, 38
    solvent extraction, 47-58
    volatilization, 58, 62
Sample pretreatment by physical methods, 13-27
  crushing and pulverizing, 13-16
    apparatus for, 13, 14
    contamination during, 14-16
  errors arising during, 13, 26, 27
  filtering of natural waters, 25
  mineral separation, 19-25
  sample splitting, 18, 19
  sieving, 16-18
  subsampling problems in, 26, 27
Sample splitting, 18, 19
  coning and quartering in, 18
  use of riffle in, 18
Sampling, aerial, 11
  of biological material, 11
  of natural waters, 7, 10
  of rocks, 4-6
  of sediments, 6, 7
  of soils, 5

of solid materials, errors in, 26, 27
  sample size in, 5, 26, 27
  target populations in, 4
Sampling constant, for geological samples, 27
Scandium, separation and analytical methods for, 39(COP), 48(IE), 119(C), 151, 155(ES), 188(AA), 222(FE), 279, 283, 287(NAA)
Secular equilibrium, in radioactive decay, 254
Sediments, atomic absorption analysis of, 201
  sampling of, 6, 7
Selectivity ratio, of membrane electrodes, 306
Selenium, environmental chemistry of, 362
  separation and analytical methods for, 35, 39(COP), 58, 60(SE), 62(VAP), 133, 134(FL), 151, 155(ES), 189, 193, 198(AA), 227, 228(AF), 279, 283, 287(NAA)
Separated flames, 170, 171, 226
Separation, by electrodeposition, 59
  by gas-liquid chromatography, 59, 61
  by ion exchange, 44-47
  by liquid-liquid chromatography, 64-66
  by solvent extraction, 56-58
  by volatilization, 58, 62
Separation and concentration techniques, 36-66
Separation of mixtures of ions, by ion exchange, 44-47
  by solvent extraction, 56-58
Separation schemes, in multielement neutron activation analysis, 281-283
Siderophile elements, 342
Sieves, standard sizes of, 16-18
Sieving, 16-18
  mesh sizes in, 16-18
Significance tests, 379-384
  chi-square test, 383, 384
  F test, 380-382
  levels of significance for, 380
  null hypothesis in, 379, 380
  Student's t test, 382, 383
Silica, vessels of, for pyrosulfate fusion, 32
Silicates, atomic absorption analysis of, 195-200
  decomposition of, 30-34, 195-200

Silicon, analytical methods for, 119(C), 151, 155(ES), 189, 191, 198, 201(AA), 287(NAA)
Silicon barrier detector, 264
Sillimanite, borate fusion of, 34
Silver, separation and analytical methods for, 35, 91(FA), 92(G), 118(C), 133(FL), 151, 155(ES), 188, 193, 196, 202(AA), 222(FE), 279, 283, 288(NAA), 318(ASV)
Single and double-sided tests in statistics, 381, 382
Sodium, separation and analytical methods for, 48(IE), 91(G), 151, 155(ES), 188, 193, 194, 198, 201(AA), 222, 224(FE), 283, 287(NAA)
Sodium borohydride, in hydride liberation, 35, 170
Sodium iodide (thallium) scintillation detector, 259-263
  resolution of, 260, 264-266
Sodium peroxide, in sample fusion, 33
Soils, crushing of, 16
  extractant solutions for analysis of, 201
  sampling of, 6
Soils and soil extracts, atomic absorption analysis of, 196-199, 201
Solid electrode voltammetry, 310
Solvent extraction, 47-58, 60, 61
  applications of, 56-58
  in atomic absorption analysis, 57, 60, 61, 187, 191, 193-195
  in colorimetry, 57, 60, 61, 109, 117
  countercurrent, 56, 57
  kinetic factors in, 56
  in neutron activation analysis, 58, 60, 61, 279, 283
Solvent extraction selectivity, 55, 56
Solvent extraction systems, 47-52, 56-58
Spaeth sedimentation glass, 23
Spark source mass spectrography, 322, 330-333
  analytical procedure for, 330, 331
  applications of, 333
  instrumentation for, 330-332
  interferences in, 331
  internal standards in, 331
  precision and sensitivity of, 331, 332
Spectral line interference, in atomic absorption, 178, 179

in flame emission, 221
Spectrofluorimeters, 130-132
Spectrophotometers, 99, 107, 108, 165-175, 218, 219
  double beam, 107, 108, 174, 175
  single beam, 107, 174
Spectrophotometry, derivative, 115
  terminology of, 100-103
Standard additions, in atomic absorption, 187, 190
  in flame emission, 220
  in ion-selective electrode measurements, 306, 307
  in polarography, 310
  in X-ray emission spectroscopy, 245
Standard rocks, 81-87
  of C.A.A.S., 85-87
  sources of, 86
  table of, 82-85
  of U.S.G.S., 26, 27, 82-87, 285
  G-1 and W-1, 26, 27, 82, 83, 87
Standards, for neutron activation analysis, 274, 278
  in X-ray emission spectroscopy, 245
Standard solutions, for atomic absorption, 184
  preparation of, 184
  storage of, 184
Statistical methods, importance of, 3
Statistical notation, 375
Statistical populations, background, 374
  distributions of, 374-376
  lognormal distributions of, 375, 376
  mixed distributions of, 377-379
  normal distributions of, 375, 376
Storage of solid samples, 7
Storage of waters, adsorption on containers during, 7-10
  deep freezing for, 10
Strontium, analytical methods for, 151, 155(ES), 189, 193-195, 199(AA), 221, 222, 225(FE), 279, 283, 288(NAA), 330(IDM)
Student's t test of significance, 382, 383
Subsampling, errors from small masses in, 26, 27
  sampling constant in, 27
Substitution rules, for trace element distribution, 341, 343
Substoichiometric extraction, 58, 280, 281, 294

Sulfides, atomic absorption analysis of, 196-200
  decomposition of, 30, 31, 200
  determination of, with ion-selective electrode, 308
Sulfur, volatilization of, 34

Tantalum, separation and analytical methods for, 48(IE), 60(SE), 64(CHR), 119(C), 134(FL), 151, 155(ES), 189(AA), 222(FE), 279, 283, 289(NAA)
Teflon, vessels of, for acid digestion, 31
Tektites, analysis of by spark source mass spectrometry, 345
  origin of, 345, 346
Tellurium, separation and analytical methods for, 31(COP), 60(SE), 58, 62(VAP), 119(C), 151, 155(ES), 189, 199(AA), 279, 283, 288(NAA)
Terbium, analytical methods for, 134(FL), 151, 155(ES), 189(AA), 222(FE), 283, 288(NAA)
Thallium, separation and analytical methods for, 48(IE), 61(SE), 119(C), 134(FL), 151, 155(ES), 189, 199(AA), 222(FE), 279, 283, 289(NAA), 318(POL)
Thermal neutrons, 275, 276
Thin layer chromatography, 63, 64
Thorium, separation and analytical methods for, 39(COP), 48(IE), 60(SE), 119(C), 134(FL), 151, 155(ES), 266-270(RAD), 279, 289(NAA)
Thorium decay series, 253
Thulium, analytical methods for, 151, 155(ES), 189(AA), 222(FE), 283, 289(NAA)
Tin, separation and analytical methods for, 35, 58, 62, 200(VAP), 48(IE), 60(SE), 119(C), 134(FL), 151, 155(ES), 189, 198, 200(AA), 222, 223(FE), 279, 288(NAA), 318(ASV)
Titanium, separation and analytical methods for, 46, 48(IE), 119(C), 151, 155(ES), 189, 191, 199(AA), 223, 225(FE), 279, 287(NAA)
Titrimetry, 77, 78, 90, 92-97, 116
Tourmaline, decomposition of, 32
Trace elements, definition of, 1
  distribution of, in extraterrestrial

material, 344, 345
  in terrestrial material, 341-344
  factors influencing the distribution of, 342-344
  in human health, 354-363
  non-uniform distribution of, in solid samples, 26, 27, 285
  in plant and animal health, 354-363
  redistribution of, by human activity, 358, 359
  uses of data on, 1, 2, 341-366
Trend surface analysis, 392-294
  autocorrelation in, 393
  residuals in, 393, 394
Tungsten, separation and analytical methods for, 48(IE), 60(SE), 119(C), 134(FL), 151, 155(ES), 189, 199(AA), 223(FE), 279, 283, 289(NAA)

Ultratrace elements, definition of, 1
Ultraviolet-visible spectrophotometry, see Absorptiometry of solutions
Uranium, decomposition of ores of, 30
  separation and analytical methods for, 39 (COP), 48(IE), 61(SE), 64, 66(CHR), 119(C), 132-134(FL), 151, 155(ES), 189(AA), 266-270(RAD), 279, 283, 285, 289(NAA), 318(POL), 329 (IDM)
Uranium decay series, 253

Vanadium, separation and analytical methods for, 39(COP), 48(IE), 57, 58, 61 (SE), 119(C), 151, 155(ES), 189, 191, 193-195, 199, 202(AA), 223 (FE), 279, 287(NAA)
Van Dorn sampler, 7, 8
Vapor discharge tubes, in atomic absorption, 167
Voltammetry, with solid electrodes, 310
  stripping, 311-316

Wet chemical analysis, 77, 78
  use for major constituents of rocks, 90
Wet oxidation, of organic matter, 36

X-ray absorption, 235-237
  absorption edges, 236, 243
X-ray emission spectroscopy, 232-248
  atom excitation processes, 232, 233
  atom relaxation processes, 232, 233
  Auger effect, 232
  X-ray emission, 232, 233
  comparison with other techniques, 157, 246
  detection limits in, 240, 243
  development of, 233
  dispersive spectrometers for, 238-240
  electron microprobe in, 246, 247
  fluorescence yield in, 237, 238
  gamma sources for, 241-244
  instrumentation for, 238-244
    detectors, 239
    dispersing crystals, 239
    sources, 233-235, 238, 241-243, 247, 248
  matrix effects in, 244-246
  non-dispersive instruments for, 240, 241
  primary X-rays for, 233-235
  proton sources for, 247, 248
  qualitative and quantitative analysis by, 244-246
  relationship between wavelength and atomic number, 234, 235

Yield determination, in radiochemical separations, 279, 280
  by reactivation, 280
Ytterbium, analytical methods for, 151, 155(ES), 189(AA), 223(FE), 283, 289(NAA)
Yttrium, separation and analytical methods for, 39(COP), 48(IE), 151, 155(ES), 189, 199(AA), 223(FE), 279, 288 (NAA)

Zinc, environmental chemistry of, 362, 363
  separation and analytical methods for, 48(IE), 96(T), 119(C), 134(FL), 151, 155(ES), 189, 193-195, 199, 201 (AA), 227(AF), 279, 283, 287(NAA), 317(POL), 318(ASV)
Zircon, decomposition of, 32-34
Zirconium, separation and analytical methods for, 39(COP), 48(IE), 58, 61(SE), 119(C), 134(FL), 151, 155(ES), 189 (AA), 223(FE), 279, 283, 288 (NAA)
  vessels of, for hydroxide and peroxide fusions, 32, 33

WITHDRAWN